QA 402.5 JET

MATHEMATICAL PROGRAMMING

PURE AND APPLIED MATHEMATICS

A Program of Monographs, Textbooks, and Lecture Notes

MONOGRAPHS AND TEXTBOOKS IN PURE AND APPLIED MATHEMATICS

1. *K. Yano*, Integral Formulas in Riemannian Geometry (1970)*(out of print)*
2. *S. Kobayashi*, Hyperbolic Manifolds and Holomorphic Mappings (1970) *(out of print)*
3. *V. S. Vladimirov*, Equations of Mathematical Physics (A. Jeffrey, editor; A. Littlewood, translator) (1970) *(out of print)*
4. *B. N. Pshenichnyi*, Necessary Conditions for an Extremum (L. Neustadt, translation editor; K. Makowski, translator) (1971)
5. *L. Narici, E. Beckenstein, and G. Bachman*, Functional Analysis and Valuation Theory (1971)
6. *D. S. Passman*, Infinite Group Rings (1971)
7. *L. Dornhoff*, Group Representation Theory (in two parts). Part A: Ordinary Representation Theory. Part B: Modular Representation Theory (1971, 1972)
8. *W. Boothby and G. L. Weiss (eds.)*, Symmetric Spaces: Short Courses Presented at Washington University (1972)
9. *Y. Matsushima*, Differentiable Manifolds (E. T. Kobayashi, translator) (1972)
10. *L. E. Ward, Jr.*, Topology: An Outline for a First Course (1972) *(out of print)*
11. *A. Babakhanian*, Cohomological Methods in Group Theory (1972)
12. *R. Gilmer*, Multiplicative Ideal Theory (1972)
13. *J. Yeh*, Stochastic Processes and the Wiener Integral (1973) *(out of print)*
14. *J. Barros-Neto*, Introduction to the Theory of Distributions (1973) *(out of print)*
15. *R. Larsen*, Functional Analysis: An Introduction (1973) *(out of print)*
16. *K. Yano and S. Ishihara*, Tangent and Cotangent Bundles: Differential Geometry (1973) *(out of print)*
17. *C. Procesi*, Rings with Polynomial Identities (1973)
18. *R. Hermann*, Geometry, Physics, and Systems (1973)
19. *N. R. Wallach*, Harmonic Analysis on Homogeneous Spaces (1973) *(out of print)*
20. *J. Dieudonné*, Introduction to the Theory of Formal Groups (1973)
21. *I. Vaisman*, Cohomology and Differential Forms (1973)
22. *B. -Y. Chen*, Geometry of Submanifolds (1973)
23. *M. Marcus*, Finite Dimensional Multilinear Algebra (in two parts) (1973, 1975)
24. *R. Larsen*, Banach Algebras: An Introduction (1973)
25. *R. O. Kujala and A. L. Vitter (eds.)*, Value Distribution Theory: Part A; Part B: Deficit and Bezout Estimates by Wilhelm Stoll (1973)
26. *K. B. Stolarsky*, Algebraic Numbers and Diophantine Approximation (1974)
27. *A. R. Magid*, The Separable Galois Theory of Commutative Rings (1974)
28. *B. R. McDonald*, Finite Rings with Identity (1974)
29. *J. Satake*, Linear Algebra (S. Koh, T. A. Akiba, and S. Ihara, translators) (1975)

30. *J. S. Golan,* Localization of Noncommutative Rings (1975)
31. *G. Klambauer,* Mathematical Analysis (1975)
32. *M. K. Agoston,* Algebraic Topology: A First Course (1976)
33. *K. R. Goodearl,* Ring Theory: Nonsingular Rings and Modules (1976)
34. *L. E. Mansfield,* Linear Algebra with Geometric Applications: Selected Topics (1976)
35. *N. J. Pullman,* Matrix Theory and Its Applications (1976)
36. *B. R. McDonald,* Geometric Algebra Over Local Rings (1976)
37. *C. W. Groetsch,* Generalized Inverses of Linear Operators: Representation and Approximation (1977)
38. *J. E. Kuczkowski and J. L. Gersting,* Abstract Algebra: A First Look (1977)
39. *C. O. Christenson and W. L. Voxman,* Aspects of Topology (1977)
40. *M. Nagata,* Field Theory (1977)
41. *R. L. Long,* Algebraic Number Theory (1977)
42. *W. F. Pfeffer,* Integrals and Measures (1977)
43. *R. L. Wheeden and A. Zygmund,* Measure and Integral: An Introduction to Real Analysis (1977)
44. *J. H. Curtiss,* Introduction to Functions of a Complex Variable (1978)
45. *K. Hrbacek and T. Jech,* Introduction to Set Theory (1978)
46. *W. S. Massey,* Homology and Cohomology Theory (1978)
47. *M. Marcus,* Introduction to Modern Algebra (1978)
48. *E. C. Young,* Vector and Tensor Analysis (1978)
49. *S. B. Nadler, Jr.,* Hyperspaces of Sets (1978)
50. *S. K. Segal,* Topics in Group Rings (1978)
51. *A. C. M. van Rooij,* Non-Archimedean Functional Analysis (1978)
54. *L. Corwin and R. Szczarba,* Calculus in Vector Spaces (1979)
53. *C. Sadosky,* Interpolation of Operators and Singular Integrals: An Introduction to Harmonic Analysis (1979)
54. *J. Cronin,* Differential Equations: Introduction and Quantitative Theory (1980)
55. *C. W. Groetsch,* Elements of Applicable Functional Analysis (1980)
56. *I. Vaisman,* Foundations of Three-Dimensional Euclidean Geometry (1980)
57. *H. I. Freedman,* Deterministic Mathematical Models in Population Ecology (1980)
58. *S. B. Chae,* Lebesgue Integration (1980)
59. *C. S. Rees, S. M. Shah, and C. V. Stanojević,* Theory and Applications of Fourier Analysis (1981)
60. *L. Nachbin,* Introduction to Functional Analysis: Banach Spaces and Differential Calculus (R. M. Aron, translator) (1981)
61. *G. Orzech and M. Orzech,* Plane Algebraic Curves: An Introduction Via Valuations (1981)
62. *R. Johnsonbaugh and W. E. Pfaffenberger,* Foundations of Mathematical Analysis (1981)
63. *W. L. Voxman and R. H. Goetschel,* Advanced Calculus: An Introduction to Modern Analysis (1981)
64. *L. J. Corwin and R. H. Szcarba,* Multivariable Calculus (1982)
65. *V. I. Istrătescu,* Introduction to Linear Operator Theory (1981)
66. *R. D. Järvinen,* Finite and Infinite Dimensional Linear Spaces: A Comparative Study in Algebraic and Analytic Settings (1981)

67. *J. K. Beem and P. E. Ehrlich*, Global Lorentzian Geometry (1981)
68. *D. L. Armacost*, The Structure of Locally Compact Abelian Groups (1981)
69. *J. W. Brewer and M. K. Smith, eds.*, Emmy Noether: A Tribute to Her Life and Work (1981)
70. *K. H. Kim*, Boolean Matrix Theory and Applications (1982)
71. *T. W. Wieting*, The Mathematical Theory of Chromatic Plane Ornaments (1982)
72. *D. B. Gauld*, Differential Topology: An Introduction (1982)
73. *R. L. Faber*, Foundations of Euclidean and Non-Euclidean Geometry (1983)
74. *M. Carmeli*, Statistical Theory and Random Matrices (1983)
75. *J. H. Carruth, J. A. Hildebrant, and R. J. Koch*, The Theory of Topological Semigroups (1983)
76. *R. L. Faber*, Differential Geometry and Relativity Theory: An Introduction (1983)
77. *S. Barnett*, Polynomials and Linear Control Systems (1983)
78. *G. Karpilovsky*, Commutative Group Algebras (1983)
79. *F. Van Oystaeyen and A. Verschoren*, Relative Invariants of Rings: The Commutative Theory (1983)
80. *I. Vaisman*, A First Course in Differential Geometry (1984)
81. *G. W. Swan*, Applications of Optimal Control Theory in Biomedicine (1984)
82. *T. Petrie and J. D. Randall*, Transformation Groups on Manifolds (1984)
83. *K. Goebel and S. Reich*, Uniform Convexity, Hyperbolic Geometry, and Nonexpansive Mappings (1984)
84. *T. Albu and C. Năstăsescu*, Relative Finiteness in Module Theory (1984)
85. *K. Hrbacek and T. Jech*, Introduction to Set Theory, Second Edition, Revised and Expanded (1984)
86. *F. Van Oystaeyen and A. Verschoren*, Relative Invariants of Rings: The Noncommutative Theory (1984)
87. *B. R. McDonald*, Linear Algebra Over Commutative Rings (1984)
88. *M. Namba*, Geometry of Projective Algebraic Curves (1984)
89. *G. F. Webb*, Theory of Nonlinear Age-Dependent Population Dynamics (1985)
90. *M. R. Bremner, R. V. Moody, and J. Patera*, Tables of Dominant Weight Multiplicities for Representations of Simple Lie Algebras (1985)
91. *A. E. Fekete*, Real Linear Algebra (1985)
92. *S. B. Chae*, Holomorphy and Calculus in Normed Spaces (1985)
93. *A. J. Jerri*, Introduction to Integral Equations with Applications (1985)
94. *G. Karpilovsky*, Projective Representations of Finite Groups (1985)
95. *L. Narici and E. Beckenstein*, Topological Vector Spaces (1985)
96. *J. Weeks*, The Shape of Space: How to Visualize Surfaces and Three-Dimensional Manifolds (1985)
97. *P. R. Gribik and K. O. Kortanek*, Extremal Methods of Operations Research (1985)
98. *J.-A. Chao and W. A. Woyczynski, eds.*, Probability Theory and Harmonic Analysis (1986)
99. *G. D. Crown, M. H. Fenrick, and R. J. Valenza*, Abstract Algebra (1986)
100. *J. H. Carruth, J. A. Hildebrant, and R. J. Koch*, The Theory of Topological Semigroups, Volume 2 (1986)

Other Volumes in Preparation

MATHEMATICAL PROGRAMMING

An Introduction to Optimization

MELVYN W. JETER

The University of Southern Mississippi
Hattiesburg, Mississippi

MARCEL DEKKER, INC.   New York and Basel

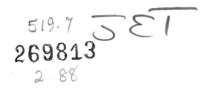
Library of Congress Cataloging-in-Publication Data

Jeter, Melvyn W., [date]
  Mathematical programming.

  (Monographs and textbooks in pure and applied
mathematics ; 102)
  Includes bibliographies and index.
  1. Programming (Mathematics) I. Title. II. Series:
Monographs and textbooks in pure and applied mathe-
matics ; v. 102.
QA402.5.J47   1986   519.7   85-29235
ISBN 0-8247-7478-7

MARCEL DEKKER, INC.
270 Madison Avenue, New York, New York 10016

Current printing (last digit):
10 9 8 7 6 5 4 3 2 1

PRINTED IN THE UNITED STATES OF AMERICA

Preface

This book is intended to be an introductory text in mathematical programming and optimization for those students having a mathematical background that includes one semester of linear algebra and a complete calculus sequence through multivariable calculus. Many of the current books on the market are either too elementary or too advanced for students with such a background. The more elementary books often spend a great deal of time presenting a linear algebra or a finite mathematics course with some optimization applications. The more advanced books may assume a background of advanced calculus or its equivalent and quite often present very few numerical examples of the concepts being presented. I found that in either case many of the available books on the market just did not fit the backgrounds and needs of my students. The student that I have in mind is most likely a junior or senior mathematics, computer science, engineering, management science, economics, or business major. Some will be first-year graduate students.

It is important for these students to develop computational as well as conceptual skills. To aid students in the development of computational skills, a respectable number of computational examples and exercises have been included. Most of the computational exercises in this book can be done without the use of a computer. However, my students were encouraged to write their own computer programs for the nonlinear optimization techniques and problems found in Chapters 9 and 10. I often instruct my students to use a particular search technique from Chapter 9 together with a particular penalty function method (or multiplier method) from Chapter 10 to solve a given constrained nonlinear programming problem. The students who have not developed their own computer programs for such techniques often find these problems tedious to do by hand. A considerable effort was made to enable the student with a minimal background (who may not be ready to read some of the more advanced texts) to read and comprehend the material at a moderately advanced level. I have tried to simplify the mathematical proofs and derivations as much as possible including primarily those which I felt were essential and within the student's ability to appreciate. If I felt that the inclusion of a proof would cause too much distraction for the student, then I left the proof out and instead gave a good reference for a complete proof. In some of the exercises the student is asked to supply a proof. Some of these are easily done using the concepts derived in the section. Others are more demanding and are meant to challenge the more advanced student.

Readers are expected to be familiar with concepts from linear algebra such as matrices, vector spaces, subspaces, bases, linear independence, elementary row operations, Gaussian elimination, Gauss-Jordan elimination, and so on. Topics not always found in a beginning linear algebra course are reviewed prior to their use in the text. Moreover, the topics just mentioned along with LU-decompositions and basic solutions of systems of equations are reviewed in Section 1 of Chapter 2. If it is desirable to begin the primal simplex procedure as soon as possible, then Chapter 3 can be started as soon as the students are familiar with the computation of the basic solutions of a system of linear equations. I personally prefer quickly to cover Chapter 1, which is a brief introduction to linear and nonlinear programming problems, to review whatever materials in Section 1 of Chapter 2 are warranted by the individual class (I do defer the review of Gaussian elimination and LU-decompositions until the revised simplex procedure is discussed in Chapter 5), and then to begin the course in earnest in the second section of Chapter 2. In that section some of the concepts from linear algebra are related to new concepts in convex analysis which are ultimately used in the third section of the chapter to describe geometrically the linear programming problem and its solutions. Even though I believe that the algebraic development, via basic solutions of systems of linear equations, of the primal simplex procedure found in Chapter 3 is more efficient than a geometric development of the procedure, I feel that mathematical programming problems are highly geometric in nature and a proper appreciation of this geometry leads to a deeper understanding of the iterative processes which are used to solve them. In any case, Sections 2 and 3 of Chapter 2 should be covered before beginning Chapter 7. Chapters 7 and 8 lay the foundation for the development in Chapters 9 and 10 of iterative procedures for solving nonlinear programming problems.

Chapters 3 through 6 are devoted entirely to the linear programming problem. Chapter 4 deals with duality. A notable feature of Chapter 4 is the inclusion of the linear complementarity problem and Lemke's complementary pivoting algorithm immediately following the complementary slackness theorem. The dual simplex procedure is also found in Chapter 4. Chapter 5 presents several revised simplex procedures as well as the primal-dual procedure. A review of LU-decompositions is needed before beginning Section 4 of Chapter 5. Chapter 6 deals with an important special type of linear programming problem known as a network flow problem. Network flow problems include the transportation, transshipment, and assignment problems. Algorithms for solving these problems are developed using three approaches. One approach is to modify the primal simplex procedure in a manner that takes advantage of the special graphical properties of these problems. Similarly, a second approach is to modify one of the other simplex procedures, in particular the primal-dual procedure, to obtain efficient algorithms for solving these problems. A third approach is the independent development of an iterative procedure directly from the special nature of the problems. Throughout the book, sensitivity and post-optimal analysis are presented whenever it is natural to do so. Exercises dealing with these topics are clearly marked whenever they appear.

Chapter 8 extends the development of convexity begun in Section 2 of Chapter 2 to include the concepts of convex and concave functions. Chapter 8 discusses the Kuhn-Tucker conditions for optimality.

The material in this book has been presented in several senior–graduate level optimization courses at the University of Southern Mississippi over the last twelve years. Based on this experience, a typical course in linear programming could include whatever material is desired from Chapters 1 and 2, Chapter 3, Chapter 4, at least Sections 2 and 5 of Chapter 5, and Chapter 6. Additional topics from Chapters 7 through 10 can be selected depending upon the needs of the students and the interest of the instructor. An emphasis on nonlinear programming can be obtained by covering Sections 2 and 3 of Chapter 2, and then Chapters 7 through 10. I always require my students to write a term paper about some recent application of these procedures. Management Science and other similar journals usually provide an excellent source of such applications.

Much of the artwork in this book was done by the Teaching, Learning, and Resource Center (TLRC) of the University of Southern Mississippi. The artists were Christopher Batchelor, David Gunter, and Linda F. Lanmon. The typing on the final as well as preliminary versions of the manuscript was done by many individuals over the last several years. I would like to express my thanks to each of these individuals and to the TLRC for their work. I would also like to thank the staff of Marcel Dekker, Inc. for making this book possible. Finally, I would like to thank my family for their patience during the preparation of the manuscript.

Melvyn W. Jeter

Contents

MATHEMATICAL
PROGRAMMING

1

An Introduction to
Mathematical Programming

The purpose of this chapter is briefly to introduce the reader to the mathematical programming problem. Both the linear programming problem and the nonlinear programming problem are introduced in Section 1. The nature of the solutions of these problems is discussed in Section 2. The concepts of local and global solutions discussed in that section are analogous to those found in calculus. Actually the concept of a local solution of an optimization problem is not used in the book before Chapter 7. Section 3 briefly introduces the reader to post optimal analysis, parametric programming, and stability of solutions. The final section gives a brief historical outline of the major optimization techniques used in this book.

1. THE MATHEMATICAL PROGRAMMING PROBLEM

Let f be a real valued function of several real variables $x_1, x_2, \ldots, x_n$. The basic *mathematical programming problem* is

$$\text{Minimize (or maximize) } f(x_1, \ldots, x_n) \text{ subject to } (x_1, \ldots, x_n) \in \Omega \quad (1)$$

where Ω is a subset of the domain of f. If $\Omega = \{(x_1, \ldots, x_n): \text{each } x_i \in \mathbf{R}\}$, then Problem (1) is said to be *unconstrained* (here $\mathbf{R}$ denotes the set of real numbers). Problem (1) is *constrained* whenever Ω is a proper subset of $\{(x_1, \ldots, x_n): \text{each } x_i \in \mathbf{R}\}$. When Problem (1) is constrained, the set Ω is usually defined by a system of equations and inequalities which are called *constraints*. the function $f(x_1, \ldots, x_n)$ is known as the *cost* (or *objective*) function, while Ω is often termed the set of *feasible* solutions.

The function $f(x_1, \ldots, x_n)$ is *linear* if it has the form

$$f(x_1, \ldots, x_n) = c_1 x_1 + \cdots + c_n x_n \quad (2)$$

1

where $c_1, \dots, c_n \in \mathbf{R}$. Thus, $c_1, \dots, c_n$ are real constants. The constant c_i is called the *cost coefficient* of the variable x_i. An equality constraint is *linear* when it has the form

$$a_1 x_1 + \cdots + a_n x_n = b \tag{3}$$

where $a_1, \dots, a_n, b \in \mathbf{R}$. A *linear inequality constraint* must have one of the following forms

$$a_1 x_1 + \cdots + a_n x_n \leq b \tag{4}$$

$$a_1 x_1 + \cdots + a_n x_n < b \tag{5}$$

$$a_1 x_1 + \cdots + a_n x_n \geq b \tag{6}$$

$$a_1 x_1 + \cdots + a_n x_n > b \tag{7}$$

where $a_1, \dots, a_n, b \in \mathbf{R}$.

When $f(x_1, \dots, x_n)$ is linear and when Ω is completely determined by a system of linear equations and linear inequalties, then the mathematical programming Problem (1) is called a *linear programming problem*. A mathematical programming problem that is not a linear programming problem is called a *nonlinear programming problem*. Generally, linear programming problems are easier to solve than are nonlinear ones.

A simple exercise (see Exercise 1.1) shows that

$$\max \{f(x_1, \dots, x_n) : (x_1, \dots, x_n) \in \Omega\} = -\min \{-f(x_1, \dots, x_n) : \\ (x_1, \dots, x_n) \in \Omega\} \tag{8}$$

Hence, any maximization problem can be restated as a minimization problem and vice versa. This means that no loss of generality occurs when studying only one type of problem. For example, any algorithm that can be used to minimize $f(x_1, \dots, x_n)$ over Ω can also be used to maximize $f(x_1, \dots x_n)$ over Ω. Simply minimize $-f(x_1, \dots, x_n)$ over Ω. If a minimum occurs when $x_1 = x_1^*, \dots, x_n = x_n^*$, then a maximum of $f(x_1, \dots, x_n)$ also occurs for the same variable values and $f(x_1^*, \dots, x_n^*) = -[-f(x_1^*, \dots, x_n^*)]$ is the corresponding maximum value.

2. EXAMPLES OF MATHEMATICAL PROGRAMMING PROBLEMS

We shall now examine several examples of linear and nonlinear programming problems. We begin with several linear ones.

EXAMPLE 1   A manufacturer produces n products $P_1, P_2, \dots P_n$ from m raw materials $M_1, \dots, M_m$. The following table indicates the amount of each basic raw material that is required for each of the products. On a given day the manufacturer has b_i units of the material M_i available (here $i = 1, \ldots m$). The total revenue is c_j dollars for each unit of the product P_j that is sold. Assuming that the manufacturer can sell all of the items produced, formulate the problem as a linear programming problem in which the total daily revenue is to be maximized.

		Products		
	P_1	P_2	$\cdots$	P_n
M_1	a_{11}	a_{12}	$\cdots$	a_{1n}
M_2	a_{21}	a_{22}	$\cdots$	a_{2n}
Materials				
M_m	a_{m1}	a_{m2}	$\cdots$	a_{mn}

Solution: Let x_j denote the number of units of product P_j to be produced. Notice that the daily revenue that is obtained from these x_j units is $c_j x_j$ dollars. The sum

$$c_1 x_1 + \cdots + c_n x_n \tag{9}$$

represents the total daily revenue that is obtained by producing x_j units of P_j for $j = 1, \ldots, n$. The function (9) must be maximized. Now consider the ith raw material M_i. Since it takes a_{ij} units of M_i to produce a single unit of P_j, then it takes $a_{ij} x_j$ units of M_i to produce x_j units of P_j. The sum

$$a_{i1} x_1 + \cdots + a_{in} x_n \tag{10}$$

represents the total number of units of M_i that is required to produce x_j units of P_j, where $j = 1, \ldots, n$. Since (10) cannot exceed the supply of M_i that is available, it follows that

$$a_{i1} x_1 + \cdots + a_{in} x_n \leq b_i$$

Thus, we have an inequality constraint of the form (4) for each $i = 1, \ldots, m$. Clearly, each $x_i \geq 0$ since we cannot produce a negative number of units of P_i. Summarizing, we must

$$\text{Maximize } c_1 x_1 + \cdots + c_n x_n$$

subject to

$$a_{11}x_1 + a_{12}x_2 + \cdots + a_{1n}x_n \leq b_1$$
$$a_{21}x_1 + a_{22}x_2 + \cdots + a_{2n}x_n \leq b_2$$

(11)

$$a_{m1}x_1 + a_{m2}x_2 + \cdots + a_{mn}x_n \leq b_m$$
$$x_1 \geq 0, x_2 \geq 0, \ldots, x_n \geq 0$$

In this example, we have assumed that the variables x_j can take on fractional as well as integral values. For example, at the end of the day the manufacturer can stop the production process leaving some products only partially finished. When a variable x_j is allowed to take on all fractional as well as integral values within its range of variation, it is called a *continuous* variable. If a variable x_j must take on values from a discrete set (i.e., a countably infinite set) of values, then the variable is said to be a *discrete* variable and the problem is said to be a *discrete* problem. Discrete optimization problems commonly have their variables restricted to the set of nonnegative integers. Such a problem is called an *integer programming problem*. This book deals primarily with optimization problems which possess continuous variables. However, the procedures developed in Chapter 6 will produce integral solutions under relatively simple conditions.

Each constraint in Problem (11) can be converted to an equality constraint [i.e., form (3)] by adding a nonnegative variable, called a *slack* variable to the appropriate side of the inequality. For example, let

$$x_{n+1} = b_1 - a_{11}x_1 - \cdots - a_{1n}x_n$$

Then the first inequality in the constraints of problem (11) can be restated as

$$a_{11}x_1 + \cdots + a_{1n}x_n + x_{n+1} = b_1$$

The new slack variable x_{n+1} is assigned a cost coefficient of $c_{n+1} = 0$. If slack variables $x_{n+2}, \ldots, x_{n+m}$ are defined in a similar fashion for the second through the mth constraints, and if the corresponding cost coefficients are $c_{n+1} = \cdots = c_{n+m} = 0$, then Problem (11) can be reformulated as

$$\text{Maximize } c_1 x_1 + \cdots + c_n x_n + c_{n+1} x_{n+1} + \cdots + c_{n+m} x_{n+m}$$
$$= c_1 x_1 + \cdots + c_n x_n$$

subject to

$$a_{11}x_1 + a_{12}x_2 + \cdots + a_{1n}x_n + x_{n+1} \qquad\qquad = b_1$$
$$a_{21}x_1 + a_{22}x_2 + \cdots + a_{2n}x_n \qquad + x_{n+2} \qquad = b_2$$

.

. (12)

.

$$a_{m1}x_1 + a_{m2}x_2 + \cdots + a_{mn}x_n \qquad + x_{n+m} \qquad = b_m$$
$$x_1 \geq 0, \ldots, x_n \geq 0, x_{n+1} \geq 0, \ldots, x_{n+m} \geq 0$$

It is shown in Section 3 of Chapter 2 that problems (11) and (12) are equivalent, i.e., they have the same solutions.

In Problems (11) or (12), the constraints $x_1 \geq 0, \ldots, x_n \geq 0$ are often referred to as *restrictions* rather than constraints. Thus, all of the variables in this problem are restricted to be nonnegative. This is a common restriction in mathematical programming problems and many of the procedures that will be developed in this book are done under the hypothesis that the variables are nonnegative. This also does not result in any loss of generality since any unrestricted variable x_i can be expressed as the difference of two nonnegative variables, i.e.,

$$x_i = y_i - z_i$$

where $y_i \geq 0$ and $z_i \geq 0$. The variable x_i is simply replaced by the difference $y_i - z_i$ in the cost function and each constraint. An alternate method of dealing with unrestricted variables can be found in Exercise 1.2. Finally, it should be noted that restrictions other than nonnegativity are common. For example, market conditions, such as consumer demand, could require that at least 10 but not more than 100 units of product P_1 be produced in Example 1. Then the variable x_1 would have the following restrictions

$$10 \leq x_1 \leq 100$$

which can also be regarded as the two restrictions $10 \leq x_1$ and $x_1 \leq 100$.

A method for solving problems of this type is found in Chapter 3. Additional methods are found in Chapters 4, 5, and 6.

EXAMPLE 2   Suppose that the latest scientific studies indicate that cattle need certain amounts $b_1, \ldots, b_m$ of nutrients $N_1, \ldots, N_m$ respectively. Moreover, these nutrients are currently found in n commercial feed materials $F_1, \ldots, F_n$ as indicated in the chart below where a_{ij} is the number of units of nutrient N_i per pound of feed material F_j. Each pound of F_j costs the rancher c_j dollars. How can a rancher supply these minimal nutrient requirements to his prize bull while minimizing his feed bill?

		Feed materials			
		F_1	F_2	$\cdots$	F_n
	N_1	a_{11}	a_{12}	$\cdots$	a_{1n}
Nutrients	N_2	a_{21}	a_{22}	$\cdots$	a_{2n}
	.	.	.	$\cdots$	.
	.	.	.	$\cdots$	.
	.	.	.	$\cdots$	.
	N_m	a_{m1}	a_{m2}	$\cdots$	a_{mn}

Solution: Most likely, the rancher can minimize his costs by mixing feeds. So for $j = 1, \ldots, n$, let x_j denote the number of pounds of F_j to be used in the mixture. Then the cost of the mixture that must be minimized is

$$\sum_{j=1}^{n} c_j x_j \tag{13}$$

where (13) is an efficient way to denote the sum $c_1 x_1 + \cdots + c_n x_n$. Since x_j pounds of F_j contains $a_{1j} x_j$ units of N_1, then the total number of units of N_1 in the mixture is

$$a_{11} x_1 + a_{12} x_2 + \cdots + a_{1n} x_n \tag{14}$$

But (14) must be no less than b_1. Hence

$$a_{11} x_1 + a_{12} x_2 + \cdots + a_{1n} x_n \geq b_1 \tag{15}$$

A constraint similar to (15) holds for each nutrient N_i, where $i = 1, \ldots, m$. Hence, the rancher must solve the following linear programming problem:

$$\text{Minimize } c_1 x_1 + \cdots + c_n x_n$$

subject to

$$a_{11} x_1 + a_{12} x_2 + \cdots + a_{1n} x_n \geq b_1$$
$$a_{21} x_1 + a_{22} x_2 + \cdots + a_{2n} x_n \geq b_2$$
$$\vdots$$
$$a_{m1} x_1 + a_{m2} x_2 + \cdots + a_{mn} x_n \geq b_m$$
$$x_1 \geq 0, x_2 \geq 0, \ldots, x_n \geq 0$$

Notice that the variables cannot be negative. (Why?) The rancher may prefer to formulate the problem using equality constraints rather than the inequality constraints (15). If that is the case, define

$$x_{n+i} = a_{i1}x_1 + a_{i2}x_2 + \cdots + a_{in}x_n - b_i$$

for each $i = 1, \ldots, m$. The variables, x_{n+i}, are called *surplus* variables. Notice that each surplus variable is nonnegative. Next let $c_{n+i} = 0$ for $i = 1, \ldots, m$. Then the problem can be expressed in the following equivalent form.

Minimize $c_1x_1 + \cdots + c_nx_n + c_{n+1}x_{n+1} + \cdots + c_{n+m}x_{n+m} = c_1x_1 + \cdots + c_nx_n$

subject to

$$
\begin{aligned}
a_{11}x_1 + a_{12}x_2 + \cdots + a_{1n}x_n \quad -x_{n+1} &= b_1 \\
a_{21}x_1 + a_{22}x_2 + \cdots + a_{2n}x_n \quad\quad\quad -x_{n+2} &= b_2 \\
&\ \ \vdots \\
a_{m1}x_1 + a_{m2}x_2 + \cdots + a_{mn}x_n \quad\quad\quad\quad\quad\quad -x_{n+m} &= b_m
\end{aligned}
$$

$$x_1 \geq 0, \ldots, x_n \geq 0, x_{n+1} \geq 0, \ldots, x_{n+m} \geq 0$$

EXAMPLE 3 Suppose that a small chemical company produces one chemical product which it stores in m storage facilities and sells in n retail outlets. The storage facilities are located in different communities. The shipping costs from storage facility F_i to retail outlet O_j is c_{ij} dollars per unit of the chemical product. On a given day, each outlet, O_j, requires exactly d_j units of the product. Moreover, each facility F_i has s_i units of the product available for shipment. The problem is to determine how many units of the product should be shipped from each storage facility to each outlet in order to minimize the total daily shipping costs.

Solution: Let x_{ij} denote the number of units of the product to be shipped from storage facility F_i to outlet O_j. Then it costs $c_{ij}x_{ij}$ to ship x_{ij} units from F_i to O_j. The total shipping cost which must be minimized is

$$\sum_{i=1}^{m}\sum_{j=1}^{n} c_{ij}x_{ij} = \sum_{i=1}^{m}\left(\sum_{j=1}^{n} c_{ij}x_{ij}\right) \tag{16}$$

where (16) is the sum of all the products $c_{ij}x_{ij}$. For a given i, the expression

$$\sum_{j=1}^{n} x_{ij} = x_{i1} + x_{i2} + \cdots + x_{in}$$

represents the total number of units of the product that is to be shipped to all of the retail outlets from F_i. Since F_i cannot ship more units than it has, it follows that

$$\sum_{j=1}^{n} x_{ij} \le s_i \tag{17}$$

Expression (17) must hold for each storage facility, i.e., for each $i = 1, \dots, m$. Also, the total number of units that outlet O_j receives from all the storage facilities is

$$x_{1j} + x_{2j} + \cdots + x_{mj} \tag{18}$$

However, outlet O_j requires exactly d_j units. Hence, it follows from expression (18) that

$$\sum_{i=1}^{m} x_{ij} = d_j \tag{19}$$

Since there are n outlets, expression (19) must hold for $j = 1, \dots, n$ Thus, the problem can be formulated as the following linear programming problem:

$$\text{Minimize} \sum_{i=1}^{m} \sum_{j=1}^{n} c_{ij} x_{ij} \text{ subject to}$$

$$\sum_{j=1}^{n} x_{ij} \le s_i, \text{ for } i = 1, \dots, m$$

$$\sum_{i=1}^{m} x_{ij} = d_j, \text{ for } i = 1, \dots, n$$

$$\text{and all } x_{ij} \ge 0$$

This example belongs to a class of linear programming problems known as *transportation* problems. A variation of the transportation problem is presented in the next example. It is called an *assignment* problem.

EXAMPLE 4 A mattress company wishes to introduce a new product to its customers. The company has four salespeople and four sales districts to be worked. Based on past sales experience the company can estimate the relative sales productivity ratings for each salesperson S_i in each of the sales districts D_j. These ratings are summarized in the following table. Assuming that each salesperson must be assigned to exactly one district, determine the best possible assignment of salespeople to sales districts from the company's point of view.

		Sales districts			
		D_1	D_2	D_3	D_4
	S_1	5	10	8	2
Salespeople	S_2	3	4	2	6
	S_3	9	5	4	5
	S_4	5	5	6	4

Solution: Define x_{ij} by

$$x_{ij} = \begin{cases} 1, & \text{if salesperson } S_i \text{ is assigned to district } D_j \\ 0, & \text{otherwise,} \end{cases} \tag{20}$$

where $i = 1, 2, 3, 4$ and $j = 1, 2, 3, 4$. The value to the company of such an assignment is

$$\begin{aligned} 5x_{11} + 10x_{12} + 8x_{13} + 2x_{14} + 3x_{21} + 4x_{22} + 2x_{23} + 6x_{24} + 9x_{31} \\ + 5x_{32} + 4x_{33} + 5x_{34} + 5x_{41} + 5x_{42} + 6x_{43} + 4x_{44} \end{aligned} \tag{21}$$

Since each salesperson must be assigned exactly one district, it follows that

$$\sum_{j=1}^{4} x_{ij} = 1, \text{ for } i = 1, 2, 3, 4 \tag{22}$$

Also, the constraints

$$\sum_{i=1}^{4} x_{ij} = 1, \text{ for } j = 1, 2, 3, 4 \tag{23}$$

insure that each district is assigned one salesperson. Thus, the company must solve the following problem:

Maximize the function (21) subject to the eight constraints found in (22) and (23), and subject to each of the variable restrictions in (20).

Clearly, this is a discrete programming problem since the variables are restricted to the two values of 0 and 1. However, the procedure given in Section 3 of Chapter 6 will produce integral solutions when the data in the constraints are integers (as is the case in this assignment problem). Thus, we can solve the continuous variable linear programming problem

$$\begin{aligned} \text{Maximize } 5_{11} + 10x_{12} + 8x_{13} + 2x_{14} + 3x_{21} + 4x_{22} + 2x_{23} + 6x_{24} + 9x_{31} \\ + 5x_{32} + 4x_{33} + 5x_{34} + 5x_{41} + 5x_{42} + 6x_{43} + 4x_{44} \end{aligned}$$

subject to

$$x_{11} + x_{12} + x_{13} + x_{14} = 1$$
$$x_{21} + x_{22} + x_{23} + x_{24} = 1$$
$$x_{31} + x_{32} + x_{33} + x_{34} = 1$$
$$x_{41} + x_{42} + x_{43} + x_{44} = 1$$
$$x_{11} + x_{21} + x_{31} + x_{41} = 1$$
$$x_{12} + x_{22} + x_{32} + x_{42} = 1$$
$$x_{13} + x_{23} + x_{33} + x_{43} = 1$$
$$x_{14} + x_{24} + x_{34} + x_{44} = 1$$
$$\text{and each } 0 \le x_{ij} \le 1$$

EXAMPLE 5   A Texas oil pipeline company has n pumping stations that are connected by pipelines. Oil enters the system at pumping station S_1, is then pumped through the system, and finally leaves the system at pumping station S_n. Every drop of oil that enters at S_1 eventually leaves the system at S_n. However, there are many routes that the oil can take as it is pumped from S_1 to S_n. Some oil will be pumped through one route while some will travel a different route. The pipeline from station S_i to station S_j will be denoted by $a(i,j)$ and its capacity by u_{ij}. Oil flows from S_i to S_j through $a(i,j)$. A different pipeline $a(j,i)$ may connect S_j to S_i so that oil can also flow in the other direction. The problem is to determine the maximum amount of oil that can be pumped through the system.

Solution: Let x_{ij} be the amount of oil to be pumped from S_i to S_j. Clearly, $0 \le x_{ij} \le u_{ij}$. For convenience, we shall assume that each pipeline $a(i,j)$ exists. When $a(i,j)$ does not in fact exist, then we shall simply let $u_{ij} = 0$ and, hence, force $x_{ij} = 0$. Since all the oil that enters at S_1 must leave at S_n, it follows that the same is true at all of the intermediate pumping stations. That is, all of the oil that enters S_j must leave S_j. Thus,

$$\sum_{i=1}^{n} (x_{ji} - x_{ij}) = 0, \text{ where } j \notin \{1, n\}$$

Let f denote the amount of oil that enters at S_1, flows through the system, and leaves at S_n. Then

$$\sum_{i=1}^{n} (x_{1i} - x_{i1}) = f \tag{24}$$

and

$$\sum_{i=1}^{n} (x_{ni} - x_{in}) = -f \tag{25}$$

Constraint (24) forces all of the oil that enters S_1 into the network of pumping stations and pipelines, while (25) ensures that all the oil in the network will eventually leave the system. We can now model our problem as a linear programming problem. Namely,

Maximize f subject to

$$\sum_{i=1}^{m} (x_{1i} - x_{i1}) = f$$

$$\sum_{i=1}^{m} (x_{ji} - x_{ij}) = 0, \text{ where } j \notin \{1,n\}$$

$$\sum_{i=1}^{n} (x_{ni} - x_{in}) = -f$$

and each $0 \leq x_{ij} \leq u_{ij}$

This is an example of a network programming problem. Chapter 6 is devoted to such problems.

EXAMPLE 6 A flour company packs several brands of flour in many sizes of containers. On a given day the company wants to produce as many ten pound bags of flour of brands A and B as possible. The company has three production lines (machines and workers) which can produce the brands A and B. The number of ten pound sacks of flour which can be produced daily by each of the production lines is summarized in the following table.

		Production lines			
		L_1	L_2	L_3	
Flour brands	A	15,000	10,000	12,000	Daily number of 10-pound sacks produced
	B	12,000	13,500	14,500	

Each production line L_i must be assigned to exactly one product brand. Formulate this problem as a mathematical programming problem.
Solution: Define variables x_1, x_2, x_3, w_1, w_2 and w_3 as follows:

$$x_i = \begin{cases} 1 & \text{if production line } L_i \text{ is assigned to brand } A \\ 0 & \text{otherwise} \end{cases}$$

and

$$w_i = \begin{cases} 1, & \text{if production line } L_i \text{ is assigned to brand } B \\ 0, & \text{otherwise} \end{cases}$$

Since each production line must be assigned to exactly one brand of flour, we shall impose the constraints

$$x_1 w_1 = 0, x_2 w_2 = 0, \text{ and } x_3 w_3 = 0 \tag{26}$$

Also, the variables are all nonnegative. Hence, the constraints in (26) can be compressed into the single constraint

$$\sum_{i=1}^{3} x_i w_i = x_1 w_1 + x_2 w_2 + x_3 w_3 = 0 \tag{27}$$

To insure that each production line is assigned to a job, the following constraints are added:

$$x_1 + w_1 = 1$$
$$x_2 + w_2 = 1$$
$$x_3 + w_3 = 1$$

For any such assignment of tasks, the total daily production of ten pound sacks of flour is

$$15000x_1 + 10000x_2 + 12000x_3 + 12000w_1 + 13500w_2 + 14500w_3$$

For example, the assignment

$$x_1 = 1, w_1 = 0$$
$$x_2 = 0, w_2 = 1$$
$$x_3 = 1, w_3 = 0$$

will produce 40,500 10-pound sacks of flour of brands A and B. The company can determine its optimal production by solving the following nonlinear programming problem:

Maximize $15000x_1 + 10000x_2 + 12000x_3 + 12000w_1 + 13500w_2 + 14500w_3$

subject to

$$x_1 + w_1 = 1$$
$$x_2 + w_2 = 1$$
$$x_3 + w_3 = 1$$

$$\sum_{i=1}^{3} x_i w_i = 0$$

$0 \le x_i \le 1$ and $0 \le w_i \le 1$ for $i = 1,2,3$

Clearly one does not need mathematical programming to solve this problem. It has an obvious solution. However, as formulated, the problem is a nonlinear programming problem due to the presence of constraint (27). Actually (27) is a special type of a nonlinear constraint called a *complementary constraint*. Each pair of variables x_i and w_i are said to be *complementary*. Complementary constraints and variables are encountered in Chapters 4 and 10.

Some other examples of nonlinear programming problems follow.

EXAMPLE 7   The following nonlinear programming problem is some-times known as a *quadratic programming problem*:

$$\text{Maximize } c_1 x_1 + \cdots + c_n x_n + \sum_{i=1}^{n} \sum_{j=1}^{n} b_{ij} x_i x_j$$

subject to

$$a_{11}x_1 + a_{12}x_2 + \cdots + a_{1n}x_n = b_1$$
$$a_{21}x_1 + a_{22}x_2 + \cdots + a_{2n}x_n = b_2 \qquad (28)$$
$$\cdot$$
$$\cdot$$
$$\cdot$$
$$a_{m1}x_1 + a_{m2}x_2 + \cdots + a_{nn}x_n = b_m$$
$$x_1 \ge 0, x_2 \ge 0, \ldots, x_n \ge 0$$

Here, each c_i, b_j and a_{ij} is a constant. Notice, in problem (28), that the objective function is nonlinear. It is said to be a *quadratic* function. Usually, each $b_{ij} = b_{ji}$ in such a function (see Section 1 of Chapter 7).

This type of programming problem arises in the *portfolio selection problem*. A portfolio is a collection of investments in securities such as stocks and bonds. In the portfolio selection problem, an investor wishes to invest a proportion x_i of his assets in a security S_i, where $i = 1, \ldots, n$. The constants c_i represent the expected rate of return on the ith investment. The expression $\sum_{i=1}^{n} \sum_{j=1}^{n} b_{ij} x_i x_j$ represents the risk involved in the portfolio of investments. The investor wishes to maximize his return while minimizing his risks. Clearly, the variables are nonnegative. Moreover,

$x_1 + \cdots + x_n = 1$. Thus the portfolio selection problem has the form

$$\text{Maximize } c_1 x_1 + \cdots + c_n x_n - \sum_{i=1}^{n} \sum_{j=1}^{n} b_{ij} x_i x_j$$

subject to

$$x_1 + \cdots + x_n = 1$$
$$x_1 \geq 0, \ldots, x_n \geq 0$$

EXAMPLE 8   Minimize $7x_1^2 + 4x_2^2 + 61x_1$ subject to $x_1, x_2 \in \mathbf{R}$. This is an example of an unconstrained nonlinear programming problem. Notice that the cost function is not linear.

EXAMPLE 9   Minimize $2x_1 + x_2 - x_3$ subject to

$$x_1 + \sin x_2 \leq 3$$
$$x_1 + x_2 + x_3 = 1$$

Here the cost function is linear, but one of the constraints is not. Thus, this is a nonlinear programming problem.

3.   GLOBAL AND LOCAL SOLUTIONS

The standard Euclidean distance between two points $x = (x_1, \ldots, x_n)$ and $\bar{x} = (\bar{x}_1, \ldots, \bar{x}_n)$ will be denoted by $\|x - \bar{x}\|$ and defined by

$$\|x - \bar{x}\| = \sqrt{\sum_{i=1}^{n} (x_i - \bar{x}_i)^2}$$

Let ε be a positive real number and define $N_\varepsilon(\bar{x})$ by

$$N_\varepsilon(\bar{x}) = \{x : \|x - \bar{x}\| < \varepsilon\}$$

Then $N(\bar{x})$ is called a *neighborhood* of the point $\bar{x}$.

A function f, which is defined over the set Ω, is said to have a *relative* (or *local*) *minimum* at $\bar{x} \in \Omega$ whenever these exists an $\varepsilon > 0$ so that $f(\bar{x}) \leq f(x)$ whenever $x \in \Omega \cap N_\varepsilon(\bar{x})$. When $f(\bar{x}) < f(x)$ for all $x \in \Omega \cap N_\varepsilon(\bar{x})$, then f has a *strict relative minimum* (or a *strict local minimum*) at $\bar{x}$. A *global minimum* of f over Ω occurs whenever $f(\bar{x}) \leq f(x)$ for all $x \in \Omega$. When this last inequality is strict, i.e., when $f(\bar{x}) < f(x)$ for all $x \in \Omega$, then we say that f has a *strict global minimum* at $\bar{x}$. Similar definitions hold for a *relative maximum*, *strict relative maximum*, *global maximum*, and a *strict global maximum*.

For example, consider the problem of finding the optimal values of the function that is represented in Figure 1.

The function clearly does not have a global maximum. It does, however, have a relative (or local) maximum when $x = 2$. Notice that the function possesses relative minimums when $x = 0$ and $x = 4$. Moreover, a global minimum occurs at $x = 4$.

When it comes to comparing the nature of the solutions of linear and nonlinear programming problems, there is one great difference. An optimal solution of a linear programming problem is always a global solution. Unfortunately, the same is not true for nonlinear programming problems. Often we will have to be content with finding only a relative optimal solution of a nonlinear problem.

Often we will make use of the following topological concepts when working with nonlinear programming problems. The *interior* of a set Ω is denoted by int (Ω) and defined by

$$\text{Int}\,(\Omega) = \{\, x \in \Omega : N_\varepsilon\,(x) \subseteq \Omega \text{ for some } \varepsilon > 0\}$$

The set Ω is *open* if and only if $\Omega = \text{Int}(\Omega)$, i.e., every point of Ω is an interior point. The *exterior*, $\text{Ext}(\Omega)$, is defined by

$$\text{Ext}\,(\Omega) = \{\, x : N_\varepsilon\,(x) \subseteq E_n \setminus \Omega \text{ for some } \varepsilon > 0\}$$

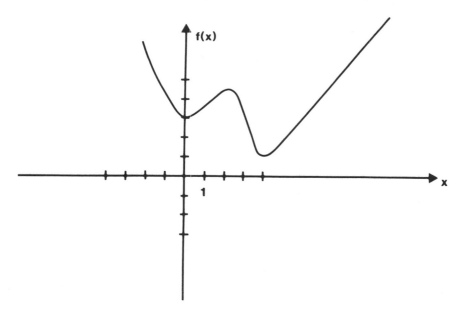

Figure 1   Global and local solutions

where $E_n = \{x = (x_1, \ldots, x_n)$: each $x_i \in \mathbf{R}\}$ and $E_n \backslash \Omega = \{x \in E_n : x \notin \Omega\}$.

A point x is a *boundary* point of Ω if and only if every neighborhood, $N_\varepsilon(x)$, of x contains a point of Ω and a point that is not in Ω. The collection of all boundary points of Ω is denoted by $Bd(\Omega)$. A set Ω is *closed* whenever it contains its boundary, i.e., whenever $Bd(\Omega) \subseteq \Omega$. A point x is an *accumulation* point of Ω if and only if every neighborhood, $N_\varepsilon(x)$, of x contains at least one point of Ω different from x. Some basic properties concerning these concepts are found in Exercises 1.9 – 1.10.

4. POST-OPTIMAL ANALYSIS, PARAMETRIC PROGRAMMING, AND STABILITY

Very often the data in a mathematical programming problem, i.e., constants such as a_{ij}, b_i and c_i, are only estimates or measurements that can be expected to change as conditions (such as market conditions, environmental conditions, etc.) change or as better data or measurements develop. For instance, in Example 1 the number of daily units b_i of material M_i that is available for production use on a given day can be reasonably expected to vary from day to day. Or in Example 2, the a_{ij} may change with an improvement in feed F_i or the replacement of F_i with a new feed. In Example 3, the cost coefficients are subject to change as transportation costs, such as oil and gas, increase. Also, additional constraints changes may appear as conditions change. In Example 2, a new feed could be added to the existing group of feeds on the market. This would amount to adding a new variable and column to the left hand side of the constraints. A new raw material in Example 1 would change or add another constraint to the system of constraints. Often it is possible to solve these new problems efficiently by properly modifying or extending the solutions of the original problems, and thus avoid the work and expense of completely resolving the new problems. The process of using an optimal solution for an existing problem to obtain an optimal solution for a modified problem is called *post-optimal* (or *sensitivity*) *analysis*.

On the other hand, expected changes sometimes can be built directly into the original model of the problem. This is usually done with the addition of an extra variable (or variables) called a parameter. To illustrate consider Example 3. If the product that is produced by the chemical company is good, then the consumer demand for the product could reasonably be expected to increase with time. If previous consumer experience indicates that consumer demand at retail outlet O_j will increase by $\bar{d}_j$, where $\bar{d}_j > 0$, units per day, then the consumer demand at O_j will be $d_j + t\bar{d}_j$ after t days. Thus, Problem (20) would become

$$\text{Minimize} \sum_{i=1}^{m} \sum_{j=1}^{n} c_{ij} x_{ij} \text{ subject to}$$

$$\sum_{j=1}^{n} x_{ij} \leq s_i, \text{ for } i = 1, \ldots, m$$

$$\sum_{i=1}^{m} x_{ij} = d_j + t\bar{d}_j \text{ for } j = 1, \ldots, n$$

$$t \geq 0 \text{ and all } x_{ij} \geq 0$$

(29)

Problem (29) is known as a *parametric programming* problem. For another example of a parametric programming problem, suppose in Example 3 that rising oil and gas prices indicate that the shipping cost from storage facility F_i to retail outlet O_j can be expected to increase at the rate of $\bar{c}_{ij}$ dollars per day. Then after t days the shipping cost from F_i to O_j will be $c_{ij} + t\bar{c}_{ij}$. So, the correct parametric formulation of the problem is

$$\text{Minimize} \sum_{i=1}^{m} \sum_{j=1}^{n} (c_{ij} + t\bar{c}_{ij}) x_{ij} \text{ subject to}$$

$$\sum_{j=1}^{n} x_{ij} \leq s_i, \text{ for } i = 1, \ldots, m$$

$$\sum_{i=1}^{m} x_{ij} = d_j, \text{ for } j = 1, \ldots, n$$

$$\text{and all } x_{ij} \geq 0$$

Rather than deal with post-optimal analysis or parametric programming in a separate chapter, this book presents these topics when it seems natural and convenient to do so. The words "post-optimal analysis" or "parametric programming" will announce the forthcoming of these topics in the body of the text and in the exercises.

A mathematical programming problem is *model stable* whenever small changes in the data, such as in a_{ij}, b_i and c_i, lead to small changes in $x_1^*, \ldots, x_n^*$ and $f(x_1^*, \ldots, x_n^*)$, where $x_1^*, \ldots x_n^*$ and $f(x_1^*, \ldots, x_n^*)$ represents an optimal solution of the problem. A problem is *model unstable* whenever it is not model stable. Small changes in the data of an unstable problem can lead to a problem that has a drastically different optimal solution – or to an infeasible problem, i.e., a problem having an empty set of feasible solutions. Frequently, problems which are model unstable possess redundant constraints. For example, if a feasible problem has constraints

$$g_i(x_1, \ldots, x_n) = b_i, \text{ for } i = 1, \ldots, m$$

and

$$\sum_{i=1}^{m} \lambda_i g_i (x_1, \dots, x_n) = \sum_{i=1}^{m} \lambda_i b_i \tag{30}$$

where $\lambda_1, \dots, \lambda_m \in \mathbf{R}$, then a slight change, ε, in the right side of the redundant constraint (30) leads to an infeasible problem having constraints

$$g_i (x_1, \dots, x_n) = b_i, \text{ for } i = 1, \dots, m$$

and

$$\sum_{i=1}^{m} \lambda_i g_i (x_1, \dots, x_n) = \varepsilon + \sum_{i=1}^{m} \lambda_i b_i$$

Since the data constants are normally numerical estimates based on observation and experience, it is clearly desirable to formulate the problem, when possible, so that it is model stable. The value of an optimal solution of a problem that is model unstable is questionable. Model stability is discussed for the linear programming problem in Section 3 of Chapter 4.

A second type of stability deals with the algorithms which are used to solve the problems. An algorithm is *numerically stable* provided the round off error produced by the algorithm stays within acceptical bounds. The algorithms presented in Sections 3 and 4 of Chapter 5 are examples of numerically (stable) superior algorithms. A quick example from linear algebra is in order. When solving a system of linear equations, the method of partial elimination is numerically superior to Gaussian elimination. (These are both discussed in Section 1 of Chapter 2.)

Exercises

1.1   Consider the two problems

Maximize $f(x_1, \dots, x_n)$ subject to $(x_1, \dots, x_n) \in \Omega$

and

Minimize $-f(x_1, \dots, x_n)$ subject to $(x_1, \dots, x_n) \in \Omega$

Show that

$$f(x_1^*, \dots, x_n^*) = \max \{f(x_1, \dots, x_n): (x_1, \dots, x_n) \in \Omega\}$$

if and only if

$$-f(x_1^*, \dots, x_n^*) = \min \{-f(x_1, \dots, x_n): (x_1, \dots, x_n) \in \Omega\}$$

1.2   Sometimes it is possible to eliminate a few unrestricted variables when they are contained in equality constraints. If possible, use the equality constraints to express the unrestricted variables in terms of the nonnegative variables. Then substitute the result into the remaining constraints and cost function. Verify that one determines a solution of the original problem by optimizing the modified cost function subject to the modified constraints. Use this technique to reformulate the following problems as problems with nonnegative variables.

(a) Minimize $x_1 + 6x_2 - x_3$ subject to $7x_1 + x_2 - x_3 = 6$, $3x_1 + x_2 + 2x_3 \le 6$, $x_1 \ge 0$ and $x_2 \ge 0$.

(b) Minimize $x_1x_2 + x_3x_4$ subject to $x_1 + 2x_2 - x_3 + x_4 = 8$, $3x_1 - x_2 + 6x_3 = 7$, $x_1 + x_2 + x_3 + x_4 \le 20$, $x_3 \ge 0$ and $x_4 \ge 0$.

1.3   Reformulate the following problem into one having only constraints of forms (4) and (6) by replacing each equality constraint with two inequality constraints.

$$\text{Minimize } x_1x_2 + x_3^2 \text{ subject to}$$
$$x_1 + x_2 - 7x_3 = 4$$
$$5x_1 - x_2 \quad\;\; = 6$$
$$x_1 \ge 0, x_2 \ge 0, x_3 \ge 0$$

1.4   Rewrite each of the problems that follow as a minimization problem with equality constraints and nonnegative variables.

(a) Maximize $x_1 + x_2 - x_3$ subject to $2x_1 + 6x_2 \ge 4$, $3x_2 + x_3 \le 8$, $x_1 \ge 0$ and $x_2 \ge 0$.

(b) Maximize $6x_1x_2 + x_3^2$ subject to $x_1 + 2x_2 \le 4$ and $x_1 + x_2 + x_3 \le 6$.

(c) Minimize $2x_1 + 3x_2 + x_4$ subject to $x_1 + x_2 + x_3 = 1$, $3x_1 + x_2 + 3x_4 \le 8$, $x_2 \ge 0$, $x_3 \ge -8$, and $x_4 \ge 0$ Hint: Treat the restriction $x_3 \ge -8$ as a constraint of form (6).

1.5   Use slack and surplus variables to convert each of the following problems into problems having all equality constraints.

(a) Minimize $7x_1 + x_2 + 2x_3$ subject to $3x_1 + x_2 \le 8$, $x_1 + x_2 + 2x_3 \ge 4$, $x_1 \ge 0$, $x_2 \ge 0$, and $x_3 \ge 0$.

(b) Maximize $x_1^2 + x_2^2 + x_3^2 + x_4^2$ subject to $x_1 + x_2 \ge 4$, $x_2 + x_3 \ge 8$, $x_3 + x_4 \ge 1$, $x_1 \ge 0$, $x_2 \ge 0$, $x_3 \ge 0$ and $x_4 \ge 0$.

(c) Minimize $x_1 + x_2 + x_3 + x_4$ subject to $x_1x_3 + x_2x_4 = 0$, $x_1^2 + x_2x_3 \ge 5$, $x_1 \ge 0$, $x_2 \ge 0$, $x_3 \ge 0$, and $x_4 \ge 0$.

1.6   Reformulate the problem

$$\text{Maximize } x_1 + 2x_2 + x_3 + 2x_4 \text{ subject to}$$

$$2x_1 + 3x_2 \qquad\qquad \geq 8$$
$$x_2 + 2x_3 \qquad \leq 9$$
$$x_1 \qquad + x_3 + 2x_4 = 10$$
$$x_1 \geq 0 \text{ and } x_4 \geq 0$$

as a minimization problem possessing only equality constraints and non-negative variables.

1.7   Express each of the following problems as a mathematical programming problem.

(a) A furniture manufacturer makes three products P_1, P_2, and P_3. The profit on P_1 is $50, on P_2 is $25, and on P_3 is $40. The three products are constructed from labor, wood, plastic, and glue. On a given day, the manufacturer has 800 hours of labor available, 4200 board feet of lumber, 150 sheets of plastic, and 150 pints of glue. The following table shows the amount of each material that is required for each product.

		Labor	Lumber	Plastic	Glue
	P_1	10 hours	61 board feet	none	2 pints
Product	P_2	5 hours	15 board feet	1 sheet	1 pint
	P_3	4 hours	10 board feet	2 sheets	2 pints

How many units of each product should be manufactured in order to maximize the profits?

(b) A mother wishes to mix two brands, brand A and brand B, of cereal to feed to her child. In doing so, she wants to insure that the child receives a daily minimum of 10 mg of vitamin B_1 and 300 mg of vitamin C, while at the same time minimizing her costs. Each ounce of brand A costs 5¢ and each ounce of brand B cost 7¢. Each ounce of brand A contains .20 mg of vitamin B_1 and 30 mg of vitamin C. Also, an ounce of brand B contains 1.5 mg of vitamin B_1 and 35 mg of vitamin C. How should she mix the two cereals?

(c) A vitamin pill company believes that there are thousands of mothers similar to the one in (b). Hence, it decides to manufacture a vitamin pill that will satisfy the daily demands of the potential customers. The company believes that the public would prefer to take such a pill rather than mix cereals! The company's problem is to determine the selling price of the new vitamin tablet. It decides to let x_1 and x_2 denote the selling price per mg of vitamin B_1 and C, respectively. Since the tablet contains 10 mg of vitamin B_1 and 150 mg of vitamin C, the tablet would then sell for $10x_1 + 150x_2$.

Based on these prices, x_1 and x_2, the value of the vitamins in an ounce of brand A cereal is $.20x_1 + 30x_2$, while the value of the vitamins in an ounce of brand B is $1.5x_1 + 35x_2$. The company believes that a smart shopper will be more likely to purchase the vitamin tablet whenever these values do not exceed the corresponding cereal prices. Express this as a mathematical programming problem.

(d) An import car company has warehouses in cities A, B and C and supplies four different dealers D_1, D_2, D_3 and D_4. The cost in dollars of transporting a car from a given warehouse to a given dealer is found in the following table.

		Dealers			
		D_1	D_2	D_3	D_4
	A	225	150	375	140
Warehouses	B	105	110	400	200
	C	200	450	310	105

Currently, D_1 needs 100, D_2 needs 50, D_3 needs 65, and D_4 needs 75 cars. Further, warehouse A has 295, B has 400, and C has 300 cars. How should the cars be shipped in order to minimize the shipping costs?

(e) A farmer in arid eastern New Mexico has a corn field on a section of land (a *section* of land is a one-mile square area of land). Unfortunately, a state highway runs through the field. The highway runs from the northeast corner to the southwest corner of the field and, thus, divides the field into a pair of identical isoceles triangular-shaped fields. The farmer wishes to irrigate the fields. Unfortunately, his irrigation equipment operates like many lawn sprinklers by throwing the water in a circular path around the sprinkler. By increasing the water pressure the farmer can make the irrigation equipment throw the water as far as needed. The problem is that the state will not allow him to spray water on the highway. Moreover, he cannot stay with the irrigation equipment. He wants to set it up and leave. Where should he set the irrigation equipment and how far should the water be thrown in order to maximize the areas to be irrigated? Experience has already taught the farmer that the equipment must be set an equal distance from the two non-highway sides of each field (Hint: Since the fields are identical, formulate the problem for one of the fields.)

(f) The value of a variable y depends on a variable x. Past observations have produced the following relationships between x and y

x_i	-3	-2	-1	0	1	2	3	4	5	6	7	8
y_i	-1.98	1.01	4.02	6.98	10.03	14	15.8	18.67	22.4	25	27.71	30.67

It is expected that y in general will vary almost linearly with respect to x, i.e., $y = mx + b$ where m and b are constants. The problem is to determine the constants m and b of this relationship in a way that minimizes the differences between the observed function values of y in the above chart and corresponding values of y produced by the function $y(x) = mx + b$. Formulate this as a mathematical programming problem in three different ways using each of the following expressions to denote the measure of the differences between the observed and the calculated values of y.

$$\text{(i)} \quad d = \sum_{i=1}^{12} (y_i - y(x_i))^2$$

$$\text{(ii)} \quad d = \sum_{i=1}^{12} |y_i - y(x_i)|$$

$$\text{(iii)} \quad d = \max \{|y_i - y(x_i)| : i = 1, \dots, 12\}$$

(g) Find the solution of the system of equations:

$$a_{11} x_1 + a_{12} x_2 + a_{13} x_3 + a_{14} x_4 = b_1$$
$$a_{21} x_1 + a_{22} x_2 + a_{23} x_3 + a_{24} x_4 = b_2$$
$$a_{31} x_1 + a_{32} x_2 + a_{33} x_3 + a_{34} x_4 = b_3$$

that is closest to the fixed point (c_1, c_2, c_3, c_4), where the distance d between the points (x_1, x_2, x_3, x_4) and (c_1, c_2, c_3, c_4) is the usual Euclidean distance function. Formulate this problem as a programming problem.

(h) A national promotion agency has just selected six individuals to win six trips to the French Riviera. The agency must fly four of the winners on a given airline, while two of the winners will travel on a luxury cruise ship. The transportation cost for each winner in each of the transportation modes is given in the following table.

Transportation costs in dollars

			Winners			
	W_1	W_2	W_3	W_4	W_5	W_6
Air fare	585	450	625	735	300	810
Ship fare	1005	700	1450	1500	685	2025

The agency alone decides how each winner will travel. How should the agency assign the modes of transportation in order to minimize its transportation costs? Express this as a mathematical programming problem.

1.8   Show that any local optimal solution of a linear programming problem is a global solution.

1.9   Establish the following results:
(a) The Int(Ω) is an open subset of Ω.
(b) The set Ω is open if and only if $\Omega \cap Bd(\Omega) = \emptyset$.
(c) The set Ω is closed if and only if its complement $En \setminus \Omega$ is open.
(d) The set Ω is closed if and only if it contains all of its accumulation points.
(e) Any union of open sets is again an open set.
(f) Any finite intersection of open sets is an open set.

1.10   Give examples of the following:
(a) A set that is neither open nor closed.
(b) A set that is both open and closed.
(c) An infinite number of open sets whose intersection is not open.
(d) A boundary point that is not an accumulation point.

5.   SOME HISTORICAL COMMENTS

One of the first mathematical programming problems to be formulated was the transportation problem. The problem was posed by L. U. Kantorovitch in 1939 [16], in 1941 by F. L. Hitchcock [14], and by T. C. Koopmans (see for example, [17]. In 1975, Kantorovitch and Koopmans received the Nobel Prize in Economic Science for their efforts.

In the 1940's, the U.S. Air Force became interested in applying mathematical techniques to military problems. It was recognized that many such problems required the minimization or the maximization of a linear function of several variables with respect to several linear constraints. In 1947, George B. Dantzig produced a method for solving these problems [3].

His technique is known as the simplex procedure (or the primal simplex procedure).

A related problem, known as the dual linear programming problem, was formulated by John Von Neuman. Von Neuman along with others like A. W. Tucker [23] and A. C. Williams [24] were instrumental in the development of a basic duality theory. Using duality, a new simplex procedure was developed by C. E. Lemke in 1954 [20]. It is known as the dual simplex procedure. The duality theory also gave rise to the primal-dual simplex procedure, which was developed by D. G. Dantzig, L. R. Ford, Jr., and D. R. Fulkerson in 1956 [5]. The complementary slackness theorem of G. B. Dantzig and A. Orden [8], allowed the linear programming problem to be expressed as a problem of solving a system of several linear equality constraints and one complementary constraint of several nonnegative complementary variables. The resulting problem is known as the linear complementarity problem. The Kuhn-Tucker optimality conditions found in Chapter 8 (see Kuhn and Tucker [18] or Kunzi and Krelle [19]) allowed the quadratic programming problem to be expressed as a linear complementarity problem. In 1968, C. E. Lemke devised a method known as the complementary pivoting algorithm for solving the linear complementarity problem [21].

Several numerically efficient forms of the primal simplex procedure, known collectively as revised simplex procedures, were developed in the earlier 1950's by G. B. Dantzig, W. Orchard-Hays, A. Orden, G. Waters, and P. Wolfe (see, for example, the papers [6], [9], and [7]). In 1969, R. H. Bartels and G. H. Golub adapted the use of LU-decompositions to these revised procedures to obtain a technique known as the elimination form of the inverse.

The pioneer work in the maximal flow problem was done by L. R. Ford, Jr. and D. R. Fulkerson [11]. G. B. Dantzig studied adaptations of the simplex procedure to the transportation problem in 1951. The current specialization of the primal simplex procedure for network programming is due largely to the paper by E. L. Johnson in 1966 [15].

The iterative techniques for optimizing unconstrained programming problems, which are found in Chapter 9, are the results of many individuals. Usually these processes are named in honor of their founders and the corresponding references can be found at the end of Chapter 10. The method of conjugate directions was originally developed by M. R. Hestenes and E. L. Stiefel in 1952 [13].

The method of penalty functions for solving a constrained nonlinear optimization problem dates back to Courant. More recently, the work of A. V. Fiacco and G. P. McCormick in 1968 has become classic [10]. The barrier function method for solving constrained programming problems is due to the work of C. W. Carroll in 1961 [2]. Finally, the quadratic penalty function method for solving these problems was independently developed by M. R. Hestenes [12] and M. J. D. Powell [22] in 1969.

REFERENCES

1. R. H. Bartels and G. H. Golub, The Simplex Method of Linear Programming Using LU Decompositions, *Comm. ACM 12, 5* (1969), pp. 266 – 268.
2. C. W. Carroll, The Created Response Surface Technique for Optimizing Nonlinear Restrained Systems, *Operations Research 9* (1961), pp. 169 – 184.
3. G. B. Dantzig, Application of the Simplex Method to a Transportation Problem, *Activity Analysis of Production and Allocation*, edited by T. C. Koopmans, John Wiley and Sons, New York (1951).
4. G. B. Dantzig, Maximization of a Linear Function of Variables Subject to Linear Inequalities, Chapter XXI, *Activity Analysis of Production and Allocation*, Cowles Commission Monograph 13, edited by T. C. Koopmans, John Wiley and Sons, New York (1951).
5. G. B. Dantzig, L. R. Ford, Jr., and D. R. Fulkerson, A Primal-Dual Algorithm, *Linear Inequalities and Related Systems, Annals of Mathematics,* Study 38, Princeton University Press, Princeton, N.J. (1956), pp. 171 – 181.
6. G. B. Dantzig and W. Orchard-Hays, Notes on Linear Programming: Part V – Alternate Algorithm for the Revised Simplex Method Using Product Form for the Inverse, Rand Report RM-1268, The Rand Corporation, Santa Monica, Ca. (1953).
7. G. B. Dantzig, W. Orchard-Hays and G. Waters, Product-Form Tableau for Revised Simplex Method, Rand Report RM-1268A, The Rand Corporation, Santa Monica, Ca. (1954).
8. G. B. Dantzig and A. Orden, Duality Theorems, Rand Report RM-1265, The Rand Corporation, Santa Monica, Ca. (1953).
9. G. B. Dantzig, A. Orden and P. Wolfe, Generalized Simplex Method for Minimizing a Linear Form Under Linear Inequality Restraints, Rand Report RM-1264, The Rand Corporation, Santa Monica, Ca. (1954).
10. A. V. Fiacco and G. P. McCormick, *Nonlinear Programming: Sequential Unconstrained Minimization Techniques*, John Wiley and Sons, New York (1968).
11. L. R. Ford, Jr. and D. R. Fulkerson, *Flows in Networks*, Princeton University Press, Princeton, N.J. (1962).
12. M. R. Hestenes, Multiplier and Gradient Methods, *Journal of Optimization Theory and Applications 4* (1969), pp. 303 – 320.
13. M. R. Hestenes and E. L. Stiefel, Methods of Conjugate Gradients for Solving Linear Systems, *Journal Research National Bureau of Standards 49* (1952), pp. 409 – 436.

14.   F. L. Hitchcock, Distribution of a Product From Several Sources to Numerous Localities, *Journal of Mathematical Physics, Vol. 20,* 1941, pp. 224 – 230.

15.   E. L. Johnson, Networks and Basic Solutions, *Operations Research 14* (1966), pp. 619 – 623.

16.   L. V. Kantorovitch, Mathematical Methods in the Organization and Planning of Production, originally published by Publication House of the Leningrad State University (1939), translated in *Management Science 6* (1958), pp. 366–422.

17.   T. C. Koopmans, Optimum Utilization of the Transportation System, *Proceedings of the International Statistical Conference,* Washington, D.C. (1947). Also, reprinted in *Econometrica 17,* supplement (1949).

18.   H. W. Kuhn and A. W. Tucker, Nonlinear Programming, *Proceedings of the Second Berkeley Symposium on Mathematical Statistics and Probability,* edited by J. Neyman, University of California Press, Berkeley, Calif. (1951).

19.   H. P. Kunzi and W. Krelle, *Nonlinear Programming,* Blaisdell Publishing Co., Waltham, Mass. (1966).

20.   C. E. Lemke, The Dual Method of Solving the Linear Programming Problem, *Naval Research Logistics Quarterly 1, 1* (1954), pp. 36 – 47.

21.   C. E. Lemke, On Complementary Pivot Theory, in *Mathematics of the Decision Sciences,* edited by G. B. Dantzig and A. F. Veinott, American Mathematical Society, Providence, R.I., (1968).

22.   M. J. D. Powell, A Method for Nonlinear Constraints in Minimization Problems, *Optimization,* edited by R. Fletcher, Academic Press, New York, (1969).

23.   A. W. Tucker, Dual Systems of Homogeneous Linear Relations, *Linear Inequalities and Related Systems,* edited by H. W. Kuhn and A. W. Tucker, Annals of Mathematics Study, No. 38, Princeton University, Princeton, N.J. (1960).

24.   A. C. Williams, Complementary Theorems for Linear Programming, *SIAM Review 12, 1* (1970), pp. 135–137.

2
Subspaces, Matrices, Affine Sets, Cones, Convex Sets, and the Linear Programming Problem

This chapter contains a brief review of linear algebra and an introduction to convex analysis and its applications to linear programming. Several topics from elementary linear algebra and matrix theory are briefly reviewed in the first section. These topics include vector spaces, matrices, elementary row operations and the Gauss-Jordan method of elimination. The section also deals with Gaussian elimination, LU-decompositions, and basic solutions of systems of equations. In that section, most results are stated without proof, and exercises are included for only the last two topics mentioned. Individuals needing more review should refer to one of the linear algebra texts mentioned in the references at the end of the chapter. In the second section, some of the concepts from linear algebra (reviewed in the first section) are related to new concepts in convex analysis. The third section illustrates the usefulness of convex analysis as a tool in describing the solutions of the linear programming problem. Additional topics and applications from convex analysis will appear in later chapters.

It is not necessary to complete this entire chapter before advancing to Chapter 3. In fact, unless the review in the first section is needed, this chapter may be temporarily skipped. In any case, if it is desirable to begin meaningful computations as soon as possible, then Chapter 3 can be started as soon as the reader is familiar with the computation of the basic solutions of a system of linear equations. This topic appears at the end of the first section and requires some knowledge of elementary row operations and block matrices. Gaussian elimination and LU-decompositions are not used until the revised simplex procedure is discussed in Chapter 5. Hence, these topics can be deferred until the reader reaches Chapter 5. Even then the section can be omitted by those who are familiar with Gaussian elimination and LU-decompositions.

The second section should be covered before beginning Chapter 8. If it is desirable to examine the linear programming problem from a geometrical point of view before introducing the algebraic procedure for solving the problem, then the second and third sections should be covered before advancing to Chapter 3. Whichever approach is selected, it is important to know that mathematical programming problems are highly geometric in nature, and many of the iterative procedures which are used to solve them result from the proper geometrical interpretations of the problems.

The use of slack variables to change a linear inequality constraint into a linear equality constraint is discussed in the third section. The individual who begins Chapter 3 before covering the second and third sections of this chapter may find it helpful to return to these sections when slack variables are encountered in Chapter 3.

1. A REVIEW OF ELEMENTARY LINEAR ALGEBRA

A $m \times n$ real *matrix* A is a rectangular array containing m rows and n columns of real numbers,

$$A = \begin{bmatrix} a_{11} & a_{12} & \cdots & a_{1n} \\ a_{21} & a_{22} & \cdots & a_{2n} \\ \cdot & \cdot & & \cdot \\ \cdot & \cdot & & \cdot \\ \cdot & \cdot & & \cdot \\ a_{m1} & a_{m2} & \cdots & a_{mn} \end{bmatrix}$$

When the number of rows and columns is understood, the above expression may be written more compactly as $A = [a_{ij}]$, where a_{ij} denotes the element in the ith row and jth column of A. The matrix above is said to be an $m \times n$ (or, m by n) matrix. The collection of all $m \times n$ matrices with real entries will be denoted by $\mathbf{R}_{m \times n}$.

The *transpose* of a $m \times n$ real matrix A is an $n \times m$ real matrix, denoted by A^T, obtained by interchanging the rows and columns of A, i.e.,

$$A^T = \begin{bmatrix} a_{11} & a_{21} & \cdots & a_{m1} \\ a_{12} & a_{22} & \cdots & a_{m2} \\ \cdot & \cdot & & \cdot \\ \cdot & \cdot & & \cdot \\ \cdot & \cdot & & \cdot \\ a_{1n} & a_{2n} & \cdots & a_{mn} \end{bmatrix}$$

Any matrix $x \in \mathbf{R}_{m \times 1}$,

$$x = \begin{bmatrix} x_1 \\ \cdot \\ \cdot \\ \cdot \\ x_m \end{bmatrix} = [x_1 \dots x_m]^T,$$

will be called a $m \times 1$ *column* matrix. A $1 \times n$ *row* matrix is a matrix of the form $x = [x_1 \dots x_n] \in \mathbf{R}_{1 \times n}$. Column matrices will also be called *vectors* and real numbers will often be called *scalars*. The collection $\mathbf{R}_{m \times 1}$ of $m \times 1$ column vectors forms a vectors space (the definition of a vector space can be found in most standard linear algebra texts) which is often called *Euclidean m-space*. Other common notations for vectors in Euclidean m-space include $x = (x_1, \dots, x_m)$ and $x = [x_1 \dots x_m]$ (a row matrix). Euclidean m-space is often denoted by E_m. However, in this book we shall usually use the notation $\mathbf{R}_{m \times 1}$. Thus, Euclidean n-space is $\mathbf{R}_{n \times 1} = \{x = [x_1 \dots x_n]^T$: each $x_i \in \mathbf{R}\}$, i.e., the set of $n \times 1$ column matrices. We shall continue our review of elementary linear algebra with this space.

If $x = [x_1 \dots x_n]^T$ and $y = [y_1 \dots y_n]^T$ both belong to $\mathbf{R}_{n \times 1}$, then we say that $x = y$ if and only if each $x_i = y_i$. *Vector addition* is defined by

$$x + y = [x_1 + y_1 \ \dots \ x_n + y_n]^T$$

For each scalar α, *scalar multiplication* is defined by

$$\alpha x = [\alpha x_1 \dots \alpha x_n]^T$$

Vector addition and scalar multiplication for vectors in $\mathbf{R}_{2 \times 1}$ can be illustrated geometrically as in Figure 1 below.

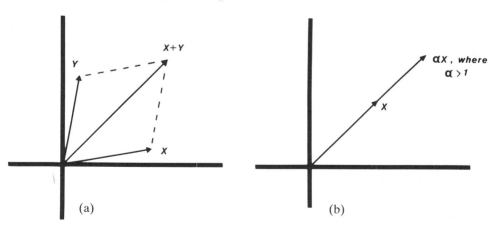

Figure 1 (a) Vector addition. (b) Scalar multiplication.

The *inner product* (or *dot product*) of x and y is defined by

$$<x|y> = \sum_{i=1}^{n} x_i y_i = x^T y$$

The following basic properties hold for inner products.

PROPOSITION 1   Let $x = [x_1 \ldots x_n]^T$, $y = [y_1 \ldots y_n]^T$ and $z = [z_1 \ldots z_n]^T$ be vectors in $\mathbf{R}_{n \times 1}$ and let both α and β be scalars. Then

(i) $<x|y> = <y|x>$
(ii) $<\alpha x + \beta y|z> = \alpha <x|z> + \beta <y|z>$
(iii) $<x|x> \geq 0$, with equality only when each $x_i = 0$.

The *norm* or the length of a vector x is denoted by $\|x\|$ and defined by

$$\|x\| = \sqrt{<x|x>} = \sqrt{x^T x}$$

A *subspace*, S, of $\mathbf{R}_{n \times 1}$ is any nonempty set of vectors for which $x \varepsilon S$, $y \varepsilon S$, $\alpha \varepsilon \mathbf{R}$ and $\beta \in \mathbf{R}$ imply that

$$\alpha x + \beta y \in S$$

For example, in $\mathbf{R}_{3 \times 1}$ the subspaces are $\{(0,0,0)\}$, straight lines passing through the origin, planes passing through the origin, and $\mathbf{R}_{3 \times 1}$ itself. The set

$$S = \{[x_1 x_2 0]^T : x_1, x_2 \in \mathbf{R}\}$$

is an example of a plane that passes through the origin. Clearly, if $x = [x_1 x_2 0]^T \in S$, $y = [y_1 y_2 0]^T \in S$, $\alpha \in \mathbf{R}$ and $\beta \in \mathbf{R}$, then

$$\alpha x + \beta y = \alpha [x_1 x_2 0]^T + \beta [y_1 y_2 0]^T = [\alpha x_1 + \beta y_1 \ \alpha x_2 + \beta y_2 \ 0]^T$$

which by definition belongs to S.

Now consider k vectors $x_1 = [x_{11} \ldots x_{1n}]^T$, $x_2 = [x_{21} \ldots x_{2n}]^T$, $\ldots$, $x_k = [x_{k1} \ldots x_{kn}]^T$ in $\mathbf{R}_{n \times 1}$. Let $\alpha_1, \ldots, \alpha_k$ be scalars. Then the expression

$$\sum_{i=1}^{k} \alpha_i x_i = \alpha_1 x_1 + \cdots + \alpha_n x_n = [\sum_{i=1}^{k} \alpha_i x_{i1} \ldots \sum_{i=1}^{k} \alpha_i x_{in}]^T$$

is called a *linear combination* of the vectors $x_1, \ldots, x_k$.

If $B \subseteq \mathbf{R}_{n \times 1}$, then we denote the set of all linear combinations of elements of B by $L(B)$, i.e.,

$$L(B) = \{\sum_{i=1}^{k} \alpha_i x_i : \text{each } \alpha_i \in \mathbf{R}, x_i \in B \text{ and } k \text{ is some natural number}\}$$

Since a subspace of $\mathbf{R}_{n \times 1}$ is a nonempty set of vectors which contains any linear combination of two vectors from itself and since any linear combination of two elements from $L(B)$ is again an element of $L(B)$, it follows that $L(B)$ is a subspace of $\mathbf{R}_{n \times 1}$. It is called the space *spanned* (or *generated*) by B. If S is a subspace and $S = L(B)$, then B is said to *span S*.

The reader should recall that any intersection of subspaces of $\mathbf{R}_{n \times 1}$ is again a subspace of $\mathbf{R}_{n \times 1}$ and, in fact, $L(B)$ is the intersection of all subspaces of $\mathbf{R}_{n \times 1}$ that contain B.

EXAMPLE 1   Let $B_1 = \{[1\,0\,0]^T, [0\,1\,0]^T, [0\,0\,1]^T\}$. Then

$$L(B_1) = \{x_1[1\,0\,0]^T + x_2[0\,1\,0]^T + x_3[0\,0\,1]^T : x_1, x_2, x_3 \in \mathbf{R}\}$$
$$= \{[x_1\,x_2\,x_3]^T : x_1, x_2, x_3 \in \mathbf{R}\} = \mathbf{R}_{3 \times 1}$$

EXAMPLE 2   Let $B_2 = \{[1\,0\,0]^T, [1\,0\,1]^T\}$. Then

$$L(B_2) = \{x_1[1\,0\,0]^T + x_2[1\,0\,1]^T : x_1, x_2 \in \mathbf{R}\}$$
$$= \{[x_1 + x_2 \;\; 0 \;\; x_2]^T : x_1, x_2 \in \mathbf{R}\}$$

which is the x_1x_3–plane.

EXAMPLE 3   Let $B_3 = \{[1\,0\,0]^T, [1\,0\,1]^T, [4\,0\,1]^T\}$. Then

$$L(B_3) = \{x_1[1\,0\,0]^T + x_2[1\,0\,1]^T + x_3[4\,0\,1]^T : x_1, x_2, x_3 \in \mathbf{R}\}$$
$$= \{[x_1 + x_2 + 4x_3 \;\; 0 \;\; x_2 + x_3]^T : x_1, x_2, x_3 \in \mathbf{R}\}$$

Actually, in the above examples $L(B_2) = L(B_3)$, since $[4\,0\,1]^T = 3[1\,0\,0]^T + [1\,0\,1]^T$.

EXAMPLE 4   Let $S = \{[x_1\,x_2\,x_3]^T : x_1 + x_2 = 0, x_1, x_2, x_3 \in \mathbf{R}\}$. Show that S is a subspace of $\mathbf{R}_{3 \times 1}$ and find a set that spans S.
Solution: Since

$$S = \{[x_1\,x_2\,x_3]^T : x_1 + x_2 = 0, x_1, x_2, x_3 \in \mathbf{R}\} = \{[x_1 \;\; -x_1 \;\; x_3]^T : x_1, x_3 \in \mathbf{R}\}$$
$$= \{x_1[1 \;\; -1 \;\; 0]^T + x_3[0\,0\,1]^T : x_1, x_3 \in \mathbf{R}\}$$
$$= L(\{[1 \;\; -1 \;\; 0]^T, [0\,0\,1]^T\})$$

then S is the subspace that is spanned by $\{[1 \;\; -1 \;\; 0]^T, [0\,0\,1]^T\}$.

Again let B be a subset of $\mathbf{R}_{n \times 1}$. Then B is a *linearly dependent* set of vectors if and only if there exists a finite collection of vectors $x_1, \ldots, x_n$ in B and scalars $\alpha_1, \ldots, \alpha_n$, where not all the $\alpha_i = 0$, such that

$$\alpha_1 x_1 + \cdots + \alpha_n x_n = 0$$

(Here 0 is the vector having all zero entries, i.e., $0 = [0 \ldots 0]^T$.)

The set of vectors B is a *linearly independent* set of vectors if and only if it is not a linearly dependent set of vectors, i.e., whenever $x_1,\ldots,x_n$ is any finite collection of vectors in B, then the expression

$$\alpha_1 x_1 + \cdots + \alpha_n x_n = 0 \quad (\text{again } 0 = [0 \ldots 0]^T)$$

always implies that each $\alpha_i = 0$.

In Example 1 and Example 2 both B_1 and B_2 are linearly independent sets. However, in Example 3, B_3 is a linearly dependent set since

$$3[1\,0\,0]^T + 1[1\,0\,1]^T - 1[4\,0\,1]^T = [0\,0\,0]^T$$

or

$$[4\,0\,1]^T = 3[1\,0\,0]^T + 1[1\,0\,1]^T$$

In Example 4, the spanning set $\{[1 \ -1 \ 0]^T, [0\,0\,1]^T\}$ is linearly independent since

$$[0\,0\,0]^T = 0 = \alpha[1 \ -1 \ 0]^T + \beta[0\,0\,1]^T = [\alpha \ -\alpha \ \beta]^T$$

implies that $\alpha = \beta = 0$.

Next let S be a subspace of $\mathbf{R}_{n \times 1}$. Then a set B of vectors in S is said to be a *basis* for S provided that

    (i) $L(B) = S$, i.e., B spans S,
    (ii) B is linearly independent

The following theorem is a basic and well known result from linear algebra. It is stated without proof.

THEOREM 1   Every subspace, S, of $\mathbf{R}_{n \times 1}$ has a basis. Any two bases of a subspace must contain the same number of elements.

As a result of this theorem we define the *dimension* of a subspace to be the number of elements in a basis for the subspace. If L is a subspace of dimension k and if both $A = \{a_1, \ldots, a_k\}$ and $B = \{b_1, \ldots, b_k\}$ are bases for L, then we say that $A = B$ provided that each $a_i = b_i$ for $i = 1, \ldots, k$.

In Example 1 B_1 is a basis for $\mathbf{R}_{3 \times 1}$. Hence dim $\mathbf{R}_{3 \times 1} = 3$. In general $\{[1\,0\ldots0]^T, [0\,1\,0\,..\,0]^T, \ldots, [0\ldots0\,1]^T\}$ is a basis for $\mathbf{R}_{n \times 1}$ and hence $\dim \mathbf{R}_{n \times 1} = n$. Another basis for $\mathbf{R}_{n \times 1}$ is $\{[0\,1\,0\,..\,0]^T, [1\,0\ldots0]^T, \ldots, [0\ldots0\,1]^T\}$.

In Example 2, B_2 is a basis for $L(B_2)$. But in Example 3, B_3 is not a basis for $L(B_3)$. In Example 4, $\{[1 \ -1 \ 0]^T, [0\,0\,1]^T\}$ is a basis for S.

From linear algebra we have the following theorems.

THEOREM 2   Every set of vectors that spans a subspace contains a basis for the subspace.

THEOREM 3   Any linearly independent subset of a subspace can be extended to a basis for the space.

One important reason for determining a basis for a subspace is that once a basis is determined then each vector in the subspace has a unique representation in terms of the basis. For example, if $B = \{b_1, \ldots, b_k\}$ is a basis for the subspace $S, x \, \varepsilon \, S$, and both

$$x = \sum_{i=1}^{k} \alpha_i b_i \quad \text{and} \quad x = \sum_{i=1}^{k} \beta_i b_i$$

are representations of x in terms of B, then

$$[0 \ldots 0]^T = 0 = x - x = \sum_{i=1}^{k} \alpha_i b_i - \sum_{i=1}^{k} \beta_i b_i = \sum_{i=1}^{k} (\alpha_i - \beta_i) b_i$$

But since B is a linearly independent set of vectors, it follows that each

$$\alpha_i - \beta_i = 0 \ , \text{i.e.,} \ \alpha_i = \beta_i$$

and, thus, the representation of x in terms of the vectors from B is unique.

Also from linear algebra we have the following theorem.

THEOREM 4   If S is a subspace of $\mathbf{R}_{n \times 1}$ with dim $S = k$, then any subset of S that contains more than k vectors is a linearly dependent set of vectors.

It follows that if S and T are both subspaces of $\mathbf{R}_{n \times 1}$, where $S \subseteq T$, then dim $S \leq$ dim T. Since dim $\mathbf{R}_{n \times 1} = n$, it follows that any subspace of $\mathbf{R}_{n \times 1}$ has dimension less than or equal to n.

Consider now an $m \times n$ real matrix A,

$$A = [a_{ij}] = \begin{bmatrix} a_{11} & \cdots & a_{1j} & \cdots & a_{1n} \\ \vdots & & \vdots & & \vdots \\ a_{i1} & \cdots & a_{ij} & \cdots & a_{in} \\ \vdots & & \vdots & & \vdots \\ a_{m1} & \cdots & a_{mj} & \cdots & a_{mn} \end{bmatrix}$$

The jth column of A will be denoted by $A^{(j)}$, i.e.,

$$A^{(j)} = \begin{bmatrix} a_{1j} \\ \vdots \\ a_{ij} \\ \vdots \\ a_{mj} \end{bmatrix} = [a_{1j} \ldots a_{ij} \ldots a_{mj}]^T$$

The ith row of A will be denoted by $A_{(i)}$, i.e.,

$$A_{(i)} = [a_{i1} \ldots a_{ij} \ldots a_{in}]$$

The space spanned by the columns of A is called the *column space* of A. The space spanned by the rows of A is called the *row space* of A. Notice that the column space of A is a subspace of $\mathbf{R}_{m \times 1}$. Also, the row space of A is a subspace of $\mathbf{R}_{n \times 1}$. The dimension of the column space of A is called the *column rank* of A. Likewise, the dimension of the row space of A is called the *row rank* of A. A basic result from linear algebra tells us that the column rank of A is equal to the row rank of A. Hence, the *rank* of A is defined to be

$$\text{rank } A = \text{column rank } A = \text{row rank } A$$

Just as with vectors, matrices may be added and multiplied by scalars. For example, if $A = [a_{ij}] \in \mathbf{R}_{m \times n}$, $B = [b_{ij}] \in \mathbf{R}_{m \times n}$ and $\alpha \in \mathbf{R}$, then

$$A + B = [a_{ij} + b_{ij}] \quad \text{and} \quad \alpha A = [\alpha a_{ij}]$$

As the following definition shows, matrices can sometimes be multiplied. If $A = [a_{ij}] \in \mathbf{R}_{m \times k}$ and $B = [b_{ij}] \in \mathbf{R}_{k \times n}$, then the product AB is defined by

$$AB = \left[\sum_{l=1}^{k} a_{il} b_{lj} \right] \in \mathbf{R}_{m \times n}$$

EXAMPLE 5   To illustrate,

$$\begin{bmatrix} 2 & 0 & 3 \\ 1 & 4 & 5 \end{bmatrix} \begin{bmatrix} 1 & 0 & 1 & 0 \\ 2 & 1 & 6 & 0 \\ 4 & 0 & 1 & 0 \end{bmatrix} = \begin{bmatrix} 14 & 0 & 5 & 0 \\ 29 & 4 & 30 & 0 \end{bmatrix}$$

EXAMPLE 6   Define $I_n = [\delta_{ij}] \in \mathbf{R}_{n \times n}$, where $\delta_{ij} = 1$ when $i = j$, and $\delta_{ij} = 0$ when $i \neq j$. Thus,

$$I_n = \begin{bmatrix} 1 & 0 & . & . & . & 0 \\ 0 & 1 & . & . & . & 0 \\ . & . & . & . & . & . \\ . & . & . & . & . & . \\ . & . & . & . & . & . \\ 0 & 0 & . & . & . & 1 \end{bmatrix}$$

Then for any $A = [a_{ij}] \in \mathbf{R}_{m \times n}$, $AI_n = A$. Likewise, $I_m A = A$. If $m = n$, then $AI_n = I_n A = A$.

A *submatrix* of A is any matrix obtained by deleting some of the rows and columns of A. If, for example,

$$A = \begin{bmatrix} 6 & 1 & 3 & 4 & 0 & 5 \\ 2 & 1 & 4 & 0 & 0 & 1 \\ 1 & 0 & 1 & -1 & 1 & 0 \end{bmatrix}$$

then

$$B = \begin{bmatrix} 4 & 0 & 1 \\ 1 & -1 & 0 \end{bmatrix}$$

is a submatrix of A obtained by deleting the first row, and the first, second and fifth columns of A. A matrix A can be *partitioned* into a rectangular array of submatrices by inserting vertical and horizontal lines between the rows and columns of A. The submatrices in a partitioned matrix are often called *blocks*. For the last example, the matrix A can be partitioned as follows

$$\left[\begin{array}{cc:cc:c:c} 6 & 1 & 3 & 4 & 0 & 5 \\ 2 & 1 & 4 & 0 & 0 & 1 \\ \hline 1 & 0 & 1 & -1 & 1 & 0 \end{array}\right] = \left[\begin{array}{c:c:c:c} B & D & O & b \\ \hline -c_B^T & -c_D^T & 1 & 0 \end{array}\right]$$

where

$$B = \begin{bmatrix} 6 & 1 \\ 2 & 1 \end{bmatrix}, D = \begin{bmatrix} 3 & 4 \\ 4 & 0 \end{bmatrix}, O = \begin{bmatrix} 0 \\ 0 \end{bmatrix}, b = \begin{bmatrix} 5 \\ 1 \end{bmatrix}$$

$$-c_B^T = \begin{bmatrix} 1 & 0 \end{bmatrix} \text{ and } -c_D^T = \begin{bmatrix} 1 & -1 \end{bmatrix}$$

The reason for using the unusual notation $-c_B^T$ and $-c_D^T$ for these last two submatrices will be apparent in Chapter 3. Partitioned matrices will be used throughout this book as a tool for simplifying the notation and the development of many of the iterative techniques for solving mathematical programming problems.

When two matrices, $A \in R_{m \times k}$ and $B \in R_{k \times n}$, are properly partitioned, the product AB can be obtained by performing the matrix multiplication using the blocks as elements. For example, if $B = [C \mathbin{\vdots} D]$, then $AB = [AC \mathbin{\vdots} AD]$. Clearly, in this case both C and D have k rows. If

$$A = \left[\frac{E}{F}\right]$$

then

$$AB = \left[\frac{EB}{FB}\right]$$

(here E and F each have k columns). For one final example, let

$$A = \begin{bmatrix} C & \vdots & D \\ \hdashline E & \vdots & F \end{bmatrix} \quad \text{and} \quad B = \begin{bmatrix} G & \vdots & H & \vdots & J \\ \hdashline K & \vdots & L & \vdots & M \end{bmatrix}$$

Then

$$AB = \begin{bmatrix} CG+DK & \vdots & CH+DL & \vdots & CJ+DM \\ \hdashline EG+FK & \vdots & EH+FL & \vdots & EJ+FM \end{bmatrix}$$

provided A and B have been partitioned so that each of the above products is possible.

Now consider a square matrix $A = [a_{ij}] \in R_{n \times n}$. If there exists another square matrix $B = [b_{ij}] \in R_{n \times n}$ such that $AB = BA = I_n$, then A is said to be *nonsingular* and B is called the *inverse* of A. When the inverse exists, it is unique and is usually denoted by A^{-1}. A matrix that is not nonsingular is said to be *singular*.

EXAMPLE 7   Notice that

$$\begin{bmatrix} 1 & 1 & -1 \\ -2 & 1 & 1 \\ 1 & 1 & 1 \end{bmatrix} \begin{bmatrix} 0 & -1/3 & 1/3 \\ 1/2 & 1/3 & 1/6 \\ -1/2 & 0 & 1/2 \end{bmatrix} = \begin{bmatrix} 0 & -1/3 & 1/3 \\ 1/2 & 1/3 & 1/6 \\ -1/2 & 0 & 1/2 \end{bmatrix} \begin{bmatrix} 1 & 1 & -1 \\ -2 & 1 & 1 \\ 1 & 1 & 1 \end{bmatrix} = \begin{bmatrix} 1 & 0 & 0 \\ 0 & 1 & 0 \\ 0 & 0 & 1 \end{bmatrix}$$

Hence

$$\begin{bmatrix} 1 & 1 & -1 \\ -2 & 1 & 1 \\ 1 & 1 & 1 \end{bmatrix} \quad \text{and} \quad \begin{bmatrix} 0 & -1/3 & 1/3 \\ 1/2 & 1/3 & 1/6 \\ -1/2 & 0 & 1/2 \end{bmatrix}$$

are both nonsingular and are inverses of one another.

As the following theorem states, an $n \times n$ matrix is nonsingular if and only if the rank of the matrix is n.

THEOREM 5   Let $A = [a_{ij}] \in \mathbf{R}_{n \times n}$. Then A is nonsingular if and only if rank $A = n$, i.e., the columns (also the rows) of A are linearly independent.

EXAMPLE 8   (a) Since its columns are linearly dependent (see Example

3), the matrix

$$\begin{bmatrix} 1 & 1 & 4 \\ 0 & 0 & 0 \\ 0 & 1 & 1 \end{bmatrix}$$

is singular.

(b) Since the columns of the matrix

$$\begin{bmatrix} 1 & 0 & 0 \\ 1 & 1 & 0 \\ 1 & 1 & 1 \end{bmatrix}$$

are linearly independent, it is nonsingular.

We shall next indicate an important process for computing the inverse of a matrix when it exists. The process may also be used to determine the rank of a singular matrix. The interested reader should consult one of the references at the end of this chapter.

There are three types of *elementary row operations* on a matrix A. These are:

1. An elementary row operation of type 1 multiplies one of the rows of A by a nonzero scalar α.
2. An elementary row operation of type 2 replaces a row of A by itself plus a scalar multiple of another row of A.
3. An elementary row operation of type 3 interchanges two rows.

EXAMPLE 9   (a) An example of a type 1 row operation is

$$\begin{bmatrix} 1 & 0 & 2 \\ 2 & 1 & 3 \end{bmatrix} \xrightarrow{4R_1} \begin{bmatrix} 4 & 0 & 8 \\ 2 & 1 & 3 \end{bmatrix}$$

where the first row is replaced by itself multiplied by the scalar 4.

(b) In the following type 2 row operation,

$$\begin{bmatrix} 1 & 0 & 1 \\ 2 & 1 & 1 \\ 0 & 1 & 0 \end{bmatrix} \xrightarrow{R_1 + 2R_3} \begin{bmatrix} 1 & 2 & 1 \\ 2 & 1 & 1 \\ 0 & 1 & 1 \end{bmatrix}$$

the first row is replaced by itself plus 2 times row 3.

(c) Finally, in the type 3 operation,

$$\begin{bmatrix} 1 & 4 \\ 5 & 1 \\ 6 & 0 \end{bmatrix} \xrightarrow{R_2, R_3} \begin{bmatrix} 1 & 4 \\ 6 & 0 \\ 5 & 1 \end{bmatrix}$$

the second and the third rows are interchanged.

Performing elementary row operations on a matrix does not change the row space of a matrix. The operations can be used to compute the inverse of a square matrix $A = [a_{ij}] \in R_{n \times n}$. This is done by forming an $n \times 2n$ matrix $[A \mid I_n]$ and then performing elementary row operations to get $[I_n \mid A^{-1}]$,

$$[A \mid I_n] \to \cdots \to [I_n \mid A^{-1}]$$

If it is impossible to go from $[A \mid I_n]$ to $[I_n \mid A^{-1}]$ via elementary row operations, then the matrix A is singular.

EXAMPLE 10   Compute the inverse of

$$A = \begin{bmatrix} 1 & 1 & -1 \\ -2 & 1 & 1 \\ 1 & 1 & 1 \end{bmatrix}$$

Solution: Since

$$\left[\begin{array}{ccc|ccc} 1 & 1 & -1 & 1 & 0 & 0 \\ -2 & 1 & 1 & 0 & 1 & 0 \\ 1 & 1 & 1 & 0 & 0 & 1 \end{array}\right] \xrightarrow[R_3 - R_1]{R_2 + 2R_1} \left[\begin{array}{ccc|ccc} 1 & 1 & -1 & 1 & 0 & 0 \\ 0 & 3 & -1 & 2 & 1 & 0 \\ 0 & 0 & 2 & -1 & 0 & 1 \end{array}\right]$$

$$\xrightarrow[(1/3)R_2]{R_1 - (1/3)R_2} \left[\begin{array}{ccc|ccc} 1 & 0 & -2/3 & 1/3 & -1/3 & 0 \\ 0 & 1 & -1/3 & 2/3 & 1/3 & 0 \\ 0 & 0 & 2 & -1 & 0 & 1 \end{array}\right]$$

$$\begin{array}{c} R_1 + (1/3)R_2 \\ R_2 + (1/6)R_3 \\ \xrightarrow{\hspace{2cm}} \\ (1/2)R_3 \end{array} \left[\begin{array}{ccc|ccc} 1 & 0 & 0 & 0 & -1/3 & 1/3 \\ 0 & 1 & 0 & 1/2 & 1/3 & 1/6 \\ 0 & 0 & 1 & -1/2 & 0 & 1/2 \end{array}\right]$$

then A is nonsingular and

$$A^{-1} = \begin{bmatrix} 0 & -1/3 & 1/3 \\ 1/2 & 1/3 & 1/6 \\ -1/2 & 0 & 1/2 \end{bmatrix}$$

(see Example 7).

In Chapters 3 and 5 we shall encounter partitioned matrices of the form

$$\begin{bmatrix} B & \vdots & O \\ \cdots & \cdots & \cdots \\ c_B^T & \vdots & 1 \end{bmatrix} \tag{1}$$

where $B \in R_{m \times m}, -c_B^T \in R_{1 \times m}$, and $O = [0 \ldots 0]^T$. When B is nonsingular, then

$$\begin{bmatrix} B^{-1} & \vdots & O \\ \cdots & \cdots & \cdots \\ c_B^T B^{-1} & \vdots & 1 \end{bmatrix} \begin{bmatrix} B & \vdots & O \\ \cdots & \cdots & \cdots \\ -c_B^T & \vdots & 1 \end{bmatrix} = \begin{bmatrix} I_m & \vdots & O \\ \cdots & \cdots & \cdots \\ 0 & \vdots & 1 \end{bmatrix} = I_{m+1}$$

and it follows that matrix (1) is nonsingular. In fact

$$\begin{bmatrix} B & \vdots & O \\ \cdots & \cdots & \cdots \\ -c_B^T & \vdots & 1 \end{bmatrix}^{-1} = \begin{bmatrix} B^{-1} & \vdots & O \\ \cdots & \cdots & \cdots \\ c_B^T B^{-1} & \vdots & 1 \end{bmatrix} \tag{2}$$

Moreover, if B^{-1} is computed by performing a sequence of elementary row operations,

$$[B \mid I_m] \quad \xrightarrow{\text{a sequence of}} \cdots \rightarrow \quad [I_m \mid B^{-1}]$$
$$\text{row operations}$$

then

$$\begin{bmatrix} B & \vdots & O \\ \cdots & \cdots & \cdots \\ -c_B^T & \vdots & 1 \end{bmatrix}^{-1}$$

can be computed by

$$\begin{bmatrix} B & \vdots O \vdots I_m \vdots O \\ \cdots & \cdots \cdots \cdots \\ -c_B^T & \vdots 1 \vdots 0 \quad \vdots 1 \end{bmatrix} \quad \xrightarrow[\text{of row operations}]{\text{the same sequence}} \cdots \rightarrow \quad \begin{bmatrix} I_m & \vdots O \vdots B^{-1} \vdots O \\ \cdots & \cdots \cdots \cdots \\ -c_B^T & \vdots 1 \vdots 0 \quad \vdots 1 \end{bmatrix}$$

$$\rightarrow \begin{bmatrix} I_m & \vdots & O & \vdots & B^{-1} & \vdots & O \\ \cdots & & \cdots & & \cdots & & \cdots \\ 0 & \vdots & 1 & \vdots & c_B^T B^{-1} & \vdots & 1 \end{bmatrix}$$

where the final row operation consists of replacing row $m + 1$ by itself plus each component of c_B^T times the corresponding row of $[I_m \quad 0 \quad B^{-1} \quad 0]$.

EXAMPLE 11   If $-c_B^T = [1 \quad 0 \ -1]$ and

$$B = \begin{bmatrix} 1 & 1 & -1 \\ -2 & 1 & 1 \\ 1 & 1 & 1 \end{bmatrix}$$

(see Example 7 for B^{-1}), then $c_B^T B^{-1} = [-1/2 \quad 1/3 \quad 1/6]^T$ and

$$
\begin{bmatrix}
1 & 1 & -1 & \vdots & 0 \\
-2 & 1 & 1 & \vdots & 0 \\
1 & 1 & 1 & \vdots & 0 \\
\hline
1 & 0 & -1 & \vdots & 1
\end{bmatrix}^{-1}
=
\begin{bmatrix}
B & \vdots & O \\
\hline
-c_B^T & \vdots & 1
\end{bmatrix}^{-1}
=
\begin{bmatrix}
B^{-1} & \vdots & O \\
\hline
c_B^T B^{-1} & \vdots & 1
\end{bmatrix}
=
\begin{bmatrix}
0 & -1/3 & 1/3 & \vdots & 0 \\
1/2 & 1/3 & 1/6 & \vdots & 0 \\
-1/2 & 0 & 1/2 & \vdots & 0 \\
\hline
-1/2 & 1/3 & 1/6 & \vdots & 1
\end{bmatrix}
$$

Moreover, this inverse can be computed by the following sequence of elementary row operations,

$$
\begin{bmatrix}
1 & 1 & -1 & \vdots & 0 & \vdots & 1 & 0 & 0 & \vdots & 0 \\
-2 & 1 & 1 & \vdots & 0 & \vdots & 0 & 1 & 0 & \vdots & 0 \\
1 & 1 & 1 & \vdots & 0 & \vdots & 0 & 0 & 1 & \vdots & 0 \\
\hline
1 & 0 & -1 & \vdots & 1 & \vdots & 0 & 0 & 0 & \vdots & 1
\end{bmatrix}
\xrightarrow[R_3 - R_1]{R_2 + 2R_1}
\begin{bmatrix}
1 & 1 & -1 & \vdots & 0 & \vdots & 1 & 0 & 0 & \vdots & 0 \\
0 & 3 & -1 & \vdots & 0 & \vdots & 2 & 1 & 0 & \vdots & 0 \\
0 & 0 & 2 & \vdots & 0 & \vdots & -1 & 0 & 1 & \vdots & 0 \\
\hline
1 & 0 & -1 & \vdots & 1 & \vdots & 0 & 0 & 0 & \vdots & 1
\end{bmatrix}
$$

$$
\xrightarrow[\;(1/3)R_2\;]{R_1 - (1/3)R_2}
\begin{bmatrix}
1 & 0 & -2/3 & \vdots & 0 & \vdots & 1/3 & -1/3 & 0 & \vdots & 0 \\
0 & 1 & -1/3 & \vdots & 0 & \vdots & 2/3 & 1/3 & 0 & \vdots & 0 \\
0 & 0 & 2 & \vdots & 0 & \vdots & -1 & 0 & 1 & \vdots & 0 \\
\hline
1 & 0 & -1 & \vdots & 1 & \vdots & 0 & 0 & 0 & \vdots & 0
\end{bmatrix}
$$

$$
\xrightarrow[\substack{R_2 + (1/6)R_3 \\ (1/2)R_3}]{R_1 + (1/3)R_3}
\begin{bmatrix}
1 & 0 & 0 & \vdots & 0 & \vdots & 0 & -1/3 & 1/3 & \vdots & 0 \\
0 & 1 & 0 & \vdots & 0 & \vdots & 1/2 & 1/3 & 1/6 & \vdots & 0 \\
0 & 0 & 1 & \vdots & 0 & \vdots & -1/2 & 0 & 1/2 & \vdots & 0 \\
\hline
1 & 0 & -1 & \vdots & 1 & \vdots & 0 & 0 & 0 & \vdots & 1
\end{bmatrix}
$$

$$
\xrightarrow{R_4 - R_1 + R_3}
\begin{bmatrix}
1 & 0 & 0 & \vdots & 0 & \vdots & 0 & -1/3 & 1/3 & \vdots & 0 \\
0 & 1 & 0 & \vdots & 0 & \vdots & 1/2 & 1/3 & 1/6 & \vdots & 0 \\
0 & 0 & 1 & \vdots & 0 & \vdots & -1/2 & 0 & 1/2 & \vdots & 0 \\
\hline
0 & 0 & 0 & \vdots & 1 & \vdots & -1/2 & 1/3 & 1/6 & \vdots & 1
\end{bmatrix}
$$

(again see Example 7). The last row operation is equivalent to first performing $R_4 - R_1$, and then $R_4 + R_3$ on the resulting matrix.

When it is necessary to solve the system of linear equations

$$a_{11}x_1 + \cdots + a_{1m}x_m = b_1$$

$$\cdot \qquad \qquad \cdot \quad \cdot$$
$$\cdot \qquad \qquad \cdot \quad \cdot$$
$$\cdot \qquad \qquad \cdot \quad \cdot$$

$$a_{m1}x_1 + \cdots + a_{mm}x_m = b_m$$

which can be expressed more compactly using matrix notation as

$$Ax = b$$

where $A = [a_{ij}] \in \mathbf{R}_{m \times m}$, $x = [x_1 \ldots x_m]^T$ and $b = [b_1 \ldots b_m]^T$, and it is necessary to compute A^{-1} (assuming that A is nonsingular), the problem may be handled very simply by modifying the previous technique involving element row operations. Consider the partitioned matrix $[A|I_m|b]$. Perform elementary row operations as in the last example

$$[A \,|\, I_m \,|\, b] \to \cdots \to [I_m \,|\, A^{-1} \,|\, A^{-1}b]$$

Recall that $Ax = b$ and A nonsingular implies that

$$Ax = b$$
$$A^{-1}Ax = A^{-1}b$$
$$x = A^{-1}b$$

EXAMPLE 12   Solve $Ax = b$ for x where

$$A = \begin{bmatrix} 1 & 1 & -1 \\ -2 & 1 & 1 \\ 1 & 1 & 1 \end{bmatrix} \text{ and } b = \begin{bmatrix} 1 \\ 2 \\ 0 \end{bmatrix}$$

Also compute A^{-1}. Then solve the new system of linear equations

$$Ax = \begin{bmatrix} 1 \\ 0 \\ 3 \end{bmatrix}$$

Solution: Since

$$\begin{bmatrix} 1 & 1 & -1 \,|\, 1 & 0 & 0 \,|\, 1 \\ -2 & 1 & 1 \,|\, 0 & 1 & 0 \,|\, 2 \\ 1 & 1 & 1 \,|\, 0 & 0 & 1 \,|\, 0 \end{bmatrix} \xrightarrow[R_3 - R_1]{R_2 + 2R_1} \begin{bmatrix} 1 & 1 & -1 \,|\, 1 & 0 & 0 \,|\, 1 \\ 0 & 3 & -1 \,|\, 2 & 1 & 0 \,|\, 4 \\ 0 & 0 & 2 \,|\, -1 & 0 & 1 \,|\, -1 \end{bmatrix}$$

$$\xrightarrow[\;(1/3)R_2\;]{R_1 - (1/3)R_2} \begin{bmatrix} 1 & 0 & -2/3 \,|\, 1/3 & -1/3 & 0 \,|\, -1/3 \\ 0 & 1 & -1/3 \,|\, 2/3 & 1/3 & 0 \,|\, 4/3 \\ 0 & 0 & 2 \,|\, -1 & 0 & 1 \,|\, -1 \end{bmatrix}$$

$$\begin{array}{c} R_1 + (1/3)R_3 \\ R_2 + (1/6)R_3 \\ \xrightarrow{} \\ (1/2)R_3 \end{array} \left[\begin{array}{ccc|ccc} 1 & 0 & 0 & 0 & -1/3 & 1/3 & -2/3 \\ 0 & 1 & 0 & 1/2 & 1/3 & 1/6 & 7/6 \\ 0 & 0 & 1 & -1/2 & 0 & 1/2 & -1/2 \end{array}\right]$$

Hence,

$$A^{-1} = \left[\begin{array}{ccc} 0 & -1/3 & 1/3 \\ 1/2 & 1/3 & 1/6 \\ -1/2 & 0 & 1/2 \end{array}\right]$$

and $x = A^{-1}b = [-2/3 \ 7/6 \ -1/2]^T$. When $Ax = [1 \ 0 \ 3]^T$, then

$$x = A^{-1}\left[\begin{array}{c} 1 \\ 0 \\ 3 \end{array}\right] = \left[\begin{array}{ccc} 0 & -1/3 & 1/3 \\ 1/2 & 1/3 & 1/6 \\ -1/2 & 0 & 1/2 \end{array}\right]\left[\begin{array}{c} 1 \\ 0 \\ 3 \end{array}\right] = \left[\begin{array}{c} 1 \\ 1 \\ 1 \end{array}\right]$$

When only the solution of $Ax = b$ is needed, we can form the partitioned matrix $[A|b]$, and then perform a sequence of elementary row operations,

$$[A|b] \rightarrow \cdots \rightarrow [I_m|A^{-1}b]$$

to obtain $[I_m|A^{-1}b]$. This last partitioned matrix represents the equivalent system of equations $x = A^{-1}b$, and the solution is apparent. The matrix $[A|b]$ is often called an *augmented* matrix. This technique for solving the system $Ax = b$ is called the *Gauss-Jordan method of elmination*. The simplex procedure of Chapter 3 for solving linear programming problems is a modification of this technique.

Rather than perform the sequence of elementary row operations, we can obtain the same result using matrix multiplication. To perform an elementary row operation on $[A|b]$ (or $Ax=b$), first perform the row operation on I_m, and then multiply the matrix $[A|b]$ (or both sides of the equation $Ax=b$) on the left by the resulting matrix.

An *elementary* matrix is a square matrix that is obtained by performing one of the elementary row operations on an identity matrix. The following matrices are examples of third order elementary matrices:

$$E_1 = \left[\begin{array}{ccc} 1 & 0 & 0 \\ 2 & 1 & 0 \\ 0 & 0 & 1 \end{array}\right], E_2 = \left[\begin{array}{ccc} 1 & 0 & 0 \\ 0 & 1 & 0 \\ -1 & 0 & 1 \end{array}\right], E_3 = \left[\begin{array}{ccc} 1 & -1/3 & 0 \\ 0 & 1 & 0 \\ 0 & 0 & 1 \end{array}\right]$$

$$E_4 = \left[\begin{array}{ccc} 1 & 0 & 0 \\ 0 & 1 & 0 \\ 0 & 0 & 1/3 \end{array}\right], E_5 = \left[\begin{array}{ccc} 1 & 0 & 1/3 \\ 0 & 1 & 0 \\ 0 & 0 & 1 \end{array}\right], E_6 = \left[\begin{array}{ccc} 1 & 0 & 0 \\ 0 & 1 & 1/6 \\ 0 & 0 & 1 \end{array}\right]$$

and

$$E_7 = \begin{bmatrix} 1 & 0 & 0 \\ 0 & 1 & 0 \\ 0 & 0 & 1/2 \end{bmatrix}$$

These seven elementary matrices represent the steps used in the solution of Example 12. It is left as an exercise for the reader to verify that when A and b are the matrices in that example, then (here $b = [1\ 2\ 0]^T$)

$$E_7(E_6(E_5(E_4(E_3(E_2(E_1A))))))x = E_7(E_6(E_5(E_4(E_3(E_2(E_1b))))))$$

reduces to the solution $x^T = [-2/3\ 7/6\ -1/2]^T$. In fact if

$$F_1 = \begin{bmatrix} 1 & 0 & 0 \\ 2 & 1 & 0 \\ -1 & 0 & 1 \end{bmatrix}, F_2 = \begin{bmatrix} 1 & -1/3 & 0 \\ 0 & 1/3 & 0 \\ 0 & 0 & 1 \end{bmatrix} \text{ and } F_3 = \begin{bmatrix} 1 & 0 & 1/3 \\ 0 & 1 & 1/6 \\ 0 & 0 & 1/2 \end{bmatrix}$$

then

$$x = F_3(F_2(F_1A))x = F_3(F_2(F_1b)) = [-2/3\ \ 7/6\ \ -1/2]^T$$

In the last solution, we performed more elementary row operations than were actually necessary. Instead let us first multiply each side of the equation $Ax = b$ on the left by F_1 to get

$$\begin{bmatrix} 1 & 1 & -1 \\ 0 & 3 & -1 \\ 0 & 0 & 2 \end{bmatrix}x = \begin{bmatrix} 1 & 0 & 0 \\ 2 & 1 & 0 \\ -1 & 0 & 1 \end{bmatrix}\begin{bmatrix} 1 & 1 & -1 \\ -2 & 1 & 1 \\ 1 & 1 & 1 \end{bmatrix}x = F_1Ax$$

$$= F_1b = \begin{bmatrix} 1 & 0 & 0 \\ 2 & 1 & 0 \\ -1 & 0 & 1 \end{bmatrix}\begin{bmatrix} 1 \\ 2 \\ 0 \end{bmatrix} = \begin{bmatrix} 1 \\ 4 \\ -1 \end{bmatrix}$$

which is equivalent to the system of equations

$$\begin{aligned} x_1 + x_2 - x_3 &= 1 \\ 3x_2 - x_3 &= 4 \\ 2x_3 &= -1 \end{aligned}$$

Clearly, $x_3 = -1/2$. Moreover, x_2 can be computed using the second equation and this value for x_3,

$$3x_2 + \frac{1}{2} = 4$$

$$3x_2 = \frac{7}{2}$$

$$x_2 = \frac{7}{6}$$

Finally, substitute these values for x_2 and x_3 into the first equation and solve for x_1,

$$x_1 + \frac{7}{6} + \frac{1}{2} = 1$$

$$x_1 = 1 - \frac{10}{6} = -\frac{2}{3}$$

Any matrix having all zero entries below the main diagonal is said to be *upper triangular*. The process of using elementary row operations on $[A|b]$ to produce an augmented matrix $[U|\bar{b}]$ (or the process of multiplying each side of the matrix equation $Ax=b$ on the left by a sequence of nonsingular matrices to obtain $Ux = \bar{b}$), where U is upper triangular is called *forward elimination*. The process of substituting the values of $x_m, x_{m-1}, \ldots, x_{m-i}$ into the $(m-i-1)$th equation and then solving for x_{m-i-1} is called *back substitution*. When forward elimination and back substitution are used together to solve $Ax=b$ the resulting technique is known as *Gaussian elimination*. For large systems, Gaussian elimination is numerically superior to Gauss-Jordan elimination. For such problems Gaussian elimination results in fewer computations and less round-off error. A more complete discussion of these advantages can be found in the references at the end of this chapter, as well as in most numerical analysis textbooks.

EXAMPLE 13   Solve the system

$$\begin{aligned} 2x_2 - x_3 &= 8 \\ x_1 + 2x_2 &= 6 \\ 3x_2 + 4x_3 &= 12 \end{aligned}$$

using Gaussian elimination. Also determine the sequence of matrices that are required to transform A into an upper triangular matrix.
Solution: If we use forward elimination on the augmented matrix $[A|b]$ we obtain

$$\begin{bmatrix} 0 & 2 & -1 & 8 \\ 1 & 2 & 0 & 6 \\ 0 & 3 & 4 & 12 \end{bmatrix} \xrightarrow{R_1, R_2} \begin{bmatrix} 1 & 2 & 0 & 6 \\ 0 & 2 & -1 & 8 \\ 0 & 3 & 4 & 12 \end{bmatrix} \xrightarrow{R_2, R_3} \begin{bmatrix} 1 & 2 & 0 & 6 \\ 0 & 3 & 4 & 12 \\ 0 & 2 & -1 & 8 \end{bmatrix}$$

$$\xrightarrow{R_3 - (2/3)R_2} \begin{bmatrix} 1 & 2 & 0 & 6 \\ 0 & 3 & 4 & 12 \\ 0 & 0 & -11/3 & 0 \end{bmatrix}$$

Using back substitution, we have $x_3=0$, $x_2=4$ and $x_1=-2$. The matrices required to produce

$$U = \begin{bmatrix} 1 & 2 & 0 \\ 0 & 3 & 4 \\ 0 & 0 & -11/13 \end{bmatrix}$$

are

$$F_1 = \begin{bmatrix} 0 & 1 & 0 \\ 1 & 0 & 0 \\ 0 & 0 & 1 \end{bmatrix}, F_2 = \begin{bmatrix} 1 & 0 & 0 \\ 0 & 0 & 1 \\ 0 & 1 & 0 \end{bmatrix} \text{ and } F_3 = \begin{bmatrix} 1 & 0 & 0 \\ 0 & 1 & 0 \\ 0 & -2/3 & 1 \end{bmatrix}$$

i.e., $F_3 F_2 F_1 A = U$. Notice that the product $F_2 F_1$ can be replaced by

$$P = \begin{bmatrix} 0 & 1 & 0 \\ 0 & 0 & 1 \\ 1 & 0 & 0 \end{bmatrix}$$

i.e., $F_3 P A = U$.

The use of elementary row operations, or elementary matrices, to introduce zero entries below (or above and below) a nonzero entry of a matrix is called a *pivot*. Moreover the entry is called the *pivot* element. The matrix F which produces this change is called a *pivot* matrix. Clearly, a pivot matrix is nonsingular and differs from an identity matrix in at most one column (see Exercise 2.8).

In Example 13, F_3 is the only pivot matrix. Notice that the second and third rows were switched before the pivot F_3 took place. The reason for the switch is that the larger the pivot element a_{ii}, the smaller the fractions a_{ji}/a_{ii} and, hence, the smaller the round-off error (see Exercise 22).

When the rows are permuted in the Gaussian elimination procedure to allow the largest possible pivot element, the technique is known as *partial* pivoting. Additional modifications used to reduce round-off error can be found in the references.

If the rows of the identity matrix I_m are permuted to obtain a matrix P, then P is called a *permutation* matrix of order m. In Example 13, $P = F_2 F_1$ is a permutation matrix. Clearly, any permutation matrix is nonsingular.

Now suppose that partial elimination has been applied to a matrix equation $Ax = b$ to obtain

$$Ux = (F_k P_k) \cdots (F_1 P_1) A x = (F_k P_k) \cdots (F_1 P_1) b = \bar{b}$$

where U is upper triangular, each F_i is a pivot matrix, and each P_i is a permutation (possibly the identity) matrix. Then $(F_k P_k)\dots(F_1 P_1)$ is nonsingular. Hence, let $L = (F_k P_k \dots F_1 P_1)^{-1} = P_1^{-1} F_1^{-1} \dots P_k^{-1} F_k^{-1}$. Then $A = LU$. This is called a *LU-decomposition* of A. Let $y = Ux$. Then

$$b = Ax = LUx = Ly$$

It follows that solving the matrix equation $Ax = b$ for x is equivalent to first solving the matrix equation $Ly = b$ for y, and then solving the matrix equation $Ux = y$ for x. Since U is upper triangular, the last equation can be solved by back substitution. The matrices F_i are all lower triangular (a *lower triangular* matrix has all zero entries above the main diagonal). Thus, if each P_i is the identity matrix, then L is lower triangular and the system $Ly = b$ can also be solved by back substitution. Even when L is not lower triangular, the solution y can be easily computed by

$$y = L^{-1}b = F_k P_k \cdots F_1 P_1 b$$

i.e., by performing the sequence of row permutations and operations on b that were performed on A to get U. Once a LU-decomposition is known for A, then any problem $Ax = b$ can be solved in this manner.

Finally it should be noted that the pivot matrices

$$F = \begin{bmatrix} 1 \dots f_{1k} \dots 0 \\ 0 \dots f_{2k} \dots 0 \\ \vdots \quad \vdots \quad \vdots \\ 0 \dots f_{mk} \dots 1 \end{bmatrix}$$

can be efficiently stored (or retained) in the form $(k, [f_{1k} \dots f_{mk}]^T)$, where k denotes the only column in which F differs from I_m and $[f_{1k} \dots f_{mk}]^T$ is that column.

We shall close this section with a brief discussion concerning the solutions of the system of equations

$$Ax = b \tag{3}$$

where $A \in \mathbf{R}_{m \times n}$ and $m < n$. We shall also assume that rank $A = m$. Since rank $A = m$, it follows that matrix A contains at least one set of m linearly independent columns. Let B be a matrix of m such columns, i.e., let B be a matrix composed of m linearly independent columns of A. Notice that B may or may not be a submatrix of A. For example if

$$A = \begin{bmatrix} 1 & 4 & 0 \\ 0 & 5 & 6 \end{bmatrix} \text{ and } B = \begin{bmatrix} 1 & 0 \\ 0 & 6 \end{bmatrix}$$

then B would be a submatrix of A. But if

$$B = \begin{bmatrix} 0 & 1 \\ 6 & 0 \end{bmatrix}$$

then B would not be a submatrix of A. Let D denote the matrix of the remaining columns of A. In the present example $D = A^{(2)}$.

To simplify our notation and discussion we shall now let $B = [A^{(1)} A^{(2)} \cdots A^{(m)}]$. Define $x_B = [x_1 \ldots x_m]^T$ and $x_D = [x_{m+1} \ldots x_n]^T$. Then (3) becomes

$$Bx_B + Dx_D = b \tag{4}$$

In augmented matrix form, (4) becomes $[B|D|b]$. If we multiply this augmented matrix on the left by B^{-1}, we have

$$B^{-1}[B|D|b] = [B^{-1}B|B^{-1}D|B^{-1}b] = [I_m|B^{-1}D|B^{-1}b] \tag{5}$$

The matrix on the far right side in (5) will be denoted by $[X_B|B^{-1}b]$, where $X_B = [I_m|B^{-1}D]$. It can also be obtained by a sequence of elementary row operations as indicated below

$$[B|D|b] \rightarrow \cdots \rightarrow [I^{(m)}|B^{-1}D|B^{-1}b] \tag{6}$$

In either case, the augmented matrix $[X_B|B^{-1}b]$ represents the system of equations

$$x_B + B^{-1}Dx_D = B^{-1}b \tag{7}$$

Systems (4) and (7) are equivalent, i.e., they have the same solutions. It is apparent from $x_B = B^{-1}b - B^{-1}Dx_D$, that (7) has infinitely many solutions (assign arbitrary values to the variables in x_D). The components of x_B are called *basic* variables. When $x_B = B^{-1}b$ and $x_D = [0 \cdots 0]^T$, the solution is said to be *basic*. In this case, the columns of B are said to be an *admissible basis* for this basic solution and B is called a *basis* matrix. Clearly, there exists at most

$$\binom{n}{m} = \frac{n!}{m!\,(n-m)!}$$

basic solutions. When none of the components of $B^{-1}b$ are zero, then the basic solution is said to be *nondegenerate*.

The concept of a basic solution for a system of linear equations is of fundamental importance in Linear programming. The connection is developed in Theorem 8 of this chapter, and in Chapter 3 as well.

It is possible to move from one basic solution of (3) to another without returning to the matrix A (in order to find a different set of m linearly independent columns). To illustrate, suppose that the first entry, denoted by d_1, of the first column $B^{-1}D^{(1)} = B^{-1}A^{(m+1)}$, where $B^{-1}D^{(1)} = [d_1 \ldots d_m]^T$, of $B^{-1}D$ is nonzero, i.e., $d_1 \neq 0$. Let $\bar{B} = [A^{(m+1)} \ A^{(2)} \ \ldots \ A^{(m)}]$. Notice that

$$B^{-1}\bar{B} = B^{-1}[A^{(m+1)} \ A^{(2)} \ \ldots \ A^{(m)}] = [B^{-1}A^{(m+1)} \ B^{-1}A^{(2)} \ \ldots \ B^{-1}A^{(m)}]$$
$$= [B^{-1}A^{(m+1)} \ I^{(2)} \ \ldots \ I^{(m)}]$$

Let

$$P = \begin{bmatrix} 1/d_1 & 0 & \ldots & 0 \\ -d_2/d_1 & 1 & & 0 \\ \vdots & \vdots & & \vdots \\ -d_m/d_1 & 0 & \ldots & 1 \end{bmatrix}$$

Then

$$(PB^{-1})\bar{B} = P[B^{-1}A^{(m+1)} \ I^{(2)} \ \ldots \ I^{(m)}] = I_m$$

This implies that $\bar{B}$ is nonsingular and $\bar{B}^{-1} = PB^{-1}$. Clearly,

$$\bar{B}^{-1}[A|b] = \bar{B}^{-1}[A^{(1)} \ A^{(2)} \ \ldots \ A^{(m)} \ A^{(m+1)} \ A^{(m+2)} \ \ldots \ A^{(n)}|b]$$
$$= [\bar{B}^{-1}A^{(1)} \ \bar{B}^{-1}A^{(2)} \ \ldots \bar{B}^{-1}A^{(m)} \ \bar{B}^{-1}A^{(m+1)} \ \bar{B}^{-1}A^{(m+2)} \ \ldots \ \bar{B}^{-1}A^{(n)}|\bar{B}^{-1}b]$$
$$= [\bar{B}^{-1}A^{(1)} \ I^{(2)} \ \ldots \ I^{(m)} \ I^{(1)} \ \bar{B}^{-1}A^{(m+2)} \ \ldots \ \bar{B}^{-1}A^{(n)}|\bar{B}^{-1}b] \qquad (8)$$

The right hand matrix in (8) can also be obtained by

$$[B|D|b] \quad \xrightarrow[\text{as (6)}]{\text{same sequence}} \quad [I_m|B^{-1}D|B^{-1}b]$$

$$\xrightarrow[\substack{(1/d_1)R_1}]{\substack{R_i - (d_i/d_1)R_1 \\ i \neq 1}} [\bar{B}^{-1}A^{(1)} \ I^{(2)} \ \ldots \ I^{(m)} \ I^{(1)} \ \bar{B}^{-1}A^{(n)}|\bar{B}^{-1}b] \qquad (9)$$

Regardless of the method used to obtain this augmented matrix, it represents the system of equations

$$x_{\bar{B}} + \bar{B}^{-1}\bar{D}x_{\bar{D}} = \bar{B}^{-1}b, \qquad (10)$$

where $x_{\bar{B}} = [x_{m+1} \ x_2 \ \ldots \ x_m]^T$, $\bar{D} = [A^{(1)} \ A^{(m+2)} \ \ldots \ A^{(n)}]$, and $x_{\bar{D}} = [x_1 \ x_{m+2} \ \ldots \ x_n]^T$. As before systems (3), (7), and (10) are equivalent. Also, $x_{\bar{B}} = \bar{B}^{-1}b$ and $x_{\bar{D}} = [0 \ \ldots \ 0]^T$ is a basic solution of (3).

EXAMPLE 14   Use the above techniques to find two basic solutions of

$$x_1 + x_2 + \qquad 6x_4 = 4$$
$$x_2 - 3x_3 + 2x_4 = 8$$

Solution: In augmented matrix form, $[A|b]$, this problem is

$$\begin{bmatrix} 1 & 1 & 0 & 6 & 4 \\ 0 & 1 & -3 & 2 & 8 \end{bmatrix}$$

Clearly, the first two columns of this matrix are linearly independent. Hence, we shall let $B = [A^{(1)} A^{(2)}]$ and $D = [A^{(3)} A^{(4)}]$. Proceeding as in (6) we have

$$\begin{bmatrix} 1 & 1 & 0 & 6 & 4 \\ 0 & 1 & -3 & 2 & 8 \end{bmatrix} \xrightarrow{R_1 - R_2} \begin{bmatrix} 1 & 0 & 3 & 4 & -4 \\ 0 & 1 & -3 & ② & 8 \end{bmatrix}$$

Hence, a basic solution is $x_1 = -4$, $x_2 = 8$, and $x_3 = x_4 = 0$. To generate another basic solution, we shall select a nonzero entry from among the entries of the third or fourth columns. The entry that we have selected is the one that has been circled in the last matrix. Continuing as in (9) gives

$$\begin{bmatrix} 1 & 0 & 3 & 4 & -4 \\ 0 & 1 & -3 & ② & 8 \end{bmatrix} \xrightarrow[\;(1/2)R_2\;]{R_1 - (4/2)R_2} \begin{bmatrix} 1 & -2 & 9 & 0 & -20 \\ 0 & 1/2 & -3/2 & 1 & 4 \end{bmatrix}$$

Thus, a second basic solution is $x_1 = -20$, $x_4 = 4$, and $x_2 = x_3 = 0$. Here, $\bar{B} = [A^{(1)} A^{(4)}]$ and $\bar{D} = [A^{(2)} A^{(3)}]$. Notice that both basic solutions are nondegenerate.

In this example

$$X_B = \begin{bmatrix} 1 & 0 & 3 & 4 \\ 0 & 1 & -3 & 2 \end{bmatrix} \quad \text{and} \quad X_{\bar{B}} = \begin{bmatrix} 1 & -2 & 9 & 0 \\ 0 & 1/2 & -3/2 & 1 \end{bmatrix}$$

If we had originally selected $B = [A^{(2)} \; A^{(1)}]$, then

$$\begin{bmatrix} 1 & 1 & 0 & 6 & 4 \\ 0 & 1 & -3 & 2 & 8 \end{bmatrix} \xrightarrow{R_2 - R_1} \begin{bmatrix} 1 & 1 & 0 & 6 & 4 \\ -1 & 0 & -3 & -4 & 4 \end{bmatrix}$$

$$\xrightarrow[\;-R_2\;]{R_1 + R_2} \begin{bmatrix} 0 & 1 & -3 & 2 & 8 \\ 1 & 0 & 3 & 4 & -4 \end{bmatrix} = X_B$$

Notice that the matrix X_B is now different from our first X_B.

Exercises

1.2   (a) Show that the product of two upper triangular matrices is an upper triangular matrix.

(b) show that the product of two lower triangular matrices is a lower triangular matrix.

2.2   Solve the system (always rounding to the nearest ten thousandth)

$$0.0001x + 1.00y = 1.00$$
$$1.00x + 1.00y = 2.00$$

using Gaussian elimination and 0.0001 as a pivot element. Then solve the system using partial pivoting. Compare the two solutions. Which is correct? [Forsythe (1967)].

2.3   Find an example of a square matrix that does not have a LU-decomposition.

2.4   Let A be a nonsingular matrix. If $A = L_1 U_1$ and $A = L_2 U_2$ are two LU-decompositions of A, and if the diagonal entries of L_1 and L_2 agree, then show that $L_1 = L_2$ and $U_1 = U_2$.

2.5   Let $A \in \mathbf{R}_{m \times n}$ have a LU-decomposition. Show that there exists a permutation matrix P so that PA has a LU-decomposition, where the factor L is lower triangular.

2.6   Show that a square matrix A can be factored as $A = LU$, where L is a lower triangular matrix with all diagonal entries $l_{ii} = 1$ and U is an upper triangular matrix, whenever none of the pivots are zero.

2.7   (a) Find an LU-decomposition of

$$A = \begin{bmatrix} 1 & 5 & -1 & 5 \\ 1 & 2 & -1 & 4 \\ 1 & 5 & 0 & 10 \\ 1 & 5 & 0 & 9 \end{bmatrix}$$

(b) Use the LU-decomposition of A to solve

$$Ax = \begin{bmatrix} 1 & -4 & 0 & 1 \end{bmatrix}^T$$

2.8   Let $A = [a_{ij}] \in \mathbf{R}_{3 \times 3}$ and let $a_{11} \neq 0$ be a pivot element. Determine the corresponding pivot matrix F and show that it is nonsingular.

2.9   If $A = LU$, where L is lower triangular and U is upper triangular, then show that $A^T = U^T L^T$ is a LU-decomposition for A^T.

2.10   Show that $A = BX_B$. In particular, each $A^{(j)} = BX_B^{(j)}$.

2.11 Determine three basic solutions of the system of equations

$$5x_1 \qquad + 7x_3 + 2x_4 \qquad + 6x_6 + x_7 \qquad = 9$$
$$6x_1 + 5x_2 \qquad - x_4 \qquad + x_7 + x_8 = 7$$
$$x_1 + 4x_2 \qquad + 3x_4 + x_5 - x_6 + x_7 \qquad = 10$$

2. AFFINE AND CONVEX SETS

Let $x_1 = [x_{11} \cdots x_{1n}]^T, \ldots, x_k = [x_{k1} \cdots x_{kn}]^T$ belong to $\mathbf{R}_{n \times 1}$. Recall that a *linear combination* of the vectors $x_1, \cdots, x_k$ is any expression of the form $\sum_{i=1}^k \alpha_i x_i$, where each $\alpha_i \in \mathbf{R}$. When $\sum_{i=1}^k \alpha_i = 1$, the linear combination $\sum_{i=1}^k \alpha_i x_i$ is called an *affine* combination of the vectors $x_1, \cdots, x_k$. When each $\alpha_i \geq 0$, the linear combination $\sum_{i=1}^k \alpha_i x_i$ is called a *conical combination* (or a *nonnegative combination*) of the vectors $x_1, \ldots, x_k$. When each $\alpha_i \geq 0$ and $\sum_{i=1}^k \alpha_i = 1$, then the linear combination $\sum_{i=1}^k \alpha_i x_i$ is called a *convex combination* of the vectors $x_1, \ldots, x_k$.

Let S be a nonempty subset of $\mathbf{R}_{n \times 1}$. Recall that the *space spanned by S* (also called the *linear hull of S*), $L(S)$, is the collection of all linear combinations of vectors of S, i.e.,

$$L(S) = \{\sum_{i=1}^k \alpha_i x_i : \text{each } x_i \in S, \text{ each } \alpha_i \in \mathbf{R}, k \text{ is a natural number}\}$$

We now define similar concepts using affine, conical and convex combinations of vectors of S.

The *affine hull*, $\text{Aff}(S)$, *of S* is defined to be the collection of all affine combinations of vectors of S, i.e.,

$$\text{Aff}(S) = \{\sum_{i=1}^k \alpha_i x_i : \text{each } x_i \in S, \text{ each } \alpha_i \in \mathbf{R}, \sum_{i=1}^k \alpha_i = 1, k \text{ is a natural number}\}$$

The *conical hull*, $\text{coni}(S)$, *of S* is defined to be the collection of all conical combinations of vectors from S, i.e.,

$$\text{coni}(S) = \{\sum_{i=1}^k \alpha_i x_i : \text{each } x_i \in S, \text{ each } \alpha_i \geq 0, k \text{ is a natural number}\}$$

The *convex hull*, $\text{conv}(S)$, *of S* is the collection of all convex combinations of vectors of S, i.e.,

$$\text{conv}(S) = \{\sum_{i=1}^k \alpha_i x_i : \text{each } x_i \in S, \text{each } \alpha_i \geq 0, \sum_{i=1}^k \alpha_i = 1, k \text{ is a natural number}\}$$

EXAMPLE   Let $S = \{[1 \quad 0 \quad 0]^T, [0 \quad 1 \quad 0]^T\}$. Determine geometrically $L(S)$, aff(S), coni(s) and conv(s).

Solution: First,

$$L(S) = \{\alpha[1 \quad 0 \quad 0]^T + \beta[0 \quad 1 \quad 0]^T : \alpha, \beta \in \mathbf{R}_1\}$$
$$= \{[\alpha \quad \beta \quad 0]^T : \alpha, \beta \in \mathbf{R}\} = x_1x_2\text{–plane}$$

(see Figure 2).

Second,

$$\text{aff}(S) = \{\alpha[1 \quad 0 \quad 0]^T + \beta[0 \quad 1 \quad 0]^T : \alpha, \beta \in \mathbf{R}, \alpha + \beta = 1\}$$
$$= \{[\alpha \quad \beta \quad 0]^T : \alpha, \beta \in \mathbf{R}, \beta = 1 - \alpha\} = \{[\alpha \quad 1-\alpha \quad 0]^T : \alpha \in \mathbf{R}\}$$

which is the straight line that passes through the vectors $[1 \quad 0 \quad 0]^T$ and $[0 \quad 1 \quad 0]^T$ (see Figure 3).

Third,

$$\text{coni}(S) = \{\alpha[1 \quad 0 \quad 0]^T + \beta[0 \quad 1 \quad 0]^T : \alpha, \beta \in \mathbf{R}, \alpha \geq 0, \beta \geq 0\}$$
$$= \{[\alpha \quad \beta \quad 0]^T : \alpha \geq 0, \beta \geq 0\}$$

which is the nonnegative orthant of the x_1x_2 – plane (see Figure 4).

Fourth,

$$\text{conv}(S) = \{\alpha[1 \quad 0 \quad 0]^T + \beta[0 \quad 1 \quad 0]^T : \alpha, \beta \in \mathbf{R}, \alpha \geq 0, \beta \geq 0, \alpha + \beta = 1\}$$
$$= \{[\alpha \quad 1-\alpha \quad 0]^T : \alpha \in \mathbf{R}, 1 \geq \alpha \geq 0\}$$

which is the straight line segment that connects the vectors $[1 \quad 0 \quad 0]^T$ and $[0 \quad 1 \quad 0]^T$ (see Figure 5).

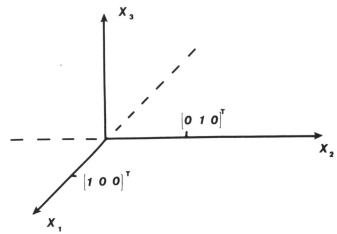

Figure 2 $L(S) = L(\{[1 \quad 0 \quad 0]^T, [0 \quad 1 \quad 0]^T\})$ is the x_1x_2 – plane.

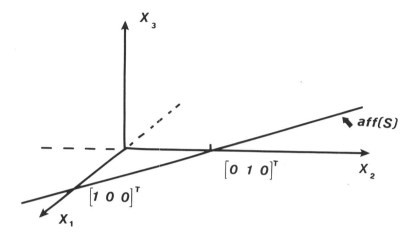

Figure 3 Aff(S) is the line passing through $[1 \quad 0 \quad 0]^T$ and $[0 \quad 1 \quad 0]^T$.

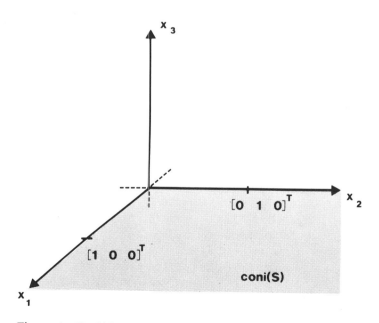

Figure 4 Coni(S) is the nonnegative orthant of the $x_1 x_2$ – plane.

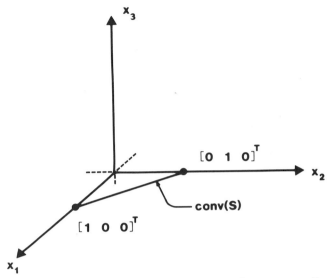

Figure 5 Conv(S) is the line segment from $[1 \quad 0 \quad 0]^T$ to $[0 \quad 1 \quad 0]^T$, inclusive.

Notice that for any nonempty subset of S of $\mathbf{R}_{n \times 1}$

$$S \subseteq \text{conv}(S) \subseteq \text{aff}(S) \subseteq L(S)$$

and

$$S \subseteq \text{conv}(S) \subseteq \text{coni}(S) \subseteq L(S)$$

Recall that S is a *subspace* of $\mathbf{R}_{n \times 1}$ if and only if $x, z \in S$ and $\alpha, \beta \in \mathbf{R}$ implies that $\alpha x + \beta z \in S$. The next result characterizes subspaces in terms of linear hulls.

PROPOSITION 1 S is a subspace of $\mathbf{R}_{n \times 1}$ if and only if $S = L(S)$.
Proof: Recall from section 1 that $L(S)$ is a subspace of $\mathbf{R}_{n \times 1}$ Hence, if $S = L(S)$, then S is a subspace of $\mathbf{R}_{n \times 1}$.

Next suppose that S is a subspace of $\mathbf{R}_{n \times 1}$. In this case we want to show that $S = L(S)$. But, $S \subseteq L(S)$. Thus, we only need to establish that $L(S) \subseteq S$. This will be done by mathematical induction. First consider any linear combination of one vector from S, i.e., $\alpha x \in L(S)$ when $\alpha \in \mathbf{R}$ and $x \in S$. Since S is a subspace it follows that $\alpha x \in S$. Next assume that any linear combination of k vectors from S is again a vector in S. Then consider a linear combination of $k + 1$ vectors from S, $\Sigma_{i=1}^{k+1} \alpha_i x_i$, where each $\alpha_i \in \mathbf{R}$ and each $x_i \in S$. Since $\Sigma_{i=1}^{k} \alpha_i x_i$ is a linear combination of k vectors from S, then by our hypothesis $\Sigma_{i=1}^{k} \alpha_i x_i \varepsilon S$. Hence,

$$\sum_{i=1}^{k+1} \alpha_i x_i = \sum_{i=1}^{k} \alpha_i x_i + \alpha_{k+1} x_{k+1}$$

is a linear combination of two vectors, $\sum_{i=1}^{k} \alpha_i x_i$ and x_{k+1}, from S. Since S is a subspace it follows that $\sum_{i=1}^{k+1} \alpha_i x_i \in S$. Hence, by the principle of mathematical induction it follows that $L(S) \subseteq S$.

We will now develop three somewhat analogous results concerning affine, conical and convex combinations.

A set S is an *affine* set if and only if $x, y \in S$ and $\alpha, \beta \in \mathbf{R}$, where $\alpha + \beta = 1$, implies that $\alpha x + \beta y \in S$. Thus, S is affine if and only if the line passing through any two distinct points of S is itself a subset of S. Examples of affine sets in $\mathbf{R}_{3 \times 1}$ are lines, planes, single points (in which case $x = y$) and $\mathbf{R}_{3 \times 1}$ itself. In $\mathbf{R}_{n \times 1}$ the affine hull of any set S, aff(S), is an affine set. To see this suppose that $\sum_{i=1}^{k} \alpha_i x_i$ and $\sum_{i=1}^{\ell} \beta_i y_i$ both belong to aff(S). Then let $\lambda, \xi \in \mathbb{R}$ such that $\lambda + \xi = 1$. Then

$$\lambda \sum_{i=1}^{k} \alpha_i x_i + \xi \sum_{i=1}^{\ell} \beta_i y_i = \sum_{i=1}^{k} (\lambda \alpha_i) x_i + \sum_{i=1}^{\ell} (\xi \beta_i) y_i$$

belongs to aff(S), since

$$\sum_{i=1}^{k} \lambda \alpha_i + \sum_{i=1}^{\ell} \xi \beta_i = \lambda \sum_{i=1}^{k} \alpha_i + \xi \sum_{i=1}^{\ell} \beta_i = \lambda + \xi = 1$$

Figure 6 gives an example of a set in $\mathbf{R}_{2 \times 1}$ that is not affine.

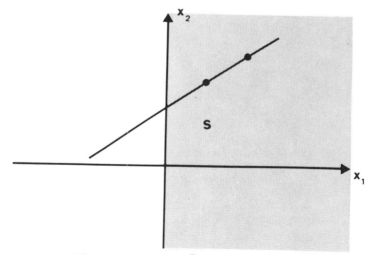

Figure 6 $S = \{[x_1 x_2]^T : x_1 \geq 0\}$ is not affine.

The set S is a *convex* set if and only if $x, y \in S$ and $a, \beta \in \mathbf{R}^+$, where $\mathbf{R}^+$ is the set of nonnegative real numbers, such that $\alpha + \beta = 1$ implies that $\alpha x + \beta y \in S$, i.e., S is convex if and only if the line segment connecting two distinct points in S is a subset of S. Figure 7 gives both an example of a convex set in $\mathbf{R}_{2 \times 1}$, and a set that is not convex. In Figure 7(b) the set is not convex because the line segment connecting points P and Q is not contained in S. Clearly, (as in the affine case) the convex hull, conv(S), of a set is a convex set. This follows since the convex combination of two convex combinations of vectors from S is again a convex combination of vectors from S. Also, any affine set is convex but some convex sets are not affine.

Finally, the set S is a *convex cone with vertex* the origin if and only if $x, y \in S$ and $\alpha, \beta \in \mathbf{R}$ such that both $\alpha \geq 0$ and $\beta \geq 0$ implies that $\alpha x + \beta y \in S$. Just as in the linear, affine and convex cases, the conical hull of any set S, coni(S), is a convex cone with vertex the origin. Geometrically, a convex cone with vertex the origin is any convex set that contains all the rays from the origin through each of the points in the set. Figures 2, 4, and 6 give examples of convex cones with vertices the origin. Neither set in Figure 7 is a convex cone with vertex the origin. The set $S = \{[x_1 x_2]^T : x_2 = |x_1|\}$ in Figure 8(a) is not a convex cone with vertex the origin since it is not a convex set. However, the set $T = \{[x_1 x_2]^T : x_2 \geq |x_1|\}$ in Figure 8(b) is such a cone.

Recall that a set S in $\mathbf{R}_{n \times 1}$ is a subspace of $\mathbf{R}_{n \times 1}$ if and only if $S = L(S)$. Analogous results also hold for affine sets, convex sets and convex cones with vertex the origin.

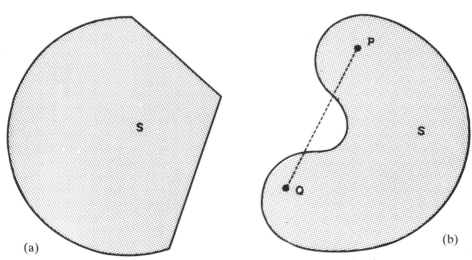

Figure 7   (a) A convex set in $\mathbf{R}_{2 \times 1}$. (b) A subset of $\mathbf{R}_{2 \times 1}$ that is not convex.

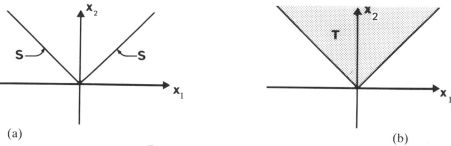

(a) **(b)**

Figure 8 (a) $S = \{[x_1\; x_2]^T : x_2 = x_1\}$ is not a convex cone with vertex the origin.
(b) $T = \{[x_1\; x_2]^T : x_2 \geq x_1\}$ is a convex cone with vertex the origin.

PROPOSITION 2 Let $S \subseteq \mathbf{R}_{n \times 1}$. Then
(i) S is an affine set if and only if $S = \text{aff}(S)$.
(ii) S is a convex cone with vertex the origin if and only if $S = \text{coni}(S)$.
(iii) S is a convex set if and only if $S = \text{conv}(S)$.
Proof: Only the proof of part (iii) will be given. The proofs of parts (i) and (ii) are left to the reader. First suppose that $S = \text{conv}(S)$, then S is convex since the conv(S) is convex.

Next suppose that S is convex. Since $S \subseteq \text{conv}(S)$, it remains to be shown that $\text{conv}(S) \subseteq S$. This will be done by induction. First the convex combination of any two elements from S is an element in S since S is convex. Next suppose that the convex combination of any k elements of S is again an element in S. Then consider any convex combination of $k + 1$ elements from S, say $\Sigma_{i=1}^{k+1} \alpha_i x_i$. Let $\alpha = \Sigma_{i=1}^{k} \alpha_i$. Since $\alpha_{k+1} = 1 - \alpha$, then

$$\sum_{i=1}^{k+1} \alpha_i x_i = \alpha \sum_{i=1}^{k} \frac{\alpha_i}{\alpha} x_i + (1 - \alpha) x_{k+1}$$

where

$$\sum_{i=1}^{k} \frac{\alpha_i}{\alpha} = \frac{1}{\alpha} \sum_{i=1}^{k} \alpha_i = \frac{1}{\alpha} \cdot \alpha = 1$$

and each $\alpha_i / \alpha \geq 0$. Thus $\Sigma_{i=1}^{k} (\alpha_i / \alpha) x_i \in S$ by the inductive hypothesis. But then $\Sigma_{i=1}^{k+1} \alpha_i x_i$ is a convex combination of two vectors from S, $\Sigma_{i=1}^{k} (\alpha_i / \alpha) x_i$ and x_{k+1}, and hence $\Sigma_{i=1}^{k+1} \alpha_i x_i \in S$. Thus, we have shown inductively that $\text{conv}(S) \subseteq S$.

Recall from linear algebra that any intersection of subspaces of $\mathbf{R}_{n \times 1}$ is again a subspace. It is an easy exercise to show that

  (i)   Any intersection of affine sets is again an affine set.

  (ii) Any intersection of convex cones (each of which has vertex the origin) is again a convex cone with vertex the origin.

 (iii) Any intersection of convex sets is again a convex set.

     We wish now to introduce the concept of dimension to subsets in $\mathbf{R}_{n\times 1}$. We begin by recalling that some examples of affine sets in $\mathbf{R}_{3\times 1}$ include lines, planes, single points and $\mathbf{R}_{3\times 1}$ itself. It appears that affine sets are merely the translates of subspaces. This is verified by the following proposition.

PROPOSITION 3   A subset S of $\mathbf{R}_{n\times 1}$ is affine if and only if $S - x_0$ is a subspace of $\mathbf{R}_{n\times 1}$, for every $x_0 \in S$ (see Figure 9).

Proof: First suppose that S is affine and that $x_0 \in S$. We want to show that $S - x_0 = \{x - x_0 : x \in S\}$ is a subspace of $\mathbf{R}_{n\times 1}$. To do this let $y,z \in S - x_0$ and $\alpha,\beta \in \mathbf{R}$. Since $y,z \in S - x_0$, then $y + x_0$, $z + x_0 \in S$. But then $x_0 \in S$, $\alpha + \beta + (1 - \alpha - \beta) = 1$ and S affine all imply that

$$\alpha(y + x_0) + \beta(z + x_0) + (1 - \alpha - \beta) x_0 \in S$$

$$\alpha y + \beta z + x_0 \in S$$

$$\alpha y + \beta z \in S - x_0$$

Hence, $S - x_0$ is a subspace of $\mathbf{R}_{n\times 1}$.

     To finish the proof we must show that if $S - x_0$ is a subspace for every $x_0 \in S$, then S is affine. To do this suppose $y,z \in S$ and $\alpha,\beta \in \mathbf{R}$ such that $\alpha + \beta = 1$. Then $y - x_0, z - x_0 \in S - x_0$. Since $S - x_0$ is a subspace and $\alpha + \beta = 1$, it follows that

$$\alpha y + \beta z - x_0 = \alpha y + \beta z - (\alpha + \beta)x_0 = \alpha(y - x_0) + \beta(z - x_0) \in S - x_0$$

i.e., $\alpha y + \beta z \in S$. Hence, S is affine.

     Thus, an affine set is just the translate of a subspace. It is an easy exercise to show that the subspace is unique. We define the *dimension of an affine set* to be the dimension of the subspace of which it is a translate. Thus, lines have dimension 1, planes have dimension 2, and points have dimension 0.

     The *dimension of any subset, T,* of $\mathbf{R}_{n\times 1}$ is defined to be the dimension of its affine hull aff(T). We say that the dimension of the empty set is -1.

EXAMPLE 15   In $\mathbf{R}_{2\times 1}$ let $S = \{[x_1\, x_2]^T : x_1 + x_2 = 1, x_1 \geq 0, x_2 \geq 0\}$. Then aff($S$) $= \{[x_1\, x_2]^T : x_1 + x_2 = 1\}$. Thus, S is the line segment that connects the points $[1\quad 0]^T$ and $[0\quad 1]^T$. Aff(S) is the line that passes through these points. Since $[1\quad 0]^T \in$ aff S, then aff(S) $- [1\quad 0]^T = \{[x_1\, x_2]^T : x_1 + x_2 = 0\}$.

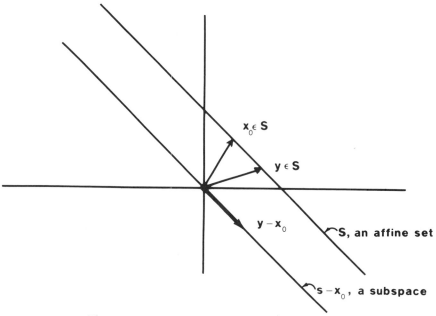

Figure 9   Affine sets are translates of subspaces.

Hence,

$$\dim S = \dim \operatorname{aff}(S) = \dim \{[x_1\, x_2]^T : x_1 + x_2 = 0\} = 1$$

(see Figure 10).

One type of affine set that is of special interest in mathematics is a generalization of the plane in $\mathbf{R}_{3\times 1}$ and the line in $\mathbf{R}_{2\times 1}$. Let $c = [c_1 \cdots c_n]^T$ be a nonzero vector in $\mathbf{R}_{n+1}$ and let $\beta \in \mathbf{R}$. Define

$$H(c,\beta) = \{x = [x_1 \ldots x_n]^T : c^T x = \beta\}$$

Then $H(c,\beta)$ is called a *hyperplane*. In $\mathbf{R}_{2\times 2}$ hyperplanes are lines. In $\mathbf{R}_{3\times 1}$ hyperplanes are planes. In general a hyperplane in $\mathbf{R}_{n\times 1}$ is a $n-1$ dimensional affine subset of $\mathbf{R}_{n\times 1}$ (see Figure 11). Let

$$H^+(c,\beta) = \{x : c^T x \geq \beta\}$$

and

$$H^-(c,\beta) = \{x : c^T x \leq \beta\}$$

Then $H^+(c,\beta)$ is called an *upper half space* and $H^-(c,\beta)$ is called a *lower half space* (see Figure 12).

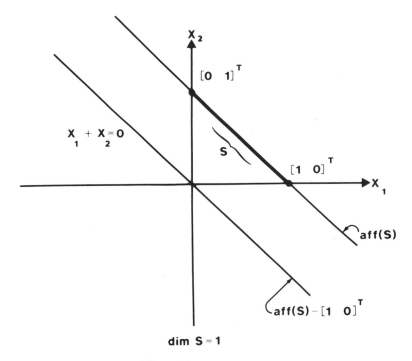

dim S = 1

Figure 10 dim S = 1.

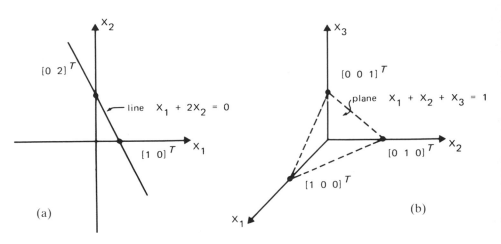

(a)   (b)

Figure 11   (a) $H([1\ 2]^T, 1) = \{[x_1 x_2]^T : x_1 + 2x_2 = 1\}$ is a hyperplane.

(b) $H([1\ 1\ 1]^T, 1) = \{[x_1 x_2 x_3]^T : x_1 + x_2 + x_3 = 1\}$ is a hyperplane.

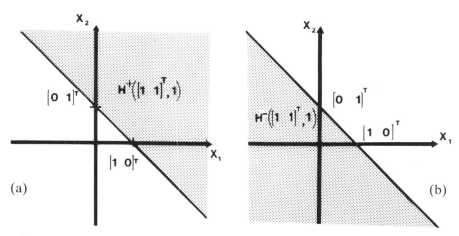

Figure 12 (a) Upper half space $H^+([1\ 1]^T, 1)$. (b) Lower half space $H^-([1\ 1]^T, 1)$.

Convex set play a very important role in mathematical programming. The convex hull of finitely many points is called a *convex polytope* (see Figure 13). In linear programming problems, convex polytopes are especially important.

The set of vectors $\{x_0, x_1 \ldots, x_k\}$ is *affinely independent* if and only if the set $\{x_1 - x_0, x_2 - x_0, \ldots, x_k - x_0\}$ is linearly independent (see Figure 14). A convex polytope is called a *k-dimensional simplex* if and only if it is the convex hull of $k + 1$ affinely independent vectors. Some examples of simplices in $\mathbf{R}_{3 \times 1}$ are present in Figure 15.

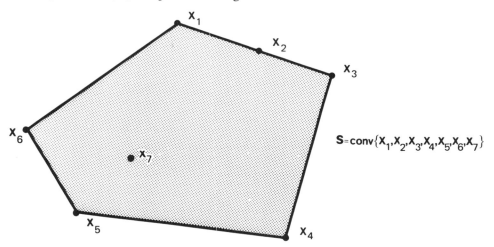

Figure 13 A convex polytope.

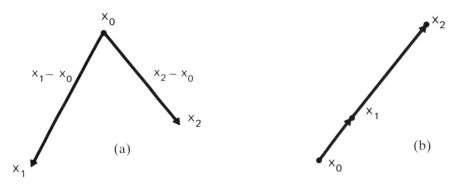

Figure 14   (a) $\{x_0, x_1, x_2\}$ is affinely independent.
(b) $\{x_0, x_1, x_2\}$ is not affinely independent.

Finally we identify a type of vector in a convex set that has great importance in linear programming problems. The importance will be seen in the next section.

Let S be a convex subset of $\mathbf{R}_{n \times 1}$ and let $z \in S$. Then z is an *extreme point* of S if and only if there does not exist distinct vectors $x, y \in S$ and $1 > \alpha > 0$ such that $z = \alpha x + (1 - \alpha)y$. The vertices of any simplex are extreme points of the simplex. The convex polytope in Figure 13 has extreme points x_1, x_3, x_4, x_5 and x_6.

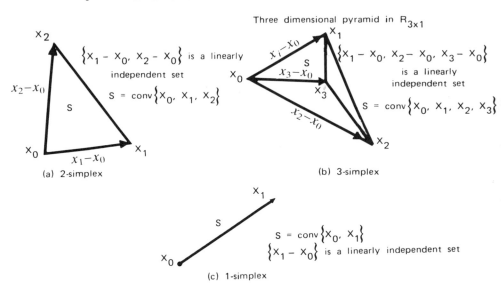

Figure 15   (a) S is a 2-simplex. (b) S is a 3-simplex. (c) S is a 1-simplex.

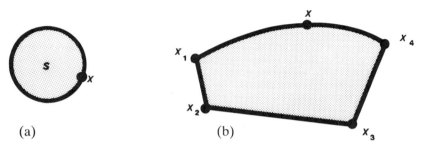

Figure 16 (a) The extreme points of S are the points x on the boundary. (b) The extreme points of S are x_1, x_2, x_3, x_4 and all the x on the curved boundary.

Clearly, the extreme points of a convex set S are those points of S that do not lie on the interior of any line segment connecting any other pair of points of S (see Figure 16).

Exercises

2.12   Graph each of the following sets and determine whether or not each set is affine, convex, a convex cone with vertex the origin, a convex polytope, a simplex, or a hyperplane. If the set is convex, determine its extreme points. If the set is not convex, determine its convex hull. Determine the dimension of each set.

(a) $S_1 = \{x = [x_1\, x_2]^T : x_1^2 + x_2^2 \le 1, x_1 - x_2 \le 1, x_2 \ge -x_1 - 1\}$

(b) $S_2 = \{x = [x_1\, x_2]^T : x_1 + x_2 \le 5, x_2 \ge 0, x_1 \ge 1\}$

(c) $S_3 = \{x = [x_1\, x_2]^T : 2x_1 - x_2 \le 4, x_1 - x_2 \ge -5, x_1 \ge 0, x_2 \ge 0\}$

(d) $S_4 = \{x = [x_1\, x_2]^T : x_2 \le (1/2)x_1 + 4,\ x_2 \ge (1/3)x_1,\ x_1 \ge 0, x_2 \ge 0\}$

(e) $S_5 = \{x = [x_1\, x_2]^T : x_2 \le |x_1|, x_1 \ge 0\}$

(f) $S_6 = \{x = [x_1\, x_2]^T : 2x_1 + x_2 = 3, x_1^2 + x_2^2 \ge 0\}$

(g) $S_7 = \{x = [x_1\, x_2\, x_3]^T : x_1^2 + x_2^2 = 1, x_3 = 3, x_1 \ge 0, x_2 \ge 0\}$

(h) $S_8 = \{x = [x_1\, x_2]^T : x_1^2 + x_2^2 = 1\} \cup \{[5\, 0]^T\}$

(i) $S_9 = \{x = [x_1\, x_2\, x_3]^T : x_2 \le |x_1|, x_2 \ge 0, -4 \le x_1 \le 4, x_3 = 3\}$

2.13   In $\mathbf{R}_{2\times1}$ let $S = \{[0\,0]^T,\ [4\,1]^T,\ [5\,0]^T,\ [3\,9]^T,\ [0\,2]^T,\ [2\,8]^T\}$. Graph $L(S)$, aff(S), conv(S) and coni(S).

2.14   Let $c = [3\,4]^T$ and $\beta = 12$. Then graph the hyperplane $H(c,\beta)$, and both half spaces $H^+(c,\beta)$ and $H^-(c,\beta)$.

2.15   Determine whether or not the set

$$S = \{[1 \quad 0 \quad 0 \quad 0 \quad 0]^T, [1 \quad 1 \quad 0 \quad 0 \quad 0]^T, [1 \quad 0 \quad 1 \quad 0 \quad 0]^T,$$
$$[1 \quad 0 \quad 0 \quad 1 \quad 0]^T, [1 \quad 0 \quad 0 \quad 0 \quad 1]^T\}$$

is an affinely independent set of vectors.

2.16   Prove that conv(S) is a convex set.

2.17   Give an example of a convex polytope that is not a simplex.

2.18   Show that a set S is a convex cone with vertex the origin if and only if
(a) $x, y \in S$ implies that $x + y \in S$.
(b) $x \in S$ and $\alpha \in \mathbf{R}$ such that $\alpha \geq 0$ implies that $\alpha x \in S$.

2.19   Show that coni(S) is a convex cone with vertex the origin.

2.20   Prove parts (i) and (ii) of Proposition 2.

2.21   Show that
(a) Any intersection of convex sets is a convex set.
(b) Any intersection of affine sets is an affine set.
(c) Any intersection of convex cones with vertices the origin is again such a cone.

2.22   Let $S \subseteq \mathbf{R}_{n \times 1}$ and show that
(a) Conv(S) is the intersection of all convex subsets of $\mathbf{R}_{n \times 1}$ that contains S.
(b) Aff(S) is the intersection of all affine subsets of $\mathbf{R}_{n \times 1}$ that contains S.
(c) Coni(S) is the intersection of all convex cones with vertex the origin that contains S.

2.23   Show that an affine set S is the translate of a unique subspace.

2.24   Prove that the following hold
(a) Any hyperplane $H(c, \beta)$ is an affine set.
(b) Both $H^+(c, \beta)$ and $H^-(c, \beta)$ are convex.
(c) Both $H^+(c, \beta) \setminus H(c, \beta)$ and $H^-(c, \beta) \setminus H(c, \beta)$ are convex (Recall that $A \setminus B = \{x \in A : x \notin B\}$.)
(d) $(H^+(c, \beta) \setminus H(c, \beta)) \cap (H^-(c, \beta) \setminus H(c, \beta)) = \emptyset$.
(e) $\mathbf{R}_{n \times 1} = H(c, \beta) \cup (H^+(c, \beta) \setminus H(c, \beta)) \cup (H^-(c, \beta) \setminus H(c, \beta))$

2.25   The set $\{x_0, x_1, \cdots, x_k\}$ is affinely independent if and only if both

$$\sum_{i=0}^{k} \alpha_i x_i = 0 \quad \text{and} \quad \sum_{i=0}^{k} \alpha_i = 0$$

imply that each $\alpha_i = 0$.

2.26   If $\{x_0, x_1, \cdots, x_k\}$ is affinely independent and $x \in$ aff $\{x_0, x_1, \cdots, x_k\}$, then there exists unique scalars $\alpha_0, \alpha_1, \cdots, \alpha_k$ such that $x = \Sigma_{i=0}^k \alpha_i x_i$ and $\Sigma_{i=0}^k \alpha_i = 1$.

2.27   In $\mathbf{R}_{n \times 1}$ show that the dimension of any hyperplane is $n - 1$.

2.28   In $\mathbf{R}_{n \times 1}$ show that the dimension of any m-dimensional simplex is m.

2.29   Let S be a convex subset of $\mathbf{R}_{n \times 1}$. Then prove that x is an extreme point of S if and only if there does not exist distinct $y, z \in S$ such that $x = (1/2)y + (1/2)z$.

2.30   Describe the convex hulls of the following sets:

   (a) $S = \{[x_1 \, x_2]^T : x_2 = |1/x_1|\}$.

   (b) $S = \{[0 \, x_2]^T : x_2 \in \mathbf{R}_1\} \cup \{[-1 \, 0]^T\}$.

2.31   If $Q = \{x = [x_1 \, x_2]^T : x_1^2 + x_2^2 \leq 4, \ x_1 \geq -1, \ x_1 - 2x_2 \leq 2\}$, then determine the extreme points of Q.

3.   THE LINEAR PROGRAMMING PROBLEM

In this section we will center our attention on the following linear programming problem and its solutions.

   Minimize the linear function

$$c_1 x_1 + \cdots + c_n x_n$$

subject to the constraints

$$a_{11} x_1 + \cdots + a_{1n} x_n \leq b_1$$
$$\cdot$$
$$\cdot$$
$$\cdot$$
$$a_{m1} x_1 + \cdots + a_{mn} x_n \leq b_m \tag{11}$$

and

$$\text{all } x_i \geq 0$$

If we let $c = [c_1 \ldots c_n]^T$, $x = [x_1 \ldots x_n]^T$, $A = [a_{ij}]_{(m, \, n)}$ and $b = [b_1 \ldots b_m]^T$, then we can restate the problem as:

$$\text{Minimize } c^T x \text{ subject to } Ax \leq b \text{ and } x \geq 0$$

If $A^{(j)}$ denotes the jth column of A, i.e., $A^{(j)} = [a_{1j} \ldots a_{mj}]^T$, then the problem can be restated in still another fashion:

$$\text{Minimize } c^T x \text{ subject to } x_1 A^{(1)} + \cdots + x_n A^{(n)} \leq b \text{ and } x \geq 0 \quad (13)$$

Let $F = \{x : Ax \leq b \text{ and } x \geq 0\}$. Then F is called the set of *feasible solutions* of problem (12). Thus, to solve Problem (12) we must determine $\min \{c^T x : x \in F\}$. The problem is *feasible* whenever $F \neq \emptyset$.

PROPOSITION 4   The set of feasible solutions, F, to the linear programming problem is a convex set.
Proof: To see this let $x, z \in F$ and $\alpha \in [0, 1]$. Then

$$A(\alpha x + (1 - \alpha)z) = \alpha Ax + (1 - \alpha)Az \leq \alpha b + (1 - \alpha)b = b$$

and

$$\alpha x + (1 - \alpha)z \geq 0$$

Hence, $\alpha x + (1 - \alpha)z \in F$ and F is convex.

Actually, F is the intersection of m lower half spaces, which are convex subsets of $\mathbf{R}_{n \times 1}$

$$F = (\bigcap_{i=1}^{m} \{x : A_{(i)}x \leq b_i\}) \cap (\bigcap_{i=1}^{n} \{x : x_i \geq 0\})$$

(here $A_{(i)} = [a_{i1} \ldots a_{in}]$ is the ith row of A) and hence F is a convex set.

Sets like F are called *convex polyhedrons*. Thus, a convex polyhedron P in $\mathbf{R}_{n \times 1}$ is the intersection of a finite number of half spaces and may be represented in the form $P = \{x \in \mathbf{R}_{n \times 1} : Mx \leq d\}$, where $M = [m_{ij}] \in \mathbf{R}_{k \times n}$ and $d = [d_1 \ldots d_k]^t$. A convex polytope is a convex polyhedron. This is true because a convex polytope can be represented as a finite intersection of half spaces. It is not true that a convex polyhedron is necessarily a convex polytope. In fact a convex polyhedron may not be bounded (convex polytopes are always bounded). Figure 17 illustrates these remarks. A bounded convex polyhedron is a convex polytope.

In Figure 17(b), F_2 is the convex hull of its extreme points, i.e., $F_2 = \text{conv} \{P_0, P_1, P_2, P_3\}$. However in Figure 17(a), F_1 is not the convex hull of its extreme points P_0 and P_1. Consider the points $[0 \quad 1]^T$ and $[2 \quad 3]^T$ in F_1. Both $[0 \quad 1]^T$ and $[2 \quad 3]^T$ lie on the ray that extends from the point $[0 \quad 1]^T$ along the line $-x_1 + x_2 = 1$. Thus, $v_0 = [2 \quad 3]^T - [0 \quad 1]^T = [2 \quad 2]^T$ may be thought of as a vector that extends from $[0 \quad 1]^T$ to $[2 \quad 3]^T$. Also since $[0 \quad 0]^T$ and $[4 \quad 0]^T$ both lie along the positive x-axis which bounds F_1, then $v_1 = [4 \quad 0]^T$ can be thought of as a vector that extends from $[0 \quad 0]^T$ through $[4 \quad 0]^T$. Figure 18 illustrates the point. Readers should convince

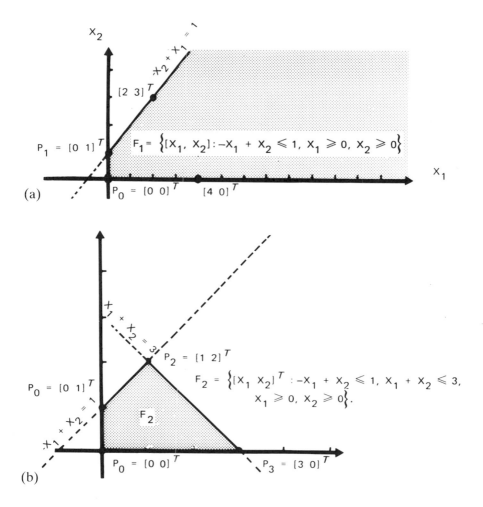

Figure 17 (a) An unbounded convex polyhedron that is not a convex polytope.
(b) A bounded convex polyhedron that is a convex polytope.

themselves that $F_2 = \text{conv}\{P_0, P_1\} + \text{coni}\{v_0, v_1\}$, i.e., each point x in F_2 can
be expressed as

$$x = \alpha_0 P_0 + \alpha_1 P_1 + \beta_0 v_0 + \beta_1 v_1,$$

where $\alpha_0 \geq 0$, $\alpha_1 \geq 0$, $\beta_0 \geq 0$, $\beta_1 \geq 0$ and $\alpha_0 + \alpha_1 = 1$ (see Figure 18).

A similar representation holds for convex polyhedrons in general. The
result is known as the finite basis theorem and is presented here without
proof (see Exercise 2.46 or [7]).

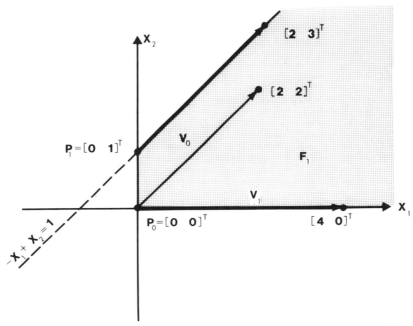

Figure 18 Vector $v_0 = [2\ 2]^T$ is parallel to and has the same length as the line
segment on $-x_1 + x_2 = 1$ that extends from $[0\ 1]^T$ to $[2\ 3]^T$.

THEOREM 6 (Finite basis theorem) Let F be a convex polyhedron in
$\mathbf{R}_{n \times 1}$ i.e., $F = \{x \in \mathbf{R}_{n \times 1} : Ax \leq b\}$ and let $\{x_1, \ldots, x_d\}$ be the set of extreme
points of F. Then there exists another set $\{y_1, \ldots, y_\ell\}$, called direction
vectors, such that

$$F = \text{conv}(\{x_1, \ldots, x_d\}) + \text{coni}(\{y_1, \ldots, y_\ell\})$$

i.e.,

$$F = \{\sum_{i=1}^{d} \alpha_i x_i + \sum_{i=1}^{l} \beta_i y_i : \text{each } \alpha_i \geq 0, \beta_i \geq 0 \text{ and } \sum_{i=1}^{d} \alpha_i = 1\}$$

The finite basis theorem is useful because it tells us when and where to
find the optimal solutions to the linear programming problem.

THEOREM 7 If the linear function $f(x) = c^T x$ is bounded from below over
the convex polyhedron $F = \{x : Ax \leq b, x \geq 0\}$, then $f(x)$ obtains a minimum
over F and the minimum occurs at one of the extreme points of F.
Proof: By the finite basis theorem, there exists sets $X = \{x_1, \ldots, x_d\}$ and
$Y = \{y_1, \ldots, y_\ell\}$ such that $F = \text{conv}\, X + \text{coni}\, Y$. Thus, if $x \in F$, then

$x = \sum_{i=1}^{d} \alpha_i x_i + \sum_{i=1}^{\ell} \beta_i y_i$, where each $\alpha_i \geq 0$, each $\beta_i \geq 0$ and $\sum_{i=1}^{d} \alpha_i = 1$. Since f is a linear function it follows that

$$f(x) = f(\sum_{i=1}^{d} \alpha_i x_i + \sum_{i=1}^{\ell} \beta_i y_i) = \sum_{i=1}^{d} \alpha_i f(x_i) + \sum_{i=1}^{\ell} \beta_i f(y_i) \qquad (14)$$

Since $f(x)$ is bounded from below over F, then there exists a real number μ such that $f(x) \geq \mu$, for every $x \in F$. Consider each $f(y_i)$. It follows that each $f(y_i) \geq 0$. For if $f(y_j) < 0$ for some j, then by letting $\beta_j \to \infty$ in equation (14) and holding all other terms constant we could make $f(x) < \mu$, which cannot happen. Let $f(x_k) = \min\{f(x_1), \ldots, f(x_d)\}$. Then

$$f(x) = \sum_{i=1}^{d} \alpha_i f(x_i) + \sum_{i=1}^{\ell} \beta_i f(y_i) \geq \sum_{i=1}^{d} \alpha_i f(x_i)$$

$$\geq \sum_{i=1}^{d} \alpha_i f(x_k) = f(x_k) \sum_{i=1}^{d} \alpha_i = f(x_k) \cdot 1 = f(x_k)$$

Thus, if f is bounded from below over F, then $f(x)$ takes on its minimum value at one of the vectors in $X = \{x_1, \ldots, x_d\}$.

Hence, when the minimum of $f(x)$ over F occurs it occurs at an extreme point. It may occur elsewhere as well. Analogously, if the function $f(x)$ is bounded over F from above then $f(x)$ obtains its maximum value at one of the extreme points of F. This is illustrated in the next example.

EXAMPLE 16   Minimize $f(x_1, x_2) = -x_1 + x_2$ subject to the constraints

$$-x_1 + 2x_2 \leq 10$$
$$5x_1 + 3x_2 \leq 41$$
$$2x_1 - 3x_2 \leq 8$$
$$x_1 \geq 0 \text{ and } x_2 \geq 0$$

Solution: If we let $c = [-1\ 1]^T, x = [x_1\ x_2]^T, b = [10\ 41\ 8]^T$ and

$$A = \begin{bmatrix} -1 & 2 \\ 5 & 3 \\ 2 & -3 \end{bmatrix}$$

then the problem may be restated as, minimize $c^T x$ subject to $Ax \leq b$ and $x \geq 0$. The set of feasible solutions $F = \{x : Ax \leq b, x \geq 0\}$ is represented graphically in Figure 19.

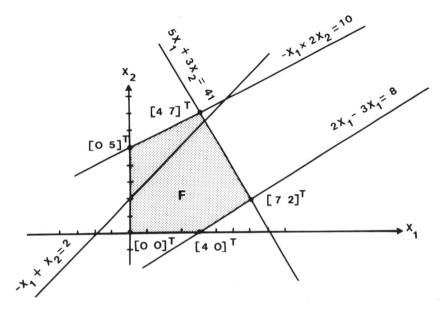

Figure 19   The set $F \cap \{[x_1\ x_2]^T : -x_1 + x_2 = u\}$, where $u = 2$

The line $-x_1 + x_2 = \mu$ shows where the objective function $f(x_1, x_2) = -x_1 + x_2$ has the value μ. For example the line $-x_1 + x_2 = 2$ in Figure 19 shows where $f(x) = 2$. As μ decreases the slope of $-x_1 + x_2 = \mu$ (or $x_2 = x_1 + \mu$) remains constantly 1 while the x_2-intercept, $(0, \mu)$, slides downward along the x_2-axis. It follows in this example that in order to determine the smallest value of μ such that $-x_1 + x_2 = \mu$ intersects F, we must slide the line $-x_1 + x_2 = \mu$ downwards until the line meets F at only the extreme point $[7\ \ 2]^T$. The minimum value of the objective function F is

$$f([7\ \ 2]^T) = 2 - 7 = -5$$

Actually, the finite basis theorem implies that we need only to check the objective function at the extreme points of F which are $[0\ \ 0]^T$, $[4\ \ 0]^T$, $[7\ \ 2]^T$, $[4\ \ 7]^T$ and $[0\ \ 5]^T$.

Since an optimal solution to the linear programming problem can always be found at one of the extreme points of the convex polyhedron of feasible solutions, we shall now turn our attention to the problem of finding these extreme points. Our first step is to modify Problem (12) by changing the inequalities $Ax \leq b$ to equalities. To do this consider each inequality of the form

$$a_{i1}x_1 + \cdots + a_{in}x_n \leq b_i$$

Define x_{n+i} by $x_{n+i} = b_i - a_{i1}x_1 - \cdots - a_{in}x_n$

Then

$$x_{n+i} \geq 0$$

and

$$a_{i1}x_1 + \cdots + a_{in}x_n + x_{n+i} = b_i$$

Thus, the constraints (11) become

$$
\begin{aligned}
a_{11}x_1 + \cdots + a_{1n}x_n + x_{n+1} &= b_1 \\
a_{21}x_1 + \cdots + a_{2n}x_n \quad\quad\quad + x_{n+2} &= b_2 \\
& \vdots \\
a_{m1}x_1 + \cdots + a_{mn}x_n \quad\quad\quad\quad + x_{n+m} &= b_m
\end{aligned}
\tag{15}
$$

where

$$x_1 \geq 0, \ldots, x_n \geq 0, x_{n+1} \geq 0, \ldots, x_{n+m} \geq 0$$

Let $\hat{c} = [c_1 \ldots c_n 0 \ldots 0]^T$ be a $n + m$ column matrix, let $\hat{A}$ be the coefficient matrix of the equations in (15), i.e.,

$$
\hat{A} =
\begin{bmatrix}
a_{11} & \cdots & a_{1n} & 1 & 0 & \cdots & 0 \\
a_{21} & \cdots & a_{2n} & 0 & 1 & \cdots & 0 \\
\vdots & & \vdots & \vdots & \vdots & & \vdots \\
& & & & & & \\
a_{m1} & \cdots & a_{mn} & 0 & 0 & . & 1
\end{bmatrix}
= [A | I_m]
$$

and let

$$
\hat{x} = [x_1 \ldots x_n x_{n+1} \ldots x_{n+m}]^T =
\begin{bmatrix}
x \\
b - Ax
\end{bmatrix}
\tag{16}
$$

Then problem (2) can be restated in the following manner.

$$\text{Minimize } \hat{c}^T \hat{x} \text{ subject to } \hat{A}\hat{x} = b \text{ and } \hat{x} \geq 0 \tag{17}$$

The variables $x_{n+1}, \ldots, x_{n+m}$ are called *slack* variables.

Next denote the set of feasible solutions of Problem (17) by $\hat{F}$, i.e.,

$$\hat{F} = \{\hat{x} \in \mathbf{R}_{(n+m)\times 1} : \hat{A}\hat{x} = b, \hat{x} \geq 0\}$$

Again $\hat{F}$ is a convex polyhedron. This follows from the fact that any linear equation can be expressed as two linear inequalities, i.e.,

$$a_{i1}x_1 + \cdots + a_{in}x_n + x_{n+i} = b_i$$

if and only if

$$a_{i1}x_1 + \cdots + a_{in}x_n + x_n \leq b_i$$

and

$$-(a_{i1}x_1 + \cdots + a_{in}x_n) \leq -b_i$$

A natural correspondence exists between the elements of F and the elements of $\hat{F}$. To see this correspondence define a function $\sigma : F \to \hat{F}$ as $\sigma(x) = \hat{x}$. The important of this correspondence can be seen in the next proposition.

PROPOSITION 5   The function σ is a one-to-one function that preserves convex combinations of vectors.
Proof: First suppose that $x, z \, \varepsilon \, F$ such that $\sigma(x) = \sigma(z)$. Then from Equation (6) we have

$$\begin{bmatrix} x \\ b - Ax \end{bmatrix} = \hat{x} = \sigma(x) = \sigma(z) = \hat{z} = \begin{bmatrix} z \\ b - Az \end{bmatrix}$$

Hence, $x = z$ and the function σ is one-to-one. Next suppose that $x, y \in F$ and $\alpha \in [0, 1]$. Then

$$\sigma(\alpha x + (1 - \alpha)y) \;\; = \begin{bmatrix} \alpha x + (1 - \alpha)y \\ b - A(\alpha x + (1 - \alpha)y) \end{bmatrix}$$

$$= \begin{bmatrix} \alpha x + (1 - \alpha)y \\ \alpha b + (1 - \alpha)b - \alpha Ax - (1 - \alpha)Ay \end{bmatrix} = \begin{bmatrix} \alpha x + (1 - \alpha)y \\ \alpha(b - Ax) + (1 - \alpha)(b - Ay) \end{bmatrix}$$

$$= \alpha \begin{bmatrix} x \\ b - Ax \end{bmatrix} + (1 - \alpha)\begin{bmatrix} y \\ b - Ay \end{bmatrix} \;\; = \alpha \, \sigma(x) + (1 - \alpha) \, \sigma(y)$$

i.e., σ preserves convex combinations.

The next proposition follows easily from the above proposition. The proof is left as an exercise (see Exercise 2.41). An illustration is given in Example 17.

PROPOSITION 6   The vector $x \in F$ is an extreme point of F if and only if $\hat{x}$ is an extreme point of $\hat{F}$.

EXAMPLE 17   Let $F = \{[x_1 x_2]^T : x_1 + x_2 \leq 1, \;\; x_1 \geq 0, \;\; x_2 \geq 0\}$. Then $\hat{F} = \{[x_1 x_2 x_3]^T : x_1 + x_2 + x_3 = 1, \;\; x_1 \geq 0, \;\; x_2 \geq 0, \;\; x_3 \geq 0\}$. Sets F and $\hat{F}$ are shown in Figure 20.

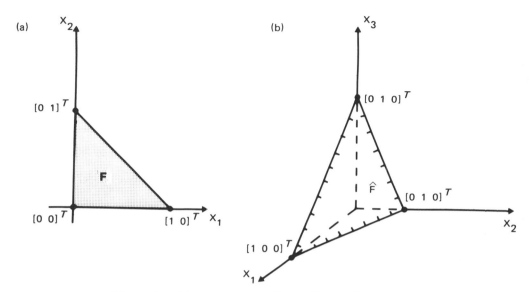

Figure 20   (a) $F = \mathrm{conv}\{[0\ 0]^T, [1\ 0]^T, [0\ 1]^T\}$.

(b) $\hat{F} = \mathrm{conv}\{[1\ 0\ 0]^T, [0\ 1\ 0]^T, [0\ 0\ 1]^T\}$.

The correspondence between the extreme points of F and $\hat{F}$ is

$$\sigma([1\ \ 0]^T) = [1\ \ 0\ \ 0]^T$$
$$\sigma([0\ \ 1]^T) = [0\ \ 1\ \ 0]^T$$
$$\sigma([0\ \ 0]^T) = [0\ \ 0\ \ 1]^T$$

Clearly, $c^T x$ is minimized at a point $z \in F$ if and only if the function $\hat{c}^T \hat{x}$ is minimized at $\hat{z} \in \hat{F}$. In fact $c^T x = \hat{c}^T \hat{x}$. As the next theorem shows, the extreme points of $\hat{F}$ are easily characterized, thus making Problem (17) easier to attack than problem (12). First recall that $\hat{A}(j)$ denotes the jth column of matrix $\hat{A}$. Also recall that A has m rows and hence rank $A \le m$. For the present time we shall assume that rank $A = m$. The case when rank $A < m$ will be handled in a later chapter by adding a suitable number of linearly independent columns to $\hat{A}$.

THEOREM 8   If $\hat{x} = [x_1 \ldots x_n x_{n+1} \ldots x_{n+m}]^T$ is an extreme point of $\hat{F}$ and $\Delta = \{j : x_j > 0\}$, then the set of columns $\{\hat{A}^{(j)} : j \in \Delta\}$ is linearly independent. Moreover, if $\{\hat{A}^{(j)} : j \in \Delta\}$ is some collection of m linearly independent columns of $\hat{A}$, if B is the matrix whose columns are $\{\hat{A}^{(j)} : j \in \Delta\}$ and if x^* is a solution of $Bx = b$, $x \ge 0$, then the vector $\hat{x}$ whose ith coordinate is 0 if $i \notin \Delta$ and x_i^* if $i \in \Delta$ is an extreme point of $\hat{F}$.

Proof: First suppose that $\hat{x}$ is an extreme point of $\hat{F}$. Without loss of generality, suppose that $\hat{x} = [x_1 \ldots x_k 0 \ldots 0]^T$, where each $x_i > 0$ for $i = 1, \ldots, k$. Since $\hat{A}\hat{x} = b$, then $b = x_1 \hat{A}^{(1)} + \cdots + x_k \hat{A}^{(k)}$. We want to show that $\{\hat{A}^{(1)}, \cdots, \hat{A}^{(k)}\}$ is linearly independent. Suppose otherwise, i.e., suppose there exists scalars $y_1, \ldots, y_k$ (notice here that each y_i is a scalar and not a vector) not all zero such that $y_1 \hat{A}^{(1)} + \cdots + y_k \hat{A}^{(k)} = 0$, where 0 is the zero column matrix. Let α be a scalar such that $\alpha = \min\{\{x_i/y_i : y_i > 0\} \cup \{-x_i/y_i : y_i < 0\}\}$. Then the vectors

$$u = [x_1 + \alpha y_1 \quad \ldots \quad x_k + \alpha y_k \quad 0 \quad \ldots \quad 0]^T$$

and

$$v = [x_1 - \alpha y_1 \quad \ldots \quad x_k - \alpha y_k \quad 0 \quad \ldots \quad 0]^T$$

are nonnegative, i.e., $u \geq 0$ and $v \geq 0$. Also,

$$\hat{A}u = A[x_1 + \alpha y_1 \quad \ldots \quad x_k + \alpha y_k]^T = \sum_{i=1}^{k} (x_i + \alpha y_i)A^{(i)}$$

$$= \sum_{i=1}^{k} x_i A^{(i)} + \alpha \sum_{i=1}^{k} y_i A^i = b + \alpha \cdot 0 = b$$

Hence, $u \in \hat{F}$. Likewise, $v \in \hat{F}$. But $\hat{x} = (1/2)u + (1/2)v$ which says that $\hat{x}$ is not an extreme point of $\hat{F}$. Since this is a contradicition, it follows that $\{\hat{A}^{(1)}, \ldots, \hat{A}^{(k)}\}$ is linearly independent.

To prove the second part of the theorem suppose that $\{\hat{A}^{(i)} : i \in \Delta\}$ is a collection of m linearly independent columns of $\hat{A}$, B is the matrix whose columns are $\{\hat{A}^{(i)} : i \in \Delta\}$, x^* is a solution to $Bx = b, x \geq 0$, and $\hat{x}$ is the vector whose ith coordinate is 0 if $i \notin \Delta$ and x_i^* if $i \in \Delta$. Again without loss of generality, suppose that $\Delta = \{1, \ldots, k\}$, i.e., $\hat{x} = [x_1 \ldots x_k 0 \ldots 0]^T$, where each $x_i \geq 0$. Clearly, $\hat{x} \in \hat{F}$. Suppose that $\hat{x}$ is not an extreme point of F. Then there must exist distinct $\hat{y}, \hat{z} \in \hat{F}$ and $\alpha \in (0, 1)$ such that $\hat{x} = \alpha \hat{y} + (1 - \alpha)\hat{z}$. Since the coordinates of both $\hat{y}$ and $\hat{z}$ are nonnegative, it follows that the only nonzero coordinates y_i of $\hat{y}$ and z_i of $\hat{z}$ must possibly occur for only $i = 1, \ldots, k$, i.e., $\hat{y} = [y_1 \ldots y_k 0 \ldots 0]^T$ and $\hat{z} = [z_1 \ldots z_k 0 \ldots 0]^T$. Also,

$$y_1 \hat{A}^{(1)} + \cdots + y_k \hat{A}^{(k)} = \hat{A}\hat{y} = b = \hat{A}\hat{z} = z_1 \hat{A}^{(1)} + \cdots + z_k \hat{A}^{(k)}$$

which implies that

$$(y_1 - z_1)\hat{A}^{(1)} + \cdots + (y_k - z_k)\hat{A}^{(k)} = 0$$

Since $\{\hat{A}^{(1)}, \ldots, \hat{A}^{(k)}\}$ is linearly independent, then it follows that each $y_i = z_i$. Thus, $\hat{y} = \hat{z}$, which is a contradiction. Hence, $\hat{x}$ is an extreme point of $\hat{F}$.

We shall now illustrate how Theorem 8 can be used to determine the extreme points of a set of feasible solutions.

EXAMPLE 18   Minimize $-x_1 + x_2$ subject to

$$-x_1 + 2x_2 \le 10$$
$$5x_1 + 3x_2 \le 41$$
$$2x_1 - 3x_2 \le 8$$
$$x_1 \ge 0 \text{ and } x_2 \ge 0$$

(see Example 16).

Reformulate the problem as

$$\text{Minimize } -x_1 + x_2 + 0 \cdot x_3 + 0 \cdot x_4 + 0 \cdot x_5 \text{ subject to}$$
$$-x_1 + 2x_2 + x_3 = 10$$
$$5x_1 + 3x_2 + x_4 = 41$$
$$2x_1 - 3x_2 + x_5 = 8$$
$$x_1 \ge 0, x_2 \ge 0, x_3 \ge 0, x_4 \ge 0, x_5 \ge 0$$

or

$$\text{minimize } -x_1 + x_2 + 0x_3 + 0x_4 + 0x_5 \text{ subject to}$$

$$\begin{bmatrix} -1 & 2 & 1 & 0 & 0 \\ 5 & 3 & 0 & 1 & 0 \\ 2 & -3 & 0 & 0 & 1 \end{bmatrix} \begin{bmatrix} x_1 \\ x_2 \\ x_3 \\ x_4 \\ x_5 \end{bmatrix} = \begin{bmatrix} 10 \\ 41 \\ 8 \end{bmatrix}$$

$$\text{and all } x_i \ge 0$$

In the above coefficient matrix there are at most $\binom{5}{3} = 10$ sets of three linearly independent columns (see Theorem 4), $\binom{5}{3} = 10$ systems of three equations in three unknowns to solve. For example selecting the first three columns (i.e., setting $x_4 = x_5 = 0$), we solve

$$\begin{bmatrix} -1 & 2 & 1 \\ 5 & 3 & 0 \\ 2 & -3 & 0 \end{bmatrix} \begin{bmatrix} x_1 \\ x_2 \\ x_3 \end{bmatrix} = \begin{bmatrix} 10 \\ 41 \\ 8 \end{bmatrix}$$

obtaining $x_1 = 7$, $x_2 = 2$, $x_3 = 13$. Since these values are all nonnegative it follows that $[7 \quad 2 \quad 13 \quad 0 \quad 0]^T$ is an extreme point of $\hat{F}$ and thus by Proposition 5, $[7 \quad 2]^T$ is an extreme point of F. Next we select columns 1,2, 4, i.e., set $x_3 = x_5 = 0$ and solve

$$\begin{bmatrix} -1 & 2 & 0 \\ 5 & 3 & 1 \\ 2 & -3 & 0 \end{bmatrix} \begin{bmatrix} x_1 \\ x_2 \\ x_4 \end{bmatrix} = \begin{bmatrix} 10 \\ 41 \\ 8 \end{bmatrix}$$

obtaining $x_1 = 46$, $x_2 = 28$ and $x_4 = -273$. Since x_4 is negative, then $[46 \quad 28 \quad 0 \quad -273 \quad 0 \quad 0]^T \notin \hat{F}$ and hence $[46 \quad 28]^T$ is not an extreme point of F. Continuing in this fashion, the reader can verify that the complete list of extreme points is $[7 \quad 2]^T$, $[4 \quad 7]^T$, $[4 \quad 0]^T$, $[0 \quad 5]^T$ and $[0 \quad 0]^T$. Recall that we want only nonnegative solutions. Upon evaluation of $f(x) = -x_1 + x_2$ at each extreme point of F we find that the minimum is $f([7 \quad 2]^T) = -5$.

Unfortunately this process is too time consuming, even for a high speed computer. In the next chapter we shall develop a procedure that will enable us to move from one extreme point $\hat{x}$ of $\hat{F}$ to another adjacent extreme point $\hat{z}$ such that $f(\hat{z}) \le f(\hat{x})$. As a result the amount of time and work required to solve the linear programming problem is greatly reduced. (The technique of converting inequalities to equalities by introducing slack variables will continue to be used.) Thus, we shall be concerned with the development of an efficient technique for solving problems of the form: Minimize $c^T x$ subject to $Ax = b$ and $x \ge 0$.

We shall now restate the results of Theorems 7 and 8 in terms of basic solutions of $\hat{A}\hat{x} = b$. Recall that any $\hat{x}$ satisfying

$$\hat{A}\hat{x} = b \text{ and } \hat{x} \ge 0 \tag{18}$$

is called a feasible solution of (18). A *basic feasible solution* (a *BFS*) of (18) is any basic solution of $\hat{A}\hat{x} = b$ that is also a feasible solution of (18). Theorem 7 says that when the objective (or cost) function is bounded from below over the set $\hat{F}$ of feasible solutions of (18), then the objective function obtains a minimum at one of the basic feasible solutions of (18). It follows from Theorem 8 that the extreme points of $\hat{F}$ are basic feasible solutions of (18).

Exercises

Write each of the following convex polyhedrons in the form $P = \{x \in \mathbf{R}_{2 \times 1} : mx \le d\}$, then graph. Determine the extreme points $x = \{x_1, \dots, x_d\}$ of P and the direction vectors $Y = \{y_1, \dots, y_\ell\}$ such that

$P = \text{conv}(X) + \text{coni}(Y)$.

(a) $P_1 = \{x = [x_1 x_2]^T : 2x_1 - x_2 \le 4, \ x_1 \ge 3, \ x_1 + x_2 \ge 1\}$.

(b) $P_2 = \{x = [x_1 x_2]^T : 6x_1 + 2x_2 \ge 6, \ x_1 + x_2 \le 4, \ x_1 \ge 0, \ x_2 \ge 0\}$.

(c) $P_3 = \{x = [x_1 x_2]^T : 6x_1 + 2x_2 \ge 6, \ x_1 + x_2 \le 4, \ x_2 = 5x_1 + 1, \ x_1 \ge 0,$
      $x_2 \ge 0\}$.

(d) $P_4 = \{x = [x_1 x_2]^T : x_1 + x_2 \ge 3, \ x_2 - 5x_1 \le 0, \ x_1 - x_2 \le 2, \ x_1 \ge 0, \ x_2 \ge 0\}$.

(e) $P_5 = \{x = [x_1 x_2]^T : x_2 \le 2x_1 + 3, \ 2x_1 + x_2 \ge 4, \ x_1 \ge 1, \ x_2 \ge 0\}$.

2.34   Let $f(x_1, x_2) = x_1 - 5x_2$. Determine whether or not f is bounded from below over each of the convex polyhedrons P_i in Problem 1. If f is bounded from below over any P_i determine the minimum value of f over that P_i and a vector in P_i for which the minimum occurs.

2.35   For the function f defined in Problem (2), determine which of the convex polyhedrons P_i from Problem (1) are such that f is bounded from above over P_i. When f is bounded from above over P_i determine the maximum of f over P_i and determine an extreme point of P_i for which the maximum occurs.

2.36   Solve graphically the following linear programming problems:

(a) Minimize $x_1 - x_2$ subject to $2x_1 - x_2 \ge -2$, $x_1 - 2x_2 \ge 3$, $x_1 + 2x_2 \le 11$, $x_1 \ge 0$ and $x_2 \ge 0$.

(b) Minimize $5x_1 - x_2$ subject to $2x_1 - x_2 \ge -2$, $x_1 - 2x_2 \ge 3$, $x_1 + 2x_2 \le 11$, $x_1 \ge 0$ and $x_2 \ge 0$.

(c) Maximize $5x_1 - x_2$ subject to $2x_1 - x_2 \ge -2$, $x_1 - 2x_2 \ge 3$, $x_1 + 2x_2 \le 11$, $x_1 \ge 0$ and $x_2 \ge 0$.

2.37   Find the extreme points of $F = \{x : Ax = b, x \ge 0\}$, where

$$x = [x_1 x_2 x_3 x_4]^T, \ b = [2 \ \ 1]^T \text{ and } A = \begin{bmatrix} 2 & 0 & -1 & 1 \\ 1 & 1 & 2 & 0 \end{bmatrix}$$

2.38   Let F be the set of vectors $x = [x_1 x_2 x_3]^T$ such that

$$2x_1 + x_2 \ge 1$$
$$x_1 - 2x_3 \le -2$$
$$x_1 + x_2 + x_3 = 3$$
$$x_1 \ge 0, x_2 \ge 0 \text{ and } x_3 \ge 0$$

Find the extreme points of F.

2.39   Restate each of the following linear programming problems as a linear programming problem with equality constraints. Then, as in Example 16, solve the original problem by generating all of its extreme points.

(a) Minimize $3x_1 - x_2$ subject to
$$2x_1 - x_2 \geq -2$$
$$x_1 - 2x_2 \leq 3$$
$$x_1 + 2x_2 \leq 11$$
$$x_1 \geq 0 \text{ and } x_2 \geq 0$$

(b) Minimize $x_1 + x_2$ subject to
$$x_1 + x_2 \geq 3$$
$$2x_1 + x_2 \geq 4$$
$$x_1 + 5x_2 \geq 10$$
$$x_1 \geq 0 \text{ and } x_2 \geq 0$$

(c) Maximize $2x_1 + 4x_2$ subject to
$$x_1 - x_2 \leq 2$$
$$x_1 + 2x_2 \geq 4$$
$$5x_1 + 8x_2 \leq 40$$
$$x_1 \geq 0 \text{ and } x_2 \geq 0$$

(d) Minimize $x_1 - x_2 + 2x_3$ subject to
$$2x_1 + x_2 \geq 1$$
$$x_1 - 2x_3 \leq -2$$
$$x_1 + x_2 + x_3 = 3$$
$$x_1 \geq 0, x_2 \geq 0 \text{ and } x_3 \geq 0$$

2.40   If the minimum of $c^T x$ occurs at two extreme points x and z of F, then show that the minimum also occurs at any point of the form $\alpha x + (1 - \alpha)z$, where $\alpha \in [0, 1]$.

2.41   Show that $\sigma(x) = \hat{x}$ preserves extreme points, i.e., prove Proposition 6.

2.42   Let $T = \{x = [x_1 \, x_2]^T : -x_1 + x_2 \leq 3, \ 5x_1 + 4x_2 \geq 20, \ x_2 \geq 1\}$.

(a) Graph T.

(b) Determine the extreme points, $\{u_i\}$, of T and direction vectors, $\{v_i\}$, such that $T = \text{conv}\{u_i\} + \text{coni}\{v_i\}$.

(c) Minimize $x_1 + x_2$ over T.

(d) Maximize $x_1 + x_2$ over T.

(e) Minimize $x_1 - x_2$ over T.

(f) Maximize $x_1 - x_2$ over T.

2.43 (Post-optimal analysis: changing the constraints.) Let $A \in \mathbf{R}_{m \times n}$ and consider the following linear programming problems:

$$\text{Minimize } c^T x \text{ subject to } x \in F = \{x : Ax = b, x \geq 0\} \qquad (19)$$

and

$$\text{Minimize } c^T x \text{ subject to } x \in \bar{F} = \{x : \bar{A}x = \bar{b}, x \geq 0\}. \qquad (20)$$

Let $\bar{F} \subseteq F$. Then prove that when there exists an $x^* \in \bar{F}$ that solves (19), then x^* solves (20) as well. In particular show that when the constraints $\bar{A}x = \bar{b}$ are obtained by joining an equation of the form

$$a_{m+11} x_1 + \cdots + a_{m+1n} x_n = b_{m+1}$$

to the constraints $Ax = b$, then any optimal solution x^* of (19) will also solve (20), whenever

$$a_{m+11} x_1^* + \cdots + a_{m+1n} x_n^* = b_{m+1}$$

2.44 Let $A \in \mathbf{R}_{m \times n}$. Prove that if there exists a feasible solution of $Ax = b$ and $x \geq 0$, then there is a basic feasible solution.

2.45 (a) Give an example of a convex polyhedron that does not have any extreme points.

(b) Prove that a convex polyhedron F does not have any extreme points if and only if F contains distinct points x and y such that $\{\alpha x + (1 - \alpha)y : \alpha \in \mathbf{R}\} \subseteq F$. (Here, F is the intersection of a finite number of closed half spaces).

2.46 A *direction* of $F = \{x : Ax = b, x \geq 0\}$ is a column vector d for which $Ad = 0$ and $d \geq 0$. An *extreme* direction of F is any direction vector d that cannot be expressed as a conical combination of two other distinct directions of F.

(a) Show that F has a finite number of both extreme points and extreme directions.

(b) Use the results of part (a) to establish the finite basis theorem.

(c) Let M be an affine set. Verify that there exists vectors $v_1, \ldots, v_d$ such that

$$M = v_0 + \{\sum_{i=1}^{d} \alpha_i v_i + \sum_{i=1}^{d} \beta_i(-v_i) : \alpha_i \geq 0, \beta_i \geq 0, \alpha_i \beta_i = 0\}$$

Using this result, can you establish the finite basis theorem? (Note, if you find that this problem is too difficult for you at this time, a proof of the finite basis theorem can be found in [7].)

REFERENCES

1. M. S. Bazaraa and C. M. Shetty, *Foundations of Optimization*, Springer-Verlag, Berlin (1976).
2. C. E. Forsythe, "Today's Computational Methods of Linear Algebra," *SIAM Review 9* (1967).
3. S. Gass, *Linear Programming*, McGraw-Hill Book Co., New York (1969).
4. L. W. Johnson and R. D. Riess, *Numerical Analysis,* Addison-Wesley, Reading, Mass. (1982).
5. A. W. Roberts and D. E. Varberg, *Convex Functions,* Academic Press, New York (1973).
6. G. Strang, *Linear Algebra and Its Applications*, Academic Press, New York (1976).
7. J. Stoer and C. Witzgall, *Convexity and Optimization in Finite Dimensions I*, Springer-Verlag, Berlin (1970).

3

The Primal Simplex Procedure

In this chapter, the procedure for generating basic solutions of a system of linear equations is extended to an efficient algorithm for generating an optimal solution of the linear programming problem. The algorithm, which is commonly known as the simplex procedure (or more precisely, the primal simplex procedure), is developed in the first section. The remaining sections contain additional generalizations of the procedure.

1. THE PRIMAL SIMPLEX PROCEDURE

Consider a linear programming problem of the form

$$\text{Minimize } c^T x \text{ subject to } Ax = b \text{ and } x \geq 0 \tag{1}$$

where $A \in \mathbf{R}_{m \times n}$, $x \in \mathbf{R}_{n \times 1}$. Notice that all the constraints are equality constraints and each variable x_i is restricted to be nonnegative. Moreover, we shall assume that rank $A = m$ (hence, $m \leq n$). It follows from the discussions in the first two chapters that any linear programming problem can be expressed in the form (1). The requirement that rank $A = m$ does not result in any real loss of generality. In fact the case where rank $A < m$ is easily resolved using the technique of artificial variables which will be developed later in this chapter.

Define z_0 by $z_0 = c^T x$. Then Problem (1) is equivalent to finding a solution $x \in \mathbf{R}_{n \times 1}$ and $z_0 \in \mathbf{R}$ of

$$
\begin{aligned}
Ax &= b \\
-c^T x + z_0 &= 0
\end{aligned}
\tag{2}
$$

for which $x \geq 0$ and z_0 is minimal. The constraints (2) can be represented by the matrix equation

$$\begin{bmatrix} A & \vdots & O \\ -c^T & \vdots & 1 \end{bmatrix} \begin{bmatrix} x \\ z_0 \end{bmatrix} = \begin{bmatrix} b \\ 0 \end{bmatrix} \tag{3}$$

or by the augumented matrix

$$\left[\begin{array}{c|c|c} A & O & b \\ \hline -c^T & 1 & 0 \end{array} \right] \tag{4}$$

where O is a column of zeros.

Since rank $A = m$, then A contains m linearly independent columns. For simplicity of notation and without loss of generality, we shall assume that the first m columns of A are linearly independent. Let $B = [A^{(1)} \dots A^{(m)}]$, i.e., the matrix of the first m columns of A, and $D = [A^{(m+1)} \dots A^{(n)}]$, i.e., the matrix of the last $n - m$ columns of A. Then $A = [B \mid D]$. Let $x_B = [x_1 \dots x_m]^T$, $x_D = [x_{m+1} \dots x_n]^T$, $c_B = [c_1 \dots c_m]^T$ and $c_D = [c_{m+1} \dots c_n]^T$. Then $x = [x_B^T \ x_D^T]^T$ and $c = [c_B^T \ c_D^T]^T$. Moreover (3) becomes

$$\left[\begin{array}{c|c|c} B & D & O \\ \hline -c_B^T & -c_D^T & 1 \end{array} \right] \begin{bmatrix} x_B \\ x_D \\ z_0 \end{bmatrix} = \begin{bmatrix} b \\ 0 \end{bmatrix} \tag{5}$$

and (4) becomes

$$\left[\begin{array}{c|c|c|c} B & D & O & b \\ \hline -c_B^T & -c_D^T & 1 & 0 \end{array} \right] \tag{6}$$

Matrix (6) will be called the *initial* matrix for problem (1).

Multiplying (6) on the left by

$$\left[\begin{array}{c|c} B & O \\ \hline -c_B^T & 1 \end{array} \right]^{-1} = \left[\begin{array}{c|c} B^{-1} & O \\ \hline c_B^T B^{-1} & 1 \end{array} \right]$$

(see Section 1 of Chapter 2) gives

$$\left[\begin{array}{c|c|c|c} I_m & B^{-1}D & O & B^{-1}b \\ \hline 0 & c_B^T B^{-1}D - c_D^T & 1 & c_B^T B^{-1}b \end{array} \right]. \tag{7}$$

Matrix (7) can also be obtained from matrix (6) by performing a sequence of elementary row operations as indicated below

$$\left[\begin{array}{c|c|c|c} B & D & O & b \\ \hline -c_B^T & -c_D^T & 1 & 0 \end{array} \right] \rightarrow \cdots \rightarrow \left[\begin{array}{c|c|c|c} I_m & B^{-1}D & O & B^{-1}b \\ \hline -c_B^T & -c_D^T & 1 & 0 \end{array} \right] \rightarrow$$

$$\left[\begin{array}{c|c|c|c} I_m & B^{-1}D & O & B^{-1}b \\ \hline 0 & c_B^T B^{-1}D - c_D^T & 1 & c_B^T B^{-1}b \end{array} \right]$$

where the final operation consists of replacing row $m + 1$ by itself plus each component of $c_B{}^T$ times the corresponding row of $[I_m \mid B^{-1}D \mid 0 \mid B^{-1}b]$.

EXAMPLE 1 If $B = [A^{(3)} A^{(2)}]$, then determine matrix (7) for the following problem:

$$\text{Minimize } -2x_1 - x_2 + 2x_3 \text{ subject to}$$
$$x_1 \qquad\ + x_3 = 4$$
$$-2x_1 + x_2 \qquad = 8$$
$$x_1 \geq 0, x_2 \geq 0, \text{ and } x_3 \geq 0$$

Solution: First form matrix (4). Then replace row 3 by itself plus twice row 1 minus row 2 to obtain

$$\begin{bmatrix} 1 & 0 & 1 & 0 & 4 \\ -2 & 1 & 0 & 0 & 8 \\ 2 & 1 & -2 & 1 & 0 \end{bmatrix} \xrightarrow{R_3 + (2R_1 - R_2)} \begin{bmatrix} 1 & 0 & 1 & 0 & 4 \\ -2 & 1 & 0 & 0 & 8 \\ 6 & 0 & 0 & 1 & 0 \end{bmatrix}.$$

In this example, $x_B = [x_3 \quad x_2]^T$, $x_D = [x_1]$, $c_B = [2 \quad -1]^T$, and $c_D = [-2]$.

EXAMPLE 2 If $B = [A^{(4)} \quad A^{(1)}]$, then determine matrix (7) for the problem:

$$\text{Minimize } x_1 + x_2 \qquad + 2x_4 \text{ subject to}$$
$$2x_1 + x_2 + x_3 + \ x_4 = 8$$
$$2x_1 \qquad\ - x_3 + 3x_4 = 7$$
$$\text{all } x_i \geq 0$$

Solution: First form matrix (4). Then use row operations to transform the fourth column of that matrix into the first column of I_3, and the first column into the second column of I_3. The work is summarized below.

$$\begin{bmatrix} 2 & 1 & 1 & 1 & 0 & 8 \\ 2 & 0 & -1 & 3 & 0 & 7 \\ -1 & -1 & 0 & -2 & 1 & 0 \end{bmatrix} \begin{array}{c} R_2 - 3R_1 \\ \xrightarrow{\hspace{1cm}} \\ R_4 + 2R_1 \\ 3 \end{array} \begin{bmatrix} 2 & 1 & 1 & 1 & 0 & 8 \\ -4 & -3 & -4 & 0 & 0 & -21 \\ 3 & 1 & 2 & 0 & 1 & 16 \end{bmatrix}$$

$$\begin{array}{c} R_1 + (1/2)R_2 \\ R_3 + (3/4)R_2 \\ \xrightarrow{\hspace{1cm}} \\ (-1/4)R_2 \end{array} \begin{bmatrix} 0 & -1/2 & -1 & 1 & 0 & -5/2 \\ 1 & 3/4 & 1 & 0 & 0 & 21/4 \\ 0 & -5/4 & -1 & 0 & 1 & 1/4 \end{bmatrix}$$

In this example $x_B = [x_4 \quad x_1]^T$, $x_D = [x_2 \quad x_3]^T$, $c_B = [2 \quad 1]^T$ and $c_D = [1 \quad 0]^T$.

EXAMPLE 3   If $B = [A^{(2)} \quad A^{(1)}]$, then determine (7) for the following problem

$$\text{Minimize} \qquad x_3 + x_4 \text{ subject to}$$

$$x_2 + 21x_3 \qquad\qquad = 53$$

$$x_1 \qquad\qquad - 19x_4 = 71$$

$$\text{all } x_i \geq 0$$

Solution: Since $A^{(2)} = I^{(1)}$, $A^{(1)} = I^{(2)}$, and $c_1 = c_2 = 0$, then (4) and (7) coincide. The required matrix is

$$\begin{bmatrix} 0 & 1 & 21 & 0 & \vdots & 0 & \vdots & 53 \\ 1 & 0 & 0 & -19 & \vdots & 0 & \vdots & 71 \\ \hline 0 & 0 & -1 & -1 & \vdots & 1 & \vdots & 0 \end{bmatrix}$$

Matrix (7) represents the system of equations below.

$$x_B + \qquad B^{-1}Dx_D \qquad\qquad = B^{-1}b \qquad\qquad (8)$$

$$(c_B{}^T B^{-1}D - c_D{}^T)x_D + z_0 = c_B{}^T B^{-1}b$$

It is well known from linear algebra that systems (2) and (8) are equivalent, i.e., they have the same solutions.

      Since $z_0 = c_B{}^T B^{-1}b - (c_B{}^T B^{-1}D - c_D{}^T)x_D$, it is clear that $x_B = B^{-1}b$ and $x_D = [0 \dots 0]^T$ is an optimal solution of (1) whenever $B^{-1}b \geq 0$ and $c_B{}^T B^{-1}D - c_D{}^T \leq 0$. This establishes the next theorem. The reader may find the alternate proof more enlightening.

THEOREM 1   If $B^{-1}b \geq 0$ and $c_B{}^T B^{-1}D - c_D{}^T \leq 0$, then $x = [x_B^T \ x_D^T]^T$, where $x_B = B^{-1}b$ and $x_D = [0 \dots 0]^T$, solves problem (1) and in fact, the minimum value of the objective function is $z_0 = c_B{}^T B^{-1}b$.

Proof: (alternate) Suppose that $A\bar{x} = b$ and $\bar{x} \geq 0$. Let $X_B = [I_m \vdots B^{-1}D]$. Notice that $BX_B = B[I_m \vdots B^{-1}D] = [B \vdots BB^{-1}D] = [B \vdots D] = A$. Thus, $b = A\bar{x} = BX_B\bar{x}$ and $b = Bx_B$. Hence, $0 = B(x_B - X_B\bar{x})$. Since B is nonsingular, it follows that $x_B - X_B\bar{x} = B^{-1}0 = 0$, i.e., $x_B = X_B\bar{x}$. Thus,

$$c_B{}^T x_B = c_B{}^T X_B\bar{x} = c_B{}^T [I_m \quad B^{-1}D]\bar{x} = [c_B{}^T \quad c_B{}^T B^{-1}D]\bar{x}$$

$$\leq [c_B{}^T \quad c_D{}^T]\bar{x} = c^T\bar{x}$$

The result then follows.

Recall (Section 1 of Chapter 2) that when $x_B = B^{-1}b$ and $x_D = [0 \ldots 0]^T$, then $x = [x_B^T x_D^T]^T$ is called a *basic solution* of $Ax = b$. (Notice that B must be nonsingular.) The matrix B is called a *basis matrix*, and the set of columns $\{A^{(1)}, \ldots, A^{(m)}\}$ is called an *admissible basis* of the basic solution. Any basic solution, x, of $Ax = b$ that is also a feasible solution of (1), i.e., $x \geq 0$, is called a *basic feasible solution* (a *BFS*). A *nondegenerate BFS* is a BFS for which $x_B = B^{-1}b > 0$. In our present setting, $F = \{x: Ax = b$ and $x \geq 0\}$ is the set of *feasible solutions* of (1). Theorem 1 gives a sufficient condition, namely $c_B^T B^{-1}D - c_D^T \leq 0$ for a BFS $x_B = B^{-1}b$, where $B^{-1}b \geq 0$, and $x_D = [0 \ldots 0]^T$ to be an optimal solution of (1).

For the remainder of this chapter, we shall assume that $B^{-1}b \geq 0$, i.e., that a BFS of F is known. In practice this is often accomplished by formulating the problem so that A contains all the columns of I_m and $b \geq 0$.

The notation $X_B = [I_m \quad B^{-1}D]$ will prove useful in our upcoming discussions. For convenience, let $X_B = [x_{ij}]$, then

$$X_B = [I_m \mid B^{-1}D] = \begin{bmatrix} 1 & 0 \ldots 0 & x_{1\,m+1} \ldots x_{1j} \ldots x_{1n} \\ 0 & 1 \ldots 0 & x_{2\,m+1} \ldots x_{2j} \ldots x_{2n} \\ \cdot & \cdot \quad \cdot & \cdot \quad \cdot \\ \cdot & \cdot \quad \cdot & \cdot \quad \cdot \\ \cdot & \cdot \quad \cdot & \cdot \quad \cdot \\ 0 & 0 \ldots 1 & x_{m\,m+1} \ldots x_{mj} \ldots x_{mn} \end{bmatrix}$$

If we define $z_j = c_B^T B^{-1}A^{(j)}$, then $z_j - c_j = 0$ when $j = 1, \ldots, m$ and $z_j - c_j$ is the jth component of $c_B^T B^{-1}D - c_D^T$ when $j = m+1, \ldots, n$. Moreover, the BFS $x_B = B^{-1}b \geq 0$ and $x_D = [0 \ldots 0]^T$ will be an optimal solution when $z_j - c_j \leq 0$ for all $j = 1, \ldots, n$.

Now suppose that some component $z_j - c_j$ of $c_B^T B^{-1}D - c_D^T$ is positive. If each corresponding x_{ij}, where $i = 1, \ldots, m$, of the jth column of $B^{-1}D$ is less than or equal to zero, then our next result shows that our problem has an unbounded minimum over the set F.

THEOREM 2   If there exists a j so that $z_j - c_j > 0$ and each $x_{ij} \leq 0$, then (1) has an unbounded minimum over F.

Proof: From (8) we have that $z_0 = c_B^T B^{-1}b - (c_B^T B^{-1}D - c_D^T)x_D$, when $x_B + B^{-1}Dx_D = B^{-1}b$. Let $x_D = [0 \ldots \theta \ldots 0]^T$, where θ is the jth component of x_D. Then

$$z_0 = c_B^T B^{-1}b - \theta(z_j - c_j)$$

Notice that $z_0 \to -\infty$ as $\theta \to \infty$. Moreover, the solution $x = [x_B^T x_D^T]^T$, where

$$x_B = B^{-1}b - \theta \begin{bmatrix} x_{1j} \\ \cdot \\ \cdot \\ \cdot \\ x_{mj} \end{bmatrix}$$

and $x_D = [0 \dots \theta \dots 0]^T$, is a feasible solution of (1), since $B^{-1}b \geq 0$ and $[x_{1j} \dots x_{mj}]^T \leq 0$ together imply that $x_B \geq 0$. So, (1) must have an unbounded minimum.

Next, consider the case where $z_j - c_j > 0$ and $x_{ij} > 0$, for some i. At the end of Section 1 of Chapter 2, a technique was developed which enabled us to generate a sequence of basic feasible solutions of F. We shall now modify that technique so that the value of the objective function, $c^T x$, will not increase as a sequence of basic feasible solutions is generated. Again let $x_D = [0 \dots \theta \dots 0]^T$, where θ is the jth entry. From (8) we have that

$$z_0 = c_B^T B^{-1} b - \theta(z_j - c_j)$$

and

$$x_B + \theta \begin{bmatrix} x_{1j} \\ \cdot \\ \cdot \\ \cdot \\ x_{mj} \end{bmatrix} = B^{-1}b$$

To simplify our notation let $B^{-1}b = [\bar{b}_1 \dots \bar{b}_m]^T$. We want $\theta \geq 0$ and

$$x_B = B^{-1}b - \theta \begin{bmatrix} x_{1j} \\ \cdot \\ \cdot \\ \cdot \\ x_{mj} \end{bmatrix} = \begin{bmatrix} \bar{b}_1 - \theta x_{1j} \\ \cdot \\ \cdot \\ \cdot \\ \bar{b}_m - \theta x_{mj} \end{bmatrix} \geq 0 \qquad (9)$$

Since each $\bar{b}_i \geq 0$, then (9) will clearly hold when $\theta \geq 0$ and $x_{ij} \leq 0$. When $x_{ij} > 0$, we must have

$$\bar{b}_i - \theta x_{ij} \geq 0$$

i.e.,

$$\theta \leq \bar{b}_i / x_{ij}$$

Thus, select the k for which

$$\bar{b}_k / x_{kj} = \min \{\bar{b}_i / x_{ij} : x_{ij} > 0\} \qquad (10)$$

and let $\theta = \bar{b}_k/x_{kj}$. Then $x_B \geq 0$ and $x_D \geq 0$. Moreover, one of the components of x_B is zero and thus at most m components of x are positive. Since $\theta \geqslant 0$ and $z_j - c_j > 0$, it follows that

$$c_B^T B^{-1} b - \theta(z_j - c_j) \leq c_B^T B^{-1} b \tag{11}$$

with strict inequality holding whenever $\theta > 0$. Thus, we have generated a new feasible solution of F that does not increase the value of z_0. To simplify our discussion, suppose that $j = m + 1$ and $k = 1$. Then the matrix $\bar{B} = [A^{(m+1)} A^{(2)} \ldots A^{(m)}]$ is nonsingular and is a basis for this new solution (see Section 1 of Chapter 2). Thus, our new solution is in fact a BFS. This proves the following result.

THEOREM 3   If there exists a j so that $z_j - c_j > 0$, and if some $x_{ij} > 0$, where x_{ij} is an entry in the jth column of the matrix X_B, then by pivoting at the element x_{kj} for which (10) holds, a new basic feasible solution is generated that does not increase the value of z_0. In fact, when $B^{-1} b > 0$, i.e., when the basic feasible solution $x_B = B^{-1} b$ and $x_D = [0 \ldots 0]^T$ is nondegenerate, then the value of z_0 decreases.

It follows from Theorem 7 and Theorem 8 of Chapter 2 that when $c^T x$ is bounded from below over the set $F = \{x : Ax = b \text{ and } x \geq 0\}$ of feasible solutions of (1), then $c^T x$ obtains a minimum over F and the minimum occurs at a BFS of F. A similar result can also be established as follows.

THEOREM 4   If $c^T x$ obtains a finite minimum over F, then the minimum occurs at a basic feasible solution of F (Note: The minimum may occur at other places as well.)
Proof: Without loss of generality, suppose that $x = [x_1 \ldots x_k 0 \ldots 0]^T$, where $k > m$, solves Problem (1). Let $B = [A^{(1)} \ldots A^{(k)}]$. Then $B \in \mathbf{R}_{m \times k}$ and is clearly singular since $k > m$ (see Theorem 4 of Chapter 2). Thus, there exists a nonzero vector y such that $By = 0$, where $0 = [0 \ldots 0]^T$. Let θ be the smallest entry in either of the sets $\{-x_i/y_i : y_i < 0\}$ or $\{x_i/y_i : y_i > 0\}$. Then

$$x_i + \theta y_i \geq 0 \text{ when } y_i < 0$$

and

$$x_i - \theta y_i \geq 0 \text{ when } y_i > 0$$

Since $x \geq 0$, it follows that $\theta \geq 0$. It is then clear that $x_i + \theta y_i \geq 0$ whenever $y_i \geq 0$, and that $x_i - \theta y_i \geq 0$ whenever $y_i \leq 0$. Thus, $[x_1 \ldots x_k]^T + \theta y \geq 0$ and $[x_1 \ldots x_k]^T - \theta y \geq 0$. If we now use the notations $x + \theta y$ and $x - \theta y$ to represent

$$x + \theta y = [x_1 + \theta y_1 \quad x_k + \theta y_k \ 0 \ldots 0]^T$$

and

$$x - \theta y = [x_1 - \theta y_1 \quad \dots \quad x_k - \theta y_k \, 0 \dots 0]^T$$

then $A(x + \theta y) = B[x_1 \dots x_k]^T + \theta By = b$. Hence, $x + \theta y \in F$. Likewise, $x - \theta y \in F$. Since x solves (1), it follows that $c^T x \le c^T(x + \theta y) = c^T x + \theta c^T y$. Hence, $\theta c^T y \ge 0$. Since $\theta \geqslant 0$, it follows that $c^T y \ge 0$. Likewise, $c^T x \le c^T(x - \theta y)$ implies that $c^T y \le 0$. Thus, $c^T y = 0$, which means that $c^T x = c^T(x + \theta y) = c^T(x - \theta y)$. Hence, the minimum also occurs at both $x + \theta y$ and $x - \theta y$. But it follows from the definition of θ that one of the vectors, $x + \theta y$ or $x - \theta y$, has at most $k - 1$ positive components. Select the one that has the fewest positive components and repeat the argument. This can be repeated at most $k - m$ times, and must terminate with a BFS which solves (1).

These four theorems enable us to solve problem (1) once we find an initial BFS of F. Using Theorem 3, we then generate a sequence of BFS's for which the cost function $z_0 = c^T x$ does not increase. The iterations stop when the conditions of Theorem 1 or Theorem 2 are satisfied. Theorem 4 ensures us that when a finite optimal value of z_0 exists, then we can find it with this process. This iterative procedure is known as the *primal simplex* procedure (or simply the simplex procedure) and was devised by G.B. Dantzig in 1947.

Clearly, when all of the BFS's are nondegenerate, the simplex procedure will move to a different BFS at each step, and none of the BFS's will be repeated. Since there is a finite number of BFS's, the process must then solve the problem in a finite number of steps. If F does have a degenerate BFS, then the possibility exists that $\theta = 0$ in some iteration. When this happens, the value of z_0 will remain constant, and not decrease, as the next BFS is computed. The possibility then exists that the simplex procedure will cycle, i.e., a repeating sequence of BFS's will be generated. This will be dealt with later in Chapter 5. In practice, when a value of $\theta = 0$ is encountered, we shall simply continue with the simplex procedure. Usually the same BFS will be repeated for a few steps (but with different basis matrices), and then a new BFS will be generated. The primal simplex procedure usually solves the problem within m to $2m$ iterations.

Matrix (7) will frequently be called a *simplex tableau*. If in the final tableau there exists $z_j - c_j = 0$, where $A^{(j)}$ is not in the admissible basis of the corresponding BFS, then another optimal solution can be determined by introducing $A^{(j)}$ into the admissible basis of another BFS of F. When this happens the problem is said to have *multiple* solutions. Certainly as in Exercise 2.40 of Section 3 of the last chapter, any convex combination of multiple solutions is again a solution of the linear programming problem. In this way we can generate solutions that are not necessarily BFS solutions, i.e., we often can generate solutions containing more than m positive components. An example will be presented shortly (see Example 7). Chart 1 summarizes our procedure at this particular point of development.

We shall now illustrate the simplex procedure with several examples.

EXAMPLE 4   Minimize $x_1 - x_2 + 2x_3 - 3x_4$ subject to

$$3x_1 \qquad + x_3 - x_4 = 1$$
$$7x_1 + x_2 \qquad - x_4 = 0$$
$$-8x_1 \qquad + x_4 \le 1$$
$$\text{all } x_i \ge 0$$

Solution: First we shall add a slack variable to the left hand side of the third constraint and rewrite the problem in matrix form

$$\text{Minimize } c^T x \text{ subject to } Ax = b \text{ and } x \ge 0$$

where $x = [x_1 \; x_2 \; x_3 \; x_4 \; x_5]^T$, $b = [1 \quad 0 \quad 1]^T$, $c^T = [1 \; -1 \; 2 \; -3 \; 0]^T$ and

$$A = \begin{bmatrix} 3 & 0 & 1 & -1 & 0 \\ 7 & 1 & 0 & -1 & 0 \\ -8 & 0 & 0 & 1 & 1 \end{bmatrix}$$

Clearly $B = [A^{(3)} A^{(2)} A^{(5)}]$ is nonsingular. Setting up matrix (4) and proceeding to matrix (7) gives

$$\begin{bmatrix} 3 & 0 & 1 & -1 & 0 & 0 & 1 \\ 7 & 1 & 0 & -1 & 0 & 0 & 0 \\ -8 & 0 & 0 & 1 & 1 & 0 & 1 \\ \hline -1 & 1 & -2 & 3 & 0 & 1 & 0 \end{bmatrix} \quad \begin{array}{c} \\ \\ R_4 - (R_2 - 2R_1) \\ \longrightarrow \end{array}$$

$$\begin{bmatrix} 3 & 0 & 1 & -1 & 0 & 0 & 1 \\ 7 & 1 & 0 & -1 & 0 & 0 & 0 \\ -8 & 0 & 0 & ① & 1 & 0 & 1 \\ \hline -2 & 0 & 0 & 2 & 0 & 1 & 2 \end{bmatrix} \quad \begin{array}{c} R_1 + R_3 \\ R_2 + R_3 \\ \longrightarrow \\ R_4 - 2R_3 \end{array}$$

In the matrix on the right $D = [A^{(1)} A^{(4)}]$ and a BFS is $x_1 = 0$, $x_2 = 0$, $x_3 = 1$, $x_4 = 0$, and $x_5 = 1$. Notice that this BFS is degenerate. Also notice that $z_4 - c_4 = 2$ and, hence, we shall try to pivot in the fourth column. In this case $\theta = \min \{x_i / x_{i4} : x_{i4} > 0\} = 1$ and the pivot element is the circled entry in the

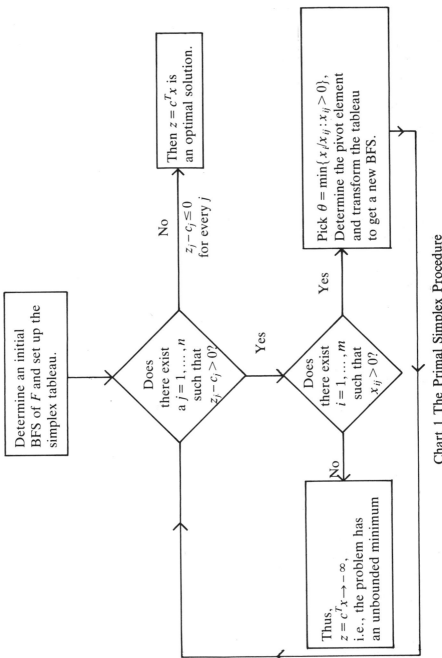

Chart 1 The Primal Simplex Procedure

matrix on the right. After performing the indicated row operations, we have the next tableau

$$\begin{bmatrix} -5 & 0 & 1 & 0 & 1 & 0 & 2 \\ -1 & 1 & 0 & 0 & 1 & 0 & 1 \\ -8 & 0 & 0 & 1 & 1 & 0 & 1 \\ \hline 14 & 0 & 0 & 0 & -2 & 1 & 0 \end{bmatrix}$$

In this tableau $B = [A^{(3)} A^{(2)} A^{(4)}]$ and $D = [A^{(1)} A^{(4)}]$. The corresponding BFS is $x_1 = x_5 = 0$, $x_2 = 1$, $x_3 = 2$ and $x_4 = 1$. Since $z_1 - c_1 = 14$, $x_{11} = -5$, $x_{21} = -1$ and $x_{31} = -8$, then Theorem 2 implies that the problem has an unbounded minimum.

EXAMPLE 5 Maximize $2x_1 + x_2 + 3x_3$ subject to

$$x_1 + 3x_2 + 2x_3 \le 6$$
$$-x_1 + x_2 - 3x_3 \ge -9$$
$$\text{all } x_i \ge 0$$

Solution: First we shall restate the problem in a more appropriate form. To do this multiply the second inequality by -1 and add slack variables to both inequalities to get

$$x_1 + 3x_2 + 2x_3 + x_4 \qquad = 6$$
$$x_1 - x_2 + 3x_3 \qquad + x_5 = 9$$
$$\text{all } x_i \ge 0$$

Next recall that max $\{c^T x : x \in F\} = -\min \{-c^T x : x \in F\}$. Hence, the problem can be restated as:

$$\text{Minimize } -2x_1 - x_2 - 3x_3 \text{ subject to}$$
$$x_1 + 3x_2 + 2x_3 + x_4 \qquad = 6$$
$$x_1 - x_2 + 3x_3 \qquad + x_5 = 9$$
$$\text{all } x_i \ge 0$$

Clearly, $\{A^{(4)}, A^{(5)}\}$ is an admissible base for the BFS $x = [0 \ 0 \ 0 \ 6 \ 9]^T$. Setting up matrix (4), we have

$$\begin{bmatrix} 1 & 3 & ② & 1 & 0 & 0 & 6 \\ 1 & -1 & 3 & 0 & 1 & 0 & 9 \\ 2 & 1 & 3 & 0 & 0 & 1 & 0 \end{bmatrix} \begin{matrix} R_2 - (3/2)/R_1 \\ R_3 - (3/2)/R_1 \\ \overline{\qquad\qquad} \\ (1/2)R_1 \end{matrix}$$

which is also matrix (7) since $c_4 = c_5 = 0$. Notice that $z_3 - c_3 = 3$ and that the column directly above contains positive entries. Since $\theta = \min\{6/2, 9/3\} = 3$, we can choose either $x_{23} = 2$ or $x_{33} = 3$ as the pivot elements. Suppose we select 2 as the pivot element. Next we will generate a new BFS that has $\{A^{(3)}, A^{(5)}\}$ as its admissible basis. First replace row 2 by itself minus 2/3 times row 1. Do the same for row 3. Then divide row 1 by 2. The results are shown in the next tableau.

$$
\left[
\begin{array}{cccccc|c|c}
1/2 & 3/2 & 1 & 1/2 & 0 & 0 & 3 \\
-1/2 & -11/2 & 0 & -3/2 & 1 & 0 & 0 \\
\hline
1/2 & -7/2 & 0 & -3/2 & 0 & 1 & -9
\end{array}
\right]
\begin{array}{l}
R_2 + R_1 \\
R_3 - R_1 \\
\;\; 2R_1
\end{array}
$$

From the last tableau we see that $[0 \;\; 0 \;\; 3 \;\; 0 \;\; 0]^T$ is a degenerate BFS. Notice that $z_1 - c_1 = 1/2$ and $x_{11} = 1/2$. As in Theorem 3 we compute $\theta = \min\{3/(1/2)\} = 6$ and let 1/2 become the new pivot element. Thus, we must replace row 2 by itself plus row 1. Next replace row 3 by itself minus row 1. Finally, we replace row 1 by itself times 2. The results are shown in the next tableau.

$$
\left[
\begin{array}{cccccc|c|c}
1 & 3 & 2 & 1 & 0 & 0 & 6 \\
0 & -4 & 1 & -1 & 1 & 0 & 3 \\
\hline
0 & -5 & -1 & -2 & 0 & 1 & -12
\end{array}
\right]
$$

From this tableau we see that $x = [6 \;\; 0 \;\; 0 \;\; 0 \;\; 3]^T$ is a BFS and $z_0 = -12$. Moreover, since the last row does not contain any positive $z_j - c_j$, then Theorem 1 says that the minimum is obtained at $[6 \;\; 0 \;\; 0 \;\; 0 \;\; 3]^T$.

Since the original problem was a maximization problem we now have $\max\{c^T x : x \in F\} = 12$ and this happens at least when $x = [6 \;\; 0 \;\; 0]^T$.

In this example the appearance of the degenerate BFS $[0 \;\; 0 \;\; 3 \;\; 0 \;\; 0]^T$ in the second tableau did not cause any difficulties. In fact the value of the objective function $-c^T x$ decreased from the second tableau to the third tableau.

EXAMPLE 6   Minimize $x_1 + x_2 + x_3 + x_4 + x_5 + x_6$ subject to

$$
\begin{aligned}
16x_1 + 2x_2 \quad\quad\quad\quad + x_5 + 10x_6 &= 4 \\
-x_1 \quad\quad\quad + x_4 \quad + \quad x_6 &= 0 \\
5x_1 + 4x_2 + x_3 \quad\quad\quad\quad\quad &= 8
\end{aligned}
$$

$$\text{all } x_i \geq 0$$

Solution: Clearly $\{A^{(5)}, A^{(4)}, A^{(3)}\}$ is an admissible basis for the degenerate BFS $[0 \;\; 0 \;\; 8 \;\; 0 \;\; 4 \;\; 0]^T$. The initial matrix (4) is first formed below.

The first simplex tableau is then computed

$$
\begin{bmatrix}
16 & 2 & 0 & 0 & 1 & 10 & 0 & 4 \\
-1 & 0 & 0 & 1 & 0 & 1 & 0 & 0 \\
5 & 4 & 1 & 0 & 0 & 0 & 0 & 8 \\
\hline
-1 & -1 & -1 & -1 & -1 & -1 & 1 & 0
\end{bmatrix}
\quad \xrightarrow{R_4 + (R_1 + R_2 + R_3)}
$$

by replacing row 4 by itself plus the sum of rows 1, 2 and 3 as indicated.

$$
\begin{bmatrix}
16 & 2 & 0 & 0 & 1 & 10 & 0 & 4 \\
-1 & 0 & 0 & 1 & 0 & ① & 0 & 0 \\
5 & 4 & 1 & 0 & 0 & 0 & 0 & 8 \\
\hline
19 & 5 & 0 & 0 & 0 & 10 & 1 & 12
\end{bmatrix}
\quad \xrightarrow[R_4 - 10R_2]{R_1 - 10R_2}
$$

Since $z_6 - c_6 = 10$ and the column above $z_6 - c_6$ contains positive elements, then we select $\theta = \min\{0/1, 4/10\} = 0$. Thus the pivot element is $x_{26} = 1$, which is circled in the first tableau. This tableau is transformed into the second tableau by replacing row 1 by itself minus 10 times row 2; and by replacing row 4 by itself minus 10 times row 2.
The second tableau is

$$
\begin{bmatrix}
26 & ② & 0 & -10 & 1 & 0 & 0 & 4 \\
-1 & 0 & 0 & 1 & 0 & 1 & 0 & 0 \\
5 & 4 & 1 & 0 & 0 & 0 & 0 & 8 \\
\hline
29 & 5 & 0 & -10 & 0 & 0 & 1 & 12
\end{bmatrix}
\quad
\begin{array}{l}
R_3 - 2R_1 \\
R_4 - (5/2)R_1 \\
\hline
(1/2)R_1
\end{array} \to
$$

Notice that the second tableau gives $[0\ 0\ 8\ 0\ 0\ 4]^T$ as a BFS, but with an admissible basis of $\{A^{(5)}, A^{(6)}, A^{(3)}\}$. Since $z_2 - c_2 = 5$ and $z_2 - c_2$ contains positive elements, let $\theta = \min\{4/2, 8/4\} = 2$. Since $\theta = 4/2$ and $\theta = 8/4$, we can select either 2 or 4 in column $A^{(2)}$ as pivot elements. In order to avoid fractions we shall select (and circle) 2 as the pivot element. Next replace row 3 by itself minus 2 times row 1. Then replace row 4 by row 4 minus 5/2 times R_1, and replace row 1 by itself times 1/2. The result is the third tableau.

$$
\begin{bmatrix}
13 & 1 & 0 & -5 & 1/2 & 0 & 0 & 2 \\
-1 & 0 & 0 & ① & 0 & 1 & 0 & 0 \\
-47 & 0 & 1 & 20 & -2 & 0 & 0 & 0 \\
\hline
-36 & 0 & 0 & 15 & -5/2 & 0 & 1 & 2
\end{bmatrix}
\quad
\begin{array}{l}
R_1 + 5R_2 \\
R_3 - 20R_2 \\
R_4 - 15R_2 \\
\hline
\end{array} \to
$$

The new BFS is $[0\ 2\ 0\ 0\ 0\ 0]^T$ and the admissible basis is

$\{A^{(2)}, A^{(6)}, A^{(3)}\}$. Notice that this is a degenerate solution. Since $z_4 - c_4 = 15$, let $\theta = \min\{0/1, 0/20\}$. Then we can choose either 1 or 20 in column $A^{(4)}$ as the pivot element. Suppose that we select and circle 1 as the pivot element. Then transforming the third tableau using the indicated row operations gives

$$\begin{bmatrix} 8 & 1 & 0 & 0 & 1/2 & 5 & 0 & 2 \\ -1 & 0 & 0 & 1 & 0 & 1 & 0 & 0 \\ -27 & 0 & 1 & 0 & -2 & -20 & 1 & 0 \\ \hline -21 & 0 & 0 & 0 & -5/2 & -15 & 1 & 2 \end{bmatrix}$$

Since each $z_j - c_j \le 0$, it follows from Theorem 1 that the minimum occurs when $x = [0 \quad 2 \quad 0 \quad 0 \quad 0 \quad 0]^T$ and the minimum is 2. The admissible basis for this extreme point is $\{A^{(2)}, A^{(4)}, A^{(3)}\}$.

EXAMPLE 7   Find several points $x = [x_1 \ x_2 \ x_3 \ x_4]^T$ that solve the problem

Minimize $3x_1 + 2x_2 + 6x_3 + x_4$ subject to

$$x_1 \qquad + x_4 = 4$$
$$x_1 + x_2 + 2x_3 = 5$$
$$\text{all } x_i \ge 0$$

Solution: Our first tableau is determined by using $\{A^{(4)}, A^{(2)}\}$ as an admissible basis.

$$\begin{bmatrix} 1 & 0 & 0 & 1 & 0 & 4 \\ 1 & 1 & 2 & 0 & 0 & 5 \\ \hline -3 & -2 & -6 & -1 & 1 & 0 \end{bmatrix} \xrightarrow{R_3 + (R_1 + 2R_2)} \begin{bmatrix} ① & 0 & 0 & 1 & 0 & 4 \\ 1 & 1 & 2 & 0 & 0 & 5 \\ \hline 0 & 0 & -2 & 0 & 1 & 14 \end{bmatrix} \xrightarrow{R_2 - R_1}$$

Notice that each $z_j - c_j \le 0$. Hence, $x = [0 \quad 5 \quad 0 \quad 4]^T$ gives a minimum of 14. Now consider $z_1 - c_1 = 0$. We shall generate another BFS by introducing $A^{(1)}$ into the admissible basis. To do this let $\theta = \min\{4/1, 5/1\} = 4/1$. Then $x_{11} = 1$ is the pivot element. Replacing row 2 by itself minus row 1 gives

$$\begin{bmatrix} 1 & 0 & 0 & 1 & 0 & 4 \\ 0 & 1 & 2 & -1 & 0 & 1 \\ \hline 0 & 0 & -2 & 0 & 1 & 14 \end{bmatrix}$$

Hence, the minimum also occurs when $x^* = [4 \quad 1 \quad 0 \quad 0]^T$. From Exercise 2.40 of Section 3 of the last chapter it follows that the minimum will occur at any of the points

$$\alpha x + (1 - \alpha)x^* = \alpha[0 \quad 5 \quad 0 \quad 4]^T + (1 - \alpha)[4 \quad 1 \quad 0 \quad 0]^T$$
$$= [(1 - \alpha)4 \quad 5\alpha + (1 - \alpha) \quad 0 \quad 4\alpha]^T$$

where $\alpha \in [0, 1]$.

The columns of the appropriate identity matrix were found in A for each of these examples. In the event that A does not contain these columns (i.e., A does not have an obvious admissible basis for a BFS), we may be able to use the technique of Example 2 to generate a BFS and the corresponding simplex tableau. Then the simplex procedure proceeds as usual. We shall develop another method of dealing with problems such as these in the next section.

A widely used rule for determining the next column of A to be introduced into the admissible basis of a new BFS is

1. Select the column that corresponds to the largest $z_j - c_j$.
2. In the event of a tie, select the column $A^{(j)}$ with the lowest index j.

When this rule is followed, the minimum is usually reached in approximately m steps. The rule was not always followed in the previous examples.

Another rule which is popular in computer software packages is the *least recently considered eligible variable rule*. To explain the rule, suppose that variable x_i was the entering variable in the previous step of the simplex procedure. Then the entering variable in the following step corresponds to the first positive entry in the sequence $z_{i+1} - c_{i+1}, z_{i+2} - c_{i+2}, \ldots, z_n - c_n,$ $z_1 - c_1, z_2 - c_2, \ldots, z_{i-1} - c_{i-1}$. In general, this rule appears to be the most efficient method that is currently available for selecting the entering variables.

Several useful points to remember when using the simplex method are:

1. The z_0 term should never increase from one tableau to the next.
2. The column $B^{-1}b$ should never contain negative components. (Recall that $B^{-1}b$ contains the components of a BFS of the set of feasible solutions.)
3. For each $A^{(j)}$ in the admissible basis, the corresponding $z_j - c_j$ must be zero.

Exercises

3.1 Let $F = \{x: Ax = b \text{ and } x \geq 0\}$, where

$$A = \begin{bmatrix} 2 & 0 & 1 & 1 & 0 \\ -3 & 0 & 0 & 1 & 4 \\ 4 & 1 & 0 & 1 & -1 \end{bmatrix} \text{ and } b = \begin{bmatrix} 14 \\ 5 \\ 6 \end{bmatrix}$$

First determine a BFS of F, and then generate two additional BFS's.

3.2   (a) Prove that whenever $F = \{x: Ax = b \text{ and } x \geq 0\} \neq \emptyset$, then F contains a BFS.

(b) If x is a nondegenerate BFS of F that solves the linear programming problem and if each $z_j - c_j < 0$ for every j where $A^{(j)}$ is not in the admissible basis of x, then prove that x is the unique solution to the problem.

3.3   Solve the following linear programming problems using the simplex procedure.

(a) Minimize $-x_1 + x_2 + x_3 + 4x_4$ subject to

$$
\begin{aligned}
x_1 - 5x_2 + x_3 + 3x_4 &= 19 \\
x_1 - 4x_2 \quad\quad + 2x_4 &\leq 5 \\
-5x_2 \quad\quad + 15x_4 &\leq 10 \\
\text{all } x_i &\geq 0
\end{aligned}
$$

(b) Minimize $2x_1 + x_2 + x_3 - x_4$ subject to

$$
\begin{aligned}
6x_1 - x_2 + x_3 \quad\quad &\leq 10 \\
x_1 + 5x_2 \quad\quad &\leq 4 \\
x_1 + x_2 \quad\quad + x_4 &= 5 \\
\text{all } x_i &\geq 0
\end{aligned}
$$

(c) Maximize $2x_1 + x_2$ subject to

$$
\begin{aligned}
2x_1 + 3x_2 &\leq 16 \\
x_1 - x_2 &\geq -2 \\
\text{all } x_i &\geq 0
\end{aligned}
$$

(d) Maximize $2x_1 - x_2 + 3x_3$ subject to

$$
\begin{aligned}
x_1 - x_2 + 4x_3 &\leq 1 \\
2x_1 + x_2 - x_3 &\leq 10 \\
-x_1 + x_2 - x_3 &\leq 8 \\
\text{all } x_i &\geq 0
\end{aligned}
$$

(e) Minimize $x_1 - x_2 + x_3$ subject to

$$
\begin{aligned}
x_1 + x_2 + x_3 &\leq 5 \\
5x_1 - x_2 \quad\quad &\leq 6 \\
x_2 + 2x_3 &\leq 4 \\
\text{all } x_i &\geq 0
\end{aligned}
$$

(f) Minimize $2x_1 + 3x_2 + 2x_3$ subject to

$$x_1 + \quad x_3 \leq 10$$
$$2x_1 + 3x_2 \leq 30$$
$$2x_2 + 5x_3 \leq 35$$
$$\text{all } x_i \geq 0$$

3.4   Determine infinitely many solutions to each of the following linear programming problems (see Exercise 2.40).

(a) Minimize $5x_1 + x_2 - x_3$ subject to $2x_1 + x_2 \leq 3$, $-x_2 + x_3 \leq 4$, $2x_1 - x_2 + x_3 \leq 1$ and all $x_i \geq 0$.

(b) Maximize $10x_1 + 4x_2$ subject to $-x_1 + 3x_2 \leq 2, x_1 - 3x_2 \leq 0, 4x_1 + 5x_2 \leq 10$ and all $x_i \geq 0$.

3.5   When one of the variables, x_i, in the linear programming problem is not required to be nonnegative the problem can be handled in the usual way by substituting $x_i = y_i - z_i$, where $y_i \geq 0$ and $z_i \geq 0$. Use this technique to solve the following problems.

(a) Minimize $7x_1 - 8x_2 - x_3$ subject to $x_1 + 2x_2 - x_3 \leq 20$, $2x_1 + x_2 \leq 10$, $2x_1 + 2x_2 + x_3 \leq 15$, $x_1 \geq 0$ and $x_2 \geq 0$.

(b) Maximize $2x_1 - 3x_2 + x_3$ subject to $x_1 + x_2 + x_3 \leq 4$, $-x_1 \leq 5$ and $x_1 + 5x_2 \leq 10$.

(c) Maximize $x_1 - x_2$ subject to

$$5x_1 + \quad x_2 \leq 8$$
$$x_1 + 3x_2 \leq 6$$
$$x_1 \text{ unrestricted and } x_2 \leq 0$$

(d) Maximize $3x_1 - x_2 + x_3 - x_4$ subject to

$$x_1 + 5x_2 + x_3 \quad \leq 10$$
$$3x_2 + 2x_3 - x_4 \geq -8.$$
$$x_1 \geq 0, x_2 \leq 0, x_3 \text{ and } x_4 \text{ unrestricted}$$

(e) Maximize $4w_1 + 8w_3$ subject to

$$16w_1 - w_2 + 5w_3 \leq 1$$
$$2w_1 \quad + 4w_3 \leq 1$$
$$10w_1 + w_2 \quad \leq 1$$
$$\text{each } w_i \leq 1$$

3.6   Minimize $2x_1 - x_2 + x_3 - x_4$ subject to $x_1 + x_3 + x_4 = 5$,

$2x_1 + 2x_2 + x_4 = 4$, $x_1 + x_2 + x_3 - x_4 = 1$, $x_1 \geq 0$, $x_2 \geq 0$ and $x_3 \geq 0$ (notice that x_4 is unconstrained) by solving $x_1 + x_2 + x_3 - x_4 = 1$ for x_4 (i.e., $x_4 = x_1 + x_2 + x_3 + 1$), and substituting this expression for x_4 throughout the problem. This will give a standard type of linear programming problem in the nonnegative variables x_1, x_2 and x_3.

3.7   A BFS for a problem of the form

$$\text{Minimize } c^T x \text{ subject to } Ax = b \text{ and } 0 \leq x \leq u$$

is obtained by setting each nonbasic variable equal to either its upper or its lower bound, and then solving the resulting system for the basic variables. Use matrices similar to (6) and (7) to represent such a solution. (Hint: Partition D into two matrices – one corresponding to the nonbasic variables assigned to their lower bounds and one corresponding to the nonbasic variables assigned to their upper bounds.) Show that a BFS is an optimal solution whenever $z_j - c_j \leq 0$ for all nonbasic variables which were assigned to their lower bounds, and whenever $z_j - c_j \geq 0$ for all basic variables which were assigned to their upper bounds. Try to determine a rule for improving a nonoptimal BFS.

3.8   (Post-optimal analysis—add a new constraint.)

(a) Join the constraint

$$x_1 + x_2 + x_3 + x_4 + x_5 + x_6 = 2$$

to the constraints in Example 6 and solve (Hint: See Exercise 2.41.)

(b) Join the constraint $x_1 - x_2 + x_3 > 1$ to the constraints in Example 5 and solve.

3.9   (a) (Post-optimal analysis – changing a cost coefficient.) In Example 6, the cost coefficients were $c_i = 1$ for $i = 1, \ldots, 6$. Change the fourth cost coefficient, $c_4 = 1$, to $\bar{c}_4 = 22$ and solve the resulting problem. Do this by recomputing the last row of the optimal tableau of Example 6, and then continuing with the simplex procedure.

(b) In Example 6 change c_6 from 1 to 9 and solve the resulting problem beginning your solution with the optimal tableau of Example 6.

3.10   (a) (Post-optimal analysis—changing a column $A^{(i)}$.) In Example 6, change the sixth column $A^{(6)} = [10 \quad 1 \quad 0]^T$ of the matrix A to the column $\bar{A}^{(6)} = [1 \quad 0 \quad 3]^T$ and solve the resulting problem beginning your solution with the optimal tableau of Example 6. (Hint, replace the appropriate items in the optimal tableau of Example 6 by $B^{-1}\bar{A}^{(6)}$ and $c_B B^{-1}\bar{A}^{(6)} - c_j$, where

$B = [A^{(2)} A^{(4)} A^{(3)}]$. Notice that

$$B^{-1} = \begin{bmatrix} 1/2 & 0 & 0 \\ 0 & 1 & 0 \\ 2 & 0 & 1 \end{bmatrix} \quad \text{(Why?)}$$

(b) (Post optimal analysis – adding a new variable.) In Example 6 add a new variable x_7 having a corresponding cost coefficient of $c_2 = 2$ and column $A^{(7)} = [8 \quad 1 \quad 6]^T$ to get the following problem

Minimize $\quad x_1 + x_2 \ + x_3 + x_4 + x_5 + \quad x_6 + 2x_7$ subject to

$$16x_1 + 2x_2 \qquad\qquad + x_5 + 10x_6 + 8x_7 = 4$$
$$-x_1 \qquad\quad + x_4 \quad + \quad x_6 + \ x_7 = 0$$
$$5x_1 + 4x_2 + x_3 \qquad\qquad\qquad + 6x_7 = 8$$
$$\text{all } x_i \geq 0$$

Solve this problem beginning with the optimal tableau of Example 6. Do this by first computing

$$z_7 - c_7 = c_B{}^T B^{-1} A^{(7)} - c_7 = w\, A^{(7)} - c_7$$

If $z_7 - c_7 > 0$, then join the column

$$\begin{bmatrix} B^{-1} A^{(7)} \\ z_7 - c_7 \end{bmatrix}$$

to the optimal tableau of Example 6 and continue with the primal simplex procedure.

2. ARTIFICIAL VARIABLES AND ARTIFICIAL COST COEFFICIENTS

To solve the linear programming problem using the simplex method we must first determine an initial BFS of F before we can set up the first tableau in the simplex procedure. This is easily done when, after possibly adding slack variables, the matrix $A = [a_{ij}]_{(m,n)}$ contains the columns of the identity matrix I_m. The next two sections are concerned with the case that A does not contain such columns.

Again consider Problem (1). Now we assume that $b \geqslant 0$. There is no loss of generality in this last assumption since any constraint for which $b_i < 0$ can be multiplied by -1. For simplicity, suppose that none of the columns of the identity matrix I_m can be found among the columns of A. Then consider the following problem.

Minimize $c_1 x_1 + \cdots + c_n x_n \quad + M x_{n+1} + \cdots + M x_{n+m}$ subject to

$$a_{11} x_1 + \cdots + a_{1n} x_n + x_{n+1} \qquad\qquad = b_1$$

$$a_{21} x_1 + \cdots + a_{2n} x_n \qquad\quad + x_{n+2} \qquad = b_2 \qquad\qquad (27)$$

$$\vdots$$

$$a_{m1} x_1 + \cdots + a_{mn} x_n \qquad\qquad\quad + x_{n+m} = b_m$$

$$x_1 \geq 0, \ldots, \; x_n \geq 0, \; x_{n+1} \geq 0, \ldots, \; x_{n+m} \geq 0$$

where M is an unspecified large positive number. The variables $x_{n+1}, \ldots,$ x_{n+m} in Problem (27) are called *artificial* variables. The set of feasible solutions of Problem (27) will be denoted by $\bar{F}$. Clearly,

$$\bar{F} = \{ x \in \mathbf{R}_{(n+m) \times 1} : [A : I_m] x = b, x \geq 0 \}$$

For convenience we denote the objective function of Problem (1) by $f(x)$, i.e.,

$$f(x) = c^T x = c_1 x_1 + \cdots + c_n x_n$$

and we denote the objective function for Problem (27) by

$$\bar{f}(x) = c_1 x_1 + \cdots + c_n x_n + M x_{n+1} + \cdots + M x_{n+m} \qquad (28)$$

where in the last case $x = [x_1 \ldots x_n \, x_{n+1} \ldots x_{n+m}]^T$. Clearly, if $x \in F$, then $\bar{x} \in \bar{F}$, where $\bar{x}$ is defined to be

$$\bar{x} = [x_1 \ldots x_n \, 0 \ldots 0]^T \qquad (29)$$

Moreover, $f(x) = \bar{f}(\bar{x})$. We now define a function $p : F \to \bar{F}$ by

$$p(x) = \bar{x} \qquad (30)$$

It is a simple exercise to show that p identifies F with the subset $p(F) = \{ p(x) : x \in F \}$ of $\bar{F}$. Also, it is clear that $\{ f(x) : x \in F \} = \{ \bar{f}(\bar{x}) : \bar{x} \in p(F) \}$ and hence,

$$\min \{ \bar{f}(x) : x \in \bar{F} \} \leq \min \{ f(x) : x \in F \} \qquad (31)$$

It is important to notice that Inequality (31) holds regardless of the magnitude of the positive number M. We shall now consider the solution of Problem (27) by the simplex procedure. Before setting up the first tableau, notice that in any tableau containing artificial variables in the BFS, each component of the last row will be of the form $aM + d$ where a and d are constants, i.e., (see Exercise 3.14)

$$z_0 = a_0 M + d_0 \text{ and } z_j - c_j = a_j M + d_j \text{ for } j = 1, \ldots, n+m$$

As a result we shall replace the last row of each tableau containing artificial variables with two rows. The $(m+2)-$row will be $[a_1, \ldots a_{n+m}\ 1\ 0\ a_0]$ and the $(m+1)-$row will be $[d_1 \ldots d_{n+m}\ 0\ 1\ d_0]$. Clearly, if we let $A^{(n+1)} = [1\ 0\ 0\ \ldots\ 0]^T$, $A_{(n+2)} = [0\ 1\ 0\ \ldots\ 0]^T$, $\ldots$, $A_{(n+m)} = [0\ 0\ 0\ \ldots\ 1]^T$, then $A^{(n+1)}, \ldots , A^{(n+m)}$ is an admissible basis for a BFS of $\bar{F}$. We shall modify matrix (4) accordingly to obtain

$$
\begin{bmatrix}
a_{11} \ldots a_{1n} & 1 & 0 \ldots & 0 & 0 & 0 & b_1 \\
a_{21} \ldots a_{2n} & 0 & 1 \ldots & 0 & 0 & 0 & b_2 \\
\cdot & & \cdot & & & & \cdot \\
\cdot & & \cdot & & & & \cdot \\
\cdot & & \cdot & & & & \cdot \\
a_{m1} \ldots a_{mn} & 0 & 0 \ldots & 1 & 0 & 0 & b_m \\
\hline
-c_1 \ldots -c_n & 0 & 0 \ldots & 0 & 1 & 0 & 0 \\
\hline
0 \ldots \quad 0 & -1 & -1 \ldots -1 & 0 & 1 & 0
\end{bmatrix}
$$

We then compute our first simplex tableau using row operations to transform columns $n+1$ through $n+m$ into the appropriate columns of I_{m+2} (here $B = [A^{(n+1)} \ldots A^{(n+m)}]$).

$$
\xrightarrow{\quad R_{m+2} + \sum_{i=1}^{m} R_i \quad}
\begin{bmatrix}
a_{11} & \ldots & a_{1n} & 1 & 0 \ldots 0 & 0 & 0 & b_1 \\
a_{21} & \ldots & a_{2n} & 0 & 1 \ldots 0 & 0 & 0 & b_2 \\
\cdot & & \cdot & & \cdot & & & \cdot \\
\cdot & & \cdot & & \cdot & & & \cdot \\
\cdot & & \cdot & & \cdot & & & \cdot \\
a_{m1} & \ldots & a_{mn} & 0 & 0 \ldots 1 & 0 & 0 & b_m \\
\hline
-c_1 & \ldots & -c_n & 0 & 0 \ldots 0 & 1 & 0 & 0 \\
\hline
\sum_{i=1}^{m} a_{i1} & \ldots & \sum_{i=1}^{m} a_{in} & 0 & 0 \ldots 0 & 0 & 1 & \sum_{i=1}^{m} b_i
\end{bmatrix}
$$

For this particular tableau, notice that

$$
z_0 = (\sum_{i=1}^{m} b_i)M + 0, \text{i.e.}, a_0 = \sum_{i=1}^{m} b_i
$$

$$
z_\ell - c_\ell = (\sum_{i=1}^{m} a_{i\ell})M - c_\ell, \text{i.e.}, a_\ell = \sum_{i=1}^{m} a_{i\ell} \text{ and } d_\ell = -c_\ell, \text{for } \ell = 1, \ldots , n
$$

and

$$
z_\ell - c_\ell = 0M + 0, \text{i.e.}, a_\ell = d_\ell = 0 \text{ for } \ell = n+1, \ldots , n+m
$$

As we generate new BFS's using the simplex procedure, we want to force each $z_j - c_j$ to be nonpositive, i.e., eventually we want

$$
z_j - c_j = a_j M + d_j \le 0
$$

for all $j = 1, \ldots, n + m$. Since M is an unspecified large positive number this means that we must first force all the $a_j \leq 0$ and then we must force each $d_j \leq 0$ whenever $a_j = 0$. Following the first rule in Section 2 for selecting the next column to enter the admissible basis of the next BFS, we first select the column of A that has the largest component in row $m + 2$ of the tableau. Once each $a_j \leq 0, j = 1, \ldots, n + m$, we then continue the process by selecting the column of A that has the largest positive entry in row $m + 1$ that is over a zero in row $m + 2$.

Suppose now that we have a tableau that gives an optimal solution to Problem (27). Consider the entry a_0^* in the

$$
\begin{bmatrix}
x_{11} & x_{12} & \ldots x_{1n} & x_{1\ n+1} & \ldots x_{1\ n+m} & 0 & 0 & b_1^* \\
\cdot & \cdot & \cdot & \cdot & \cdot & \cdot & \cdot & \cdot \\
\cdot & \cdot & \cdot & \cdot & \cdot & \cdot & \cdot & \cdot \\
\cdot & \cdot & \cdot & \cdot & \cdot & \cdot & \cdot & \cdot \\
x_{m1} & x_{m2} & \ldots x_{mn} & x_{m\ n+1} & \ldots x_{m\ n+n} & 0 & 0 & b_m^* \\
\hline
d_1^* & d_2^* & \ldots d_n^* & d_{n+1}^* & \ldots d_{n+m}^* & 1 & 0 & d_0^* \\
\hline
a_1^* & a_2^* & \ldots a_n^* & a_{n+1}^* & \ldots a_{n+m}^* & 0 & 1 & a_0^*
\end{bmatrix}
\tag{33}
$$

last row of matrix (33). If $a_0^* \neq 0$, then the admissible basis for the optimal solution contains columns that correspond to artificial variables. Notice in particular that $a_0^* \not< 0$. (Why?) Thus, either $a_0^* = 0$ or $a_0^* > 0$.

First consider the case where $a^* > 0$. Then Inequality (31) gives

$$a^* M + d_0^* = \min \{\bar{f}(x) : x \in \bar{F}\} \leq \min \{f(x) : x \in F\}$$

regardless of size of the positive number M. This easily implies that $F = \emptyset$, i.e., Problem (1) is not feasible (see Exercise 3.12).

Before illustrating this with our next example, we note that we need only add as many artificial variables as is necessary so that our coefficient matrix contains all the columns of the appropriate identity matrix. As before, the last two rows of each tableau can be transformed by elementary row operations.

EXAMPLE 8   Minimize $2x_1 + x_2 + 3x_3$ subject to

$$
\begin{aligned}
x_1 \quad - 2x_3 &= 2 \\
x_1 \quad + x_3 &= 1 \\
x_2 + x_3 &= 1 \\
\text{all } x_i &\geq 0
\end{aligned}
$$

Solution: The coefficient matrix for the above system of constraints contains the third column $[0 \quad 0 \quad 1]^T$ of I_3. Hence we need only to add two artificial variables. Thus, we shall consider the problem

$$\text{Minimize } 2x_1 + x_2 + 3x_3 + Mx_4 + Mx_5 \text{ subject to}$$

$$x_1 \quad\ -2x_3 + x_4 \qquad = 2$$
$$x_1 \ + \ x_3 \qquad\ + x_5 = 1$$
$$x_2 + \ x_3 \qquad\qquad = 1$$
$$\text{all } x_i \geq 0$$

The initial matrix and the first two iterations of the simplex procedure are

$$\begin{bmatrix} 1 & 0 & -2 & 1 & 0 & 0 & 0 & 2 \\ 1 & 0 & 1 & 0 & 1 & 0 & 0 & 1 \\ 0 & 1 & 1 & 0 & 0 & 0 & 0 & 1 \\ -2 & -1 & -3 & 0 & 0 & 1 & 0 & 0 \\ 0 & 0 & 0 & -1 & -1 & 0 & 1 & 0 \end{bmatrix} \begin{array}{l} R_4 + R_3 \\ \\ R_5 + (R_1 + R_2) \\ \longrightarrow \end{array}$$

$$\theta = \min\{2/1, 1/1\} = 1$$

$$\begin{bmatrix} 1 & 0 & -2 & 1 & 0 & 0 & 0 & 2 \\ ① & 0 & 1 & 0 & 1 & 0 & 0 & 1 \\ 0 & 1 & 1 & 0 & 0 & 0 & 0 & 1 \\ -2 & 0 & -2 & 0 & 0 & 1 & 0 & 1 \\ 2 & 0 & -1 & 0 & 0 & 0 & 1 & 3 \end{bmatrix} \begin{array}{l} R_1 - R_2 \\ R_4 + 2R_2 \\ \longrightarrow \\ R_5 - 2R_2 \end{array}$$

$$\begin{bmatrix} 0 & 0 & -3 & 1 & -1 & 0 & 0 & 1 \\ 1 & 0 & 1 & 0 & 1 & 0 & 0 & 1 \\ 0 & 1 & 1 & 0 & 0 & 0 & 0 & 1 \\ 0 & 0 & 0 & 0 & 2 & 1 & 0 & 3 \\ 0 & 0 & -3 & 0 & -2 & 0 & 1 & 1 \end{bmatrix}$$

In the last row, $a_0^* > 0$ and all other entries are nonpositive. Hence, the original problem is not feasible.

Next consider the case where $a_0^* = 0$ and the admissible basis for the optional BFS contains columns that correspond to an artificial variable. Since $a_0^* = 0$, any such artificial variable must itself be equal to zero. Thus dropping all the artificial variables, which must all be zero, in the optimal solution vector $\bar{x}$ gives a vector $x \in F$ such that

$$f(x) = \bar{f}(\bar{x}) = \min\{\bar{f}(x) : x \in \bar{F}\} \leq \min\{f(x) : x \in F\}$$

and hence, x solves the original problem.

EXAMPLE 9   Minimize $2x_1 + x_2 + x_3 + 3x_4$ subject to

$$-\frac{1}{3}x_1 \quad\quad + x_3 + x_4 = 1$$

$$x_2 + x_3 \quad\quad = 1$$

$$-x_1 + x_2 + x_3 \quad\quad = 1$$

$$\text{all } x_i \geq 0$$

Solution: Adding two artificial variables changes the problem to

Minimize $2x_1 + x_2 + x_3 + 3x_4 + Mx_5 + Mx_6$ subject to

$$-\frac{1}{3}x_1 \quad\quad + x_3 + \quad x_4 \quad\quad\quad = 1$$

$$x_2 + x_3 \quad\quad + x_5 \quad\quad = 1$$

$$-x_1 + x_2 + x_3 \quad\quad\quad + x_6 = 1$$

$$\text{all } x_i \geq 0$$

Setting up the initial matrix, where $B = [A^{(4)} \quad A^{(5)} \quad A^{(6)}]$, and solving gives

$$
\left[
\begin{array}{cccccc|c|c|c}
-1/3 & 0 & 1 & 1 & 0 & 0 & 0 & 0 & 1 \\
0 & 1 & 1 & 0 & 1 & 0 & 0 & 0 & 1 \\
-1 & 1 & 1 & 0 & 0 & 1 & 0 & 0 & 1 \\
\hline
-2 & -1 & -1 & -3 & 0 & 0 & 1 & 0 & 0 \\
\hline
0 & 0 & 0 & 0 & -1 & -1 & 0 & 1 & 0
\end{array}
\right]
\begin{array}{l}
\\ R_4 + 3R_1 \\ \\ R_5 + (R_2 + R_3) \\ \\ \longrightarrow
\end{array}
$$

$$
\left[
\begin{array}{cccccc|c|c|c}
-1/3 & 0 & 1 & 1 & 0 & 0 & 0 & 0 & 1 \\
0 & ① & 1 & 0 & 1 & 0 & 0 & 0 & 1 \\
-1 & 1 & 1 & 0 & 0 & 1 & 0 & 0 & 1 \\
\hline
-3 & -1 & 2 & 0 & 0 & 0 & 1 & 0 & 3 \\
\hline
-1 & 2 & 2 & 0 & 0 & 0 & 0 & 1 & 2
\end{array}
\right]
\begin{array}{l}
\\ R_3 - R_2 \\ R_4 + R_2 \\ \longrightarrow \\ R_5 - 2R_2
\end{array}
$$

$$
\left[
\begin{array}{cccccc|c|c|c}
-1/3 & 0 & 1 & 1 & 0 & 0 & 0 & 0 & 1 \\
0 & 1 & ① & 0 & 1 & 0 & 0 & 0 & 1 \\
-1 & 0 & 0 & 0 & -1 & 1 & 0 & 0 & 0 \\
\hline
-3 & 0 & 3 & 0 & 1 & 0 & 1 & 0 & 4 \\
\hline
-1 & 0 & 0 & 0 & -2 & 0 & 0 & 1 & 0
\end{array}
\right]
\begin{array}{l}
\\ R_1 - R_2 \\ \\ R_4 - 3R_2 \\ \longrightarrow
\end{array}
$$

$$\begin{bmatrix} -1/3 & -1 & 0 & 1 & -1 & 0 & 0 & 0 & 0 \\ 0 & 1 & 1 & 0 & 1 & 0 & 0 & 0 & 1 \\ -1 & 0 & 0 & 0 & -1 & 1 & 0 & 0 & 0 \\ -3 & -3 & 0 & 0 & -2 & 0 & 1 & 0 & 1 \\ -1 & 0 & 0 & 0 & -2 & 0 & 0 & 1 & 0 \end{bmatrix}$$

Hence, we have a degenerate BFS $[0 \quad 0 \quad 1 \quad 0]^T$ to the original problem. The minimum is 1. Actually, the second constraint in the original problem is redundant.

Finally, consider the case where $a_0^* = 0$ and the admissible basis for the BFS does not contain any columns that correspond to an artificial variable. In this case each artificial variable is zero and hence dropping the artificial variables gives a BFS of the original problem.

EXAMPLE 10   Minimize $4x_1 + x_2 + x_3 + x_4$ subject to

$$x_1 \qquad\qquad + x_4 = 4$$
$$x_1 \qquad + x_3 \qquad = 5$$
$$x_1 + x_2 \qquad + x_4 = 9$$
$$\text{all } x_i \geq 0$$

Solution: We shall add one artificial variable to the first constraint and solve the following problem:

$$\text{Minimize } 4x_1 + x_2 + x_3 + x_4 + Mx_5 \text{ subject to}$$
$$x_1 \qquad\qquad + x_4 + x_5 = 4$$
$$x_1 \qquad + x_3 \qquad = 5$$
$$x_1 + x_2 \qquad + x_4 \qquad = 9$$
$$\text{all } x_i \geq 0$$

where M is some large positive number. The initial matrix and first simplex tableau are given below. Notice that $B = [A^{(5)} A^{(3)} A^{(2)}]$.

$$\begin{bmatrix} 1 & 0 & 0 & 1 & 1 & 0 & 0 & 4 \\ 1 & 0 & 1 & 0 & 0 & 0 & 0 & 5 \\ 1 & 1 & 0 & 1 & 0 & 0 & 0 & 9 \\ -4 & -1 & -1 & -1 & 0 & 1 & 0 & 0 \\ 0 & 0 & 0 & 0 & -1 & 0 & 1 & 0 \end{bmatrix} \begin{matrix} \\ \\ R_4 + (R_2 + R_3) \\ R_5 + R_1 \\ \longrightarrow \end{matrix} \begin{bmatrix} 1 & 0 & 0 & ① & 1 & 0 & 0 & 4 \\ 1 & 0 & 1 & 0 & 0 & 0 & 0 & 5 \\ 1 & 1 & 0 & 1 & 0 & 0 & 0 & 9 \\ -2 & 0 & 0 & 0 & 0 & 1 & 0 & 14 \\ 1 & 0 & 0 & 1 & 0 & 0 & 1 & 4 \end{bmatrix}$$

From this tableau we see that $[0 \quad 9 \quad 5 \quad 0 \quad 4]^T$ is a BFS of the set of feasible solutions of the second problem. Selecting $A^{(4)}$ to enter the admissible basis and performing the indicated row operations gives the next tableau:

$$
\begin{array}{c}
R_3 - R_1 \\
R_5 - R_1 \\
\longrightarrow
\end{array}
\left[
\begin{array}{ccccc|c|c|c}
1 & 0 & 0 & 1 & 1 & 0 & 0 & 4 \\
1 & 0 & 1 & 0 & 0 & 0 & 0 & 5 \\
0 & 1 & 0 & 0 & -1 & 0 & 0 & 5 \\
\hline
-2 & 0 & 0 & 0 & 0 & 1 & 0 & 14 \\
\hline
0 & 0 & 0 & 0 & -1 & 0 & 1 & 0
\end{array}
\right]
$$

The second tableau indicates that $\bar{x}^* = [0 \quad 5 \quad 5 \quad 4 \quad 0]^T$ is a BFS of $\bar{F}$ and $\bar{f}(\bar{x}^*) = 0 \cdot M + 14 = 14$. Since $x_5 = 0$, this indicates that $x^* = [0 \quad 5 \quad 5 \quad 4]^T$ is a BFS of F and $f(x^*) = 14$. But inequality (31) tells us that

$$f(x^*) = f(\bar{x}^*) = \min \{\bar{f}(\bar{x}): \bar{x} \in \bar{F}\} \leq \min \{f(x): x \in F\}$$

Hence, x^* solves the original problem and the minimum is 14.

Now consider Problem (27) again. Suppose that one of the columns $A^{(1)}, \ldots, A^{(n)}$ is selected to replace one of the columns $A^{(n+1)}, \ldots, A^{(n+m)}$ in the admissible basis for the next BFS to be generated. For simplicity suppose that $A^{(1)}$ replaces $A^{(n+1)}$. Then in the new BFS $[x_1 \ldots x_n x_{n+1} \ldots x_{n+m}]^T$ to be generated it follows that $x_{n+1} = 0$. Thus, the point $x = [x_1 \ldots x_n \\ x_{n+2} \ldots x_{n+m}]^T$ is a feasible solution of the following problem:

$$
\begin{array}{ll}
\text{Minimize } c_1 x_1 + \cdots + c_n x_n + M x_{n+2} + \cdots + M x_{n+m} \text{ subject to} \\
\quad a_{11} x_1 + \cdots + a_{1n} x_n \qquad\qquad = b_1 \\
\quad a_{21} x_1 + \cdots + a_{2n} x_n + x_{n+2} \qquad = b_2 & (34) \\
\qquad\qquad\qquad\quad \cdot \qquad\qquad\qquad \cdot \\
\qquad\qquad\qquad\quad \cdot \qquad\qquad\qquad \cdot \\
\qquad\qquad\qquad\quad \cdot \qquad\qquad\qquad \cdot \\
\quad a_{m1} x_1 + \cdots + a_{mn} x_n \qquad + x_{n+m} = b_m \\
\qquad\qquad \text{all } x_i \geq 0
\end{array}
$$

If we redefine $\bar{f}(x)$ and $\bar{F}$, respectively, to be the objective function and the set of feasible solutions for Problem (34), then as before Inequality (31) holds. We then continue as usual. Thus, once a column corresponding to an artificial variable is eliminated from an admissible basis it is never again selected to enter the admissible basis of another BFS. In fact any such column can be dropped from each tableau. If all the artificial variables can be eliminated, then in row $m + 2$, $a_0 = 0$, $a_1 = 0, \ldots, a_n = 0$. In this case we can then drop row $m + 2$.

EXAMPLE 11   Maximize $x_1 + x_2 + x_3$ subject to

$$x_1 + x_2 \quad\quad = 2$$
$$x_2 + x_3 \geq 1$$
$$x_1 \quad\quad + x_3 \leq 5$$
$$\text{all } x_i \geq 0$$

Solution: First we must express the constraints in the form $Ax = b$ where $x \geq 0$ and $b \geq 0$. To do this first let $x_4 \geq 0$ and $x_5 \geq 0$ so that the second and fourth constraints become

$$x_2 + x_3 - x_4 = 1 \text{ and } x_1 + x_3 + x_5 = 5$$

Then let x_6 and x_7 be nonnegative artificial variables so that

$$x_1 + x_2 + x_6 = 2 \text{ and } x_2 + x_3 - x_4 + x_7 = 1$$

Finally, changing the problem to a minimization problem gives

$$\text{Minimize} -x_1 - x_2 - x_3 + 0x_4 + 0x_5 + Mx_6 + Mx_7 \text{ (where } M > 0)$$

subject to

$$x_1 + x_2 \quad\quad\quad + x_6 \quad\quad = 2$$
$$x_2 + x_3 - x_4 \quad\quad + x_7 = 1$$
$$x_1 \quad + x_3 \quad + x_5 \quad\quad = 5$$
$$\text{all } x_i \geq 0$$

Clearly columns $A^{(6)}, A^{(7)}$ and $A^{(5)}$ form an admissible basis for a first BFS. Solving by the simplex procedure we have the following.

$$\begin{bmatrix} 1 & 1 & 0 & 0 & 0 & 1 & 0 & \vdots & 0 & \vdots & 0 & \vdots & 2 \\ 0 & 1 & 1 & -1 & 0 & 0 & 1 & \vdots & 0 & \vdots & 0 & \vdots & 1 \\ 1 & 0 & 1 & 0 & 1 & 0 & 0 & \vdots & 0 & \vdots & 0 & \vdots & 5 \\ \hdashline 1 & 1 & 1 & 0 & 0 & 0 & 0 & \vdots & 1 & \vdots & 0 & \vdots & 0 \\ \hdashline 0 & 0 & 0 & 0 & 0 & -1 & -1 & \vdots & 0 & \vdots & 1 & \vdots & 0 \end{bmatrix} \quad \begin{array}{l} R_5 + (R_1 + R_2) \\ \longrightarrow \end{array}$$

$$\begin{bmatrix} 1 & 1 & 0 & 0 & 0 & 1 & 0 & \vdots & 0 & \vdots & 0 & \vdots & 2 \\ 0 & \textcircled{1} & 1 & -1 & 0 & 0 & 1 & \vdots & 0 & \vdots & 0 & \vdots & 1 \\ 1 & 0 & 1 & 0 & 1 & 0 & 0 & \vdots & 0 & \vdots & 0 & \vdots & 5 \\ \hdashline 1 & 1 & 1 & 0 & 0 & 0 & 0 & \vdots & 1 & \vdots & 0 & \vdots & 0 \\ \hdashline 1 & 2 & 1 & -1 & 0 & 0 & 0 & \vdots & 0 & \vdots & 1 & \vdots & 3 \end{bmatrix} \quad \begin{array}{l} R_1 - R_2 \\ R_4 - R_2 \\ \longrightarrow \\ R_5 - 2R_2 \end{array}$$

$$
\left[\begin{array}{ccccccc|c|c|c}
1 & 0 & -1 & ① & 0 & 1 & 0 & 0 & 1 \\
0 & 1 & 1 & -1 & 0 & 0 & 0 & 0 & 1 \\
1 & 0 & 1 & 0 & 1 & 0 & 0 & 0 & 5 \\
\hline
1 & 0 & 0 & 1 & 0 & 0 & 1 & 0 & -1 \\
\hline
1 & 0 & -1 & 1 & 0 & 0 & 0 & 1 & 1
\end{array}\right]
\begin{array}{l}
R_2 + R_1 \\
R_4 - R_1 \\
\xrightarrow{\hspace{1.5cm}} \\
R_5 - R_1
\end{array}
$$

$$
\left[\begin{array}{ccccc|c|c}
1 & 0 & -1 & 1 & 0 & 0 & 1 \\
1 & 1 & 0 & 0 & 0 & 0 & 2 \\
1 & 0 & ① & 0 & 1 & 0 & 5 \\
\hline
0 & 0 & 1 & 0 & 0 & 1 & -2
\end{array}\right]
\begin{array}{l}
R_1 + R_3 \\
\xrightarrow{\hspace{1.5cm}} \\
R_4 - R_3
\end{array}
$$

$$
\left[\begin{array}{ccccc|c|c}
2 & 0 & 0 & 1 & 1 & 0 & 6 \\
1 & 1 & 0 & 0 & 0 & 0 & 2 \\
1 & 0 & 1 & 0 & 1 & 0 & 5 \\
\hline
-1 & 0 & 0 & 0 & -1 & 1 & -7
\end{array}\right]
$$

Hence, $x = [0 \quad 2 \quad 5]^T$ (recall that x_4 is a slack variable) solves the original problem and the maximum is 7 (recall that $\max\{c^T x : x \in F\} = -\min\{-c^T x : x \in F\}$).

When working with M artificial variables, as in Problem (27), one can usually expect the optional solution in approximately $2m$ iterations.

Exercises

3.11   Solve the following problems by the simplex method.

(a) Minimize $3x_1 - x_2 + x_3$ subject to

$$
\begin{aligned}
x_1 + x_2 &= 3 \\
2x_1 + 4x_2 - x_3 &\ge 2 \\
\text{all } x_i &\ge 0
\end{aligned}
$$

(Hint: Subtract a nonnegative slack variable from the left side of the second constraint, then add artificial variables.)

(b) Maximize $4x_1 - x_2 + x_3$ subject to

$$
\begin{aligned}
x_1 + x_2 + 2x_3 &\le 6 \\
2x_1 + x_2 - x_3 &\ge 4 \\
\text{all } x_i &\ge 0
\end{aligned}
$$

(c) Maximize $x_1 - x_2 + 3x_3 - x_4$ subject to

$$2x_1 \quad + \quad x_3 + x_4 \geq 4$$
$$x_1 - x_2 - \quad x_3 \quad \geq 2$$
$$\text{all } x_i \geq 0$$

(d) Minimize $2x_1 - x_2 + x_3$ subject to

$$x_1 + x_2 - x_3 = 1$$
$$x_1 - x_2 + x_3 = 1$$
$$-x_1 + x_2 + x_3 = 1$$
$$\text{all } x_i \geq 0$$

(e) Minimize $x_1 + x_2 - x_3 + 2x_4$ subject to

$$2x_1 + 7x_2 \quad + \quad x_4 \leq 8$$
$$x_2 + 2x_3 - \quad x_4 \geq 2$$
$$6x_1 + 5x_2 \quad + \quad x_4 = 7$$
$$\text{all } x_i \geq 0$$

(f) Minimize $6x_1 + x_2 - x_3 + x_4$ subject to

$$3x_1 \quad + x_3 + 5x_4 \geq 10$$
$$x_1 + 6x_2 \quad - \quad x_4 \geq -2$$
$$2x_1 + x_2 - x_3 \quad = 7$$
$$\text{all } x_i \geq 0$$

3.12 Prove that in any linear programming problem with artificial variables, if the $m+2$ row of any tableau has the property that $a_0 > 0$ and all other $a_j \leq 0$, then the original problem is not feasible.

3.13 Show that the function p in Equation (30) is injective (one to one), i.e., $p(x) = p(z)$ implies that $x = z$.

3.14 Prove that when working with artificial variables as in Problem (27), both row $m+1$ and row $m+2$ can be transformed by elementary row operations. Moreover, $z_0 = a_0 M + d_0$ and each $z_j - c_j = a_j M + d_j$.

3. ARTIFICIAL VARIABLES AND THE TWO-PHASE METHOD

In this section we shall again be interested in the problem at the beginning of Section 2, namely, how we can handle Problem (1) whenever A does not contain all the columns of I_m (recall that A is $m \times n$). Again our approach will be to add artificial variables. However, we shall modify the approach used in the last section. Rather than assign each artificial variable the cost coefficient M, where M is some arbitrarily large number, we shall first assign a cost coefficient of 1 to each artificial variable and reassign a cost coefficient of 0 to the other variables. We then proceed as in the last section to

$$\text{Minimize } 0 \cdot x_1 + \cdots + 0 \cdot x_n + x_{n+1} + \cdots + x_{n+m}$$

subject to

$$a_{11}x_1 + \cdots + a_{1n}x_n + x_{n+1} \qquad\qquad = b_1$$
$$a_{21}x_1 + \cdots + a_{2n}x_n \qquad + x_{n+2} \qquad = b_2$$
$$\vdots$$
$$a_{m1}x_1 + \cdots + a_{mn}x_n \qquad\qquad + x_{n+m} = b_m$$
$$x_1 \geq 0, \ldots, \quad x_n \geq 0, \quad x_{n+1} \geq 0, \ldots, \quad x_{n+m} \geq 0$$

while eliminating the artificial variables. This is called Phase I of the two-phase method. We added m artificial variables for notational simplicity. In practice we only add enough artificial variables to guarantee that the coefficient matrix of linear constraints contains the columns of the identity matrix I_m.

Clearly, the objective function in Phase I is nonnegative and has a minimum valve of zero whenever the original problem is feasible. As with artificial cost coefficients once an artificial variable is removed from an admissible basis it is discarded. Phase I ends when all $z_j - c_j \leq 0$. If Phase I ends and the Phase I objective function is positive (i.e., $z_0 > 0$), then clearly the original problem was not feasible. This is illustrated in the following simple example.

EXAMPLE 12   Show that the following problem is not feasible

$$\text{Minimize } x_1 + x_2 - x_3 \text{ subject to}$$
$$x_1 + x_2 + x_3 \leq 1$$
$$2x_1 + 3x_2 + 2x_3 \geq 6$$
$$\text{all } x_i \geq 0$$

Solution: Adding both artificial and slack variables and using Phase I cost coefficients gives the following Phase I problem:

$$\text{Minimize } 0 \cdot x_1 + 0 \cdot x_2 + 0 \cdot x_3 + 0 \cdot x_4 + 0 \cdot x_5 + 1 \cdot x_6 \text{ subject to}$$
$$x_1 + x_2 + x_3 + x_4 \qquad\qquad = 1$$
$$2x_1 + 3x_2 + 2x_3 \qquad - x_5 + x_6 = 6$$
$$\text{all } x_i \geq 0$$

Thus,

$$\begin{bmatrix} 1 & 1 & 1 & 1 & 0 & 0 & 0 & 1 \\ 2 & 3 & 2 & 0 & -1 & 1 & 0 & 6 \\ 0 & 0 & 0 & 0 & 0 & -1 & 1 & 0 \end{bmatrix} \xrightarrow{R_3 + R_2} \begin{bmatrix} 1 & ① & 1 & 1 & 0 & 0 & 0 & 1 \\ 2 & 3 & 2 & 0 & -1 & 1 & 0 & 6 \\ 2 & 3 & 2 & 0 & -1 & 0 & 1 & 6 \end{bmatrix}$$

$$\xrightarrow[\substack{R_3 - 3R_1}]{R_2 - 3R_1} \begin{bmatrix} 1 & 1 & 1 & 1 & 0 & 0 & 0 & 1 \\ -1 & 0 & -1 & -3 & -1 & 1 & 0 & 3 \\ -1 & 0 & -1 & -3 & -1 & 0 & 1 & 3 \end{bmatrix}$$

Hence, Phase I ends with $z_0 = 3$ and hence, the original problem is not feasible.

If Phase I ends with $z_0 = 0$ and all the artificial variables have been eliminated, then we have a BFS of the set of feasible solutions of the original problem. When this happens we restore the original cost coefficients and proceed with the usual simplex procedure. This is known as Phase II.

EXAMPLE 13   Minimize $4x_1 + x_2 - 2x_3$ subject to

$$2x_1 \quad\quad - x_3 = 4$$
$$2x_2 + x_3 = 7$$
$$\text{all } x_i \geq 0$$

Solution: First we add two artificial variables, change to Phase I cost coefficients and perform the Phase I calculations on the following problem.

$$\text{Minimize } 0 \cdot x_1 + 0 \cdot x_2 + 0 \cdot x_3 + x_4 + x_5 \text{ subject to}$$
$$2x_1 \quad\quad - x_3 + x_4 \quad\quad = 4$$
$$2x_2 + x_3 \quad\quad + x_5 = 7$$
$$\text{all } x_i \geq 0$$

$$\begin{bmatrix} 2 & 0 & -1 & 1 & 0 & 0 & 4 \\ 0 & 2 & 1 & 0 & 1 & 0 & 7 \\ 0 & 0 & 0 & -1 & -1 & 1 & 0 \end{bmatrix} \xrightarrow{R_3 + (R_1 + R_2)} \begin{bmatrix} ② & 0 & -1 & 1 & 0 & 0 & 4 \\ 0 & 2 & 1 & 0 & 1 & 0 & 7 \\ 2 & 2 & 0 & 0 & 0 & 1 & 11 \end{bmatrix}$$

$$\xrightarrow[\substack{(1/2)R_1}]{R_3 - R_1} \begin{bmatrix} 1 & 0 & -1/2 & 0 & 0 & 2 \\ 0 & ② & 1 & 1 & 0 & 7 \\ 0 & 2 & 1 & 0 & 1 & 7 \end{bmatrix} \quad \text{(Here } x_4 \text{ was deleted.)}$$

$$\xrightarrow[\substack{(1/2)R_2}]{R_3 - R_2} \begin{bmatrix} 1 & 0 & -1/2 & 0 & 2 \\ 0 & 1 & 1/2 & 0 & 7/2 \\ 0 & 0 & 0 & 1 & 0 \end{bmatrix} \quad \begin{array}{l} \text{(End of Phase I)} \\ \text{(Here } x_5 \text{ was deleted.)} \end{array}$$

In this case Phase I ends with $z_0 = 0$ and all the artificial variables removed. Notice that $[x_1 \ x_2 \ x_3]^T = [2 \ 7/2 \ 0]^T$ must be a BFS of the set of feasible solutions of the original problem. We now restore the original cost coefficients, recompute the last row and perform the Phase II calculations.

$$
\begin{bmatrix}
1 & 0 & -1/2 & 0 & 2 \\
0 & 1 & 1/2 & 0 & 7/2 \\
\hline
-4 & -1 & 2 & 1 & 0
\end{bmatrix}
\begin{array}{l} \\ \\ \xrightarrow{R_3 + (4R_1 + R_2)} \end{array}
\begin{bmatrix}
1 & 0 & -1/2 & 0 & 2 \\
0 & 1 & \boxed{1/2} & 0 & 7/2 \\
\hline
0 & 0 & 1/2 & 1 & 23/2
\end{bmatrix}
\begin{array}{l} R_1 + R_2 \\ R_3 - R_2 \\ \hline \xrightarrow{} \\ 2R_2 \end{array}
$$

$$
\begin{bmatrix}
1 & 1 & 0 & 0 & 11/2 \\
0 & 2 & 1 & 0 & 7 \\
\hline
0 & -1 & 0 & 1 & 8
\end{bmatrix} \qquad \text{(End of Phase II)}
$$

Thus, the minimum is 8 when $x_1 = 11/2$, $x_2 = 0$ and $x_3 = 7$.

Phase I can end with $z_0 = 0$ before all the artificial variables are eliminated. When this happens the artificial variables in the BFS of the Phase I problem are equal to zero. This case is more complicated than the other two cases and we must take special steps to deal with it. There are several known techniques that can be used in this situation. Our approach will be to continue with Phase II as before while not allowing any remaining artificial variables to become positive. This insures that each extreme point solution obtained will be an extreme point of the set of feasible solutions of the original problem.

To begin Phase II we first return the original cost coefficients to the original variables and set all cost coefficients of remaining artificial variables equal to zero. We then recompute each $z_j - c_j$ term. Suppose for simplicity that the admissible basis for the BFS obtained at the end of Phase I is $A^{(1)}$, $A^{(2)}, \ldots, A^{(k)}$, $A^{(n+k+1)}, \ldots, A^{(n+m)}$ where $x_1, x_2, \ldots, x_k$ are original variables and $x_{n+k+1, \ldots}, x_{n+m}$ are the remaining artificial variables, and let matrix (35) represent the first simplex tableau in Phase II.

Suppose that $z_j - c_j > 0$ and we that intend to construct a new BFS by allowing $A^{(j)}$ to enter the admissible basis. Clearly, if any $x_{ij} > 0$ for $i = k+1, \ldots, n+m$ then we can pivot at x_{ik} (here $\theta = 0/x_{ij}$) and get a new solution in which the remaining artificial variables are zero. Also, if all $x_{ij} = 0$ for $i = k+1, \ldots, n+m$, then regardless of where we pivot the remaining artificial variables will be zero. However, if any $x_{gj} < 0$, for $g = k+1,\ldots,n+m$, then the artificial variable x_g can become positive when we perform a row operation to change x_{gj} to 0. This is what we wish to avoid. To do this we simply pivot at x_{gj} rather than where the minimum ratio test would have us pivot. Thus, $A^{(j)}$ enters the admissible basis while $A^{(g)}$ is removed. We then discard the artificial variable x_g and continue.

nonbasic columns

$$
\begin{bmatrix}
1 & \cdots & 0 & x_{1\,k+1} & \cdots x_{1j} & \cdots & x_{1n} & 0 & \cdots & 0 & \cdots & 0 & \mid & 0 & \mid & b_1 \\
\cdot & & \cdot & \cdot & \cdot & & \cdot & & \cdot & & \cdot & & \mid & \cdot & \mid & \cdot \\
0 & \cdots & 1 & x_{k\,k+1} & \cdots x_{kj} & \cdots & x_{kn} & 0 & \cdots & 0 & \cdots & 0 & \mid & 0 & \mid & b_k \\
0 & \cdots & 0 & x_{k+1\,k+1} & \cdots x_{k+1\,j} & \cdots x_{k\,+1n} & & 1 & \cdots & 0 & \cdots & 0 & \mid & 0 & \mid & 0 \\
\cdot & & \cdot & \cdot & \cdot & & \cdot & & \cdot & & \cdot & & \mid & \cdot & \mid & \\
0 & \cdots & 0 & x_{i\,k+1} & \cdots x_{ij} & \cdots x_{in} & & 0 & \cdots & 1 & \cdots & 0 & \mid & 0 & \mid & 0 \\
\cdot & & \cdot & \cdot & \cdot & & & \cdot & & \cdot & & & \mid & \cdot & \mid & \\
0 & \cdots & 0 & x_{m\,k+1} & \cdots x_{mj} & & x_{mn} & 0 & \cdots & 0 & \cdots & 1 & \mid & 0 & \mid & 0 \\
0 & \cdots & 0 & z_{k+1}-c_{k+1} & \cdots z_j-c_j & \cdots z_n-c_n & & 0 & \cdots & 0 & \cdots & 0 & \mid & 1 & \mid & z_0
\end{bmatrix} \quad (35)
$$

Our next example illustrates these last three situations. In each part of the example we shall deal with only one pivot in the Phase II calculations.

EXAMPLE 14   (a) Suppose we have the following Phase II tableau where x_4 and x_5 are artificial variables. Since

$$
\begin{bmatrix}
4 & 1 & 6 & 0 & 0 & \vdots & 0 & \vdots & 1 \\
1 & 0 & \textcircled{1} & 1 & 0 & \vdots & 0 & \vdots & 0 \\
0 & 0 & 0 & 0 & 1 & \vdots & 0 & \vdots & 0 \\
\hdashline
0 & 0 & 1 & 0 & 0 & \vdots & 1 & \vdots & 1
\end{bmatrix}
\begin{array}{l}
R_1 - 6R_2 \\
R_4 - R_2 \\
\\
\xrightarrow{\hspace{1cm}}
\end{array}
\quad \text{(Phase II)}
$$

$z_3 - c_3 > 0$ and $x_{23} > 0$, we pivot at $x_{23} = 1$, and drop the fourth column to obtain

$$
\begin{bmatrix}
-2 & 1 & 0 & 0 & \vdots & 0 & \vdots & 1 \\
1 & 0 & 1 & 0 & \vdots & 0 & \vdots & 0 \\
0 & 0 & 0 & 1 & \vdots & 0 & \vdots & 0 \\
\hdashline
-1 & 0 & 0 & 0 & \vdots & 1 & \vdots & 1
\end{bmatrix}
$$

Notice that the artificial variable x_5 remains equal to zero.

(b) Again suppose we have a Phase II tableau where x_4 and x_5 are artificial variables.

$$\left[\begin{array}{ccccc|c|c}
1 & -6 & ③ & 0 & 0 & 0 & 9 \\
0 & 12 & 0 & 1 & 0 & 0 & 0 \\
0 & 3 & 0 & 0 & 1 & 0 & 0 \\
\hline
0 & -41 & 19 & 0 & 0 & 1 & 63
\end{array}\right]
\begin{array}{l}
R_4 - (19/3)R_1 \text{ (Phase II)} \\
\overrightarrow{} \\
(1/3)R_1
\end{array}$$

In this case $z_3 - c_3 > 0$ and both $x_{23} = 0$ and $x_{33} = 0$. Thus, we pivot at $x_{13} = 3$ obtaining

$$\left[\begin{array}{ccccc|c|c}
1/3 & -2 & 1 & 0 & 0 & 0 & 3 \\
0 & 12 & 0 & 1 & 0 & 0 & 0 \\
0 & 3 & 0 & 0 & 1 & 0 & 0 \\
\hline
-19/3 & -3 & 0 & 0 & 0 & 1 & 6
\end{array}\right]$$

Notice that while $A^{(4)}$ and $A^{(5)}$ remain in the admissible basis both $x_4 = 0$ and $x_5 = 0$.

(c) Finally suppose that x_4 and x_5 are artificial variables in the following tableau

$$\left[\begin{array}{ccccc|c|c}
1 & 2 & -1 & 0 & 0 & 0 & 2 \\
0 & ⊖ & 1 & 1 & 0 & 0 & 0 \\
0 & 0 & 4 & 0 & 1 & 0 & 0 \\
\hline
0 & 7 & -5 & 0 & 0 & 1 & 6
\end{array}\right]
\begin{array}{l}
R_1 + 2R_2 \qquad \text{(Phase II)} \\
R_4 + 7R_2 \\
\overrightarrow{} \\
-R_2
\end{array}$$

Since $z_2 - c_2 > 0$ and $x_{22} < 0$ we must pivot at x_{22} rather than at the usual place which would be x_{12}. After deleting the fourth column we have

$$\left[\begin{array}{cccc|c|c}
1 & 0 & 1 & 0 & 0 & 2 \\
0 & 1 & -1 & 0 & 0 & 0 \\
0 & 0 & 4 & 1 & 0 & 0 \\
\hline
0 & 0 & 2 & 0 & 1 & 6
\end{array}\right]$$

Notice that the remaining artificial variable x_5 is still equal to zero.

In Sections 3 and 4 we have presented two different but similar methods that can be used when dealing with artificial variables. The simplicity of the cost coefficients in Phase I of the two-phase method often simplifies calculations. However, as we have just noted, the two-phase method requires very special attention when artificial variables, each equal to zero, remain at the end of Phase I. This is avoided in the first technique by using the unspecified large positive artificial cost coefficient M. Many people, when using the big $- M$ method of Section 3, assign a large positive value to M and thus need only one row to specify the values of z_0 and $z_j - c_j$. This causes two problems. First, it is difficult to determine beforehand just how large M should be selected in order to drive the artificial variables out of the admissible basis. Second, a very large value of M very often leads to serious round-off error. However, both of these problems are avoided by using the two rows, as in Section 3, to specify the values of a_j and d_j in the expressions $z_j - c_j = a_j M + d_j$. (M is never specified.) The two-phase method can be useful in determining the presence of redundant constraints. This is done at the end of Phase I and is explored in the following exercises.

Exercises

3.15 Solve the following problems using the two-phase method.

(a) Problem 3.11 (9).

(b) Problem 3.11 (6).

(c) Problem 3.11 (c).

(d) Problem 3.11 (d).

(e) Problem 3.11 (e).

(f) Problem 3.11 (f).

3.16 For simplicity suppose that Phase I ends with artificial variables present at zero levels in the following tableau:

$$
\begin{bmatrix}
1 \dots 0 & x_{1k+1} & & \dots x_{1n} & 0 \dots 0 & 0 & \bar{b}_1 \\
\cdot & \cdot & & \cdot & \cdot & \cdot & \cdot \\
\cdot & \cdot & & \cdot & \cdot & \cdot & \cdot \\
\cdot & \cdot & & \cdot & \cdot & \cdot & \cdot \\
0 \dots 1 & x_{k\,k+1} & & \dots x_{kn} & 0 \quad 0 & 0 & \bar{b}_k \\
0 \dots 0 & & & & 1 \dots 0 & 0 & 0 \\
\cdot & \cdot & & & \cdot & \cdot & \cdot \\
\cdot & \cdot & & Q & \cdot & \cdot & \cdot \\
\cdot & \cdot & & & \cdot & \cdot & \cdot \\
0 \dots 0 & & & & 0 \quad 1 & 0 & 0 \\
\hline
0 \dots 0 & z_{k+1} - c_{k+1} & & \dots z_n - c_n & 0 \dots 0 & 1 & 0
\end{bmatrix}
$$

where

$$Q = \begin{bmatrix} x_{n+1\,k+1} & \cdots x_{n+1\,n} \\ & \cdot & \\ & \cdot & \\ & \cdot & \\ x_{n+n-k\,k+1} & \cdots x_{n+n-k\,n} \end{bmatrix}$$

Show that when any row of Q is the zero row matrix, say $Q_{(i)} = [0 \ldots 0]$, then the i-th constraint is redundant and can be removed from the original set of constraints.

(b) Use part (a) to discover the redundant constraint in the following problem:

$$\text{Minimize } 10x_1 + x_2 + 17x_3 \text{ subject to}$$
$$x_1 + 3x_2 + 4x_3 = 16$$
$$x_1 + x_2 + x_3 = 6$$

$$x_2 + \frac{3}{2}x_3 = 5$$

$$x_3 \leq 2$$
$$\text{all } x_i \geq 0$$

3.17   Solve the following problems:

(a) Maximize $x_1 - 3x_2 + x_3$ subject to
$$3x_1 + 2x_2 \qquad = 6$$
$$4x_1 + x_2 + 4x_3 = 12$$
$$x_1 \text{ unrestricted}, x_2 \geq 0, x_3 \leq 0$$

(Recall that x_1 can be expressed as the difference of two nonnegative numbers.)

(b) Minimize $6w_1 - 9w_2$ subject to
$$w_1 - w_2 \geq 2$$
$$3w_1 + w_2 \geq 1$$
$$2w_1 - 3w_2 \geq 3$$
$$w_1 \geq 0, w_2 \leq 0$$

REFERENCES

1. M. Bazaraa and J. Jarvis, *Linear Programming and Network Flows*, John Wiley and Sons, New York (1977).
2. G. B. Dantzig, Maximization of a Linear Function of Variables Subject to Linear Inequalities, *Econometrica* (1949), 200 – 211.
3. G. B. Dantzig, *Linear Programming and Extensions*, Princeton University Press, Princeton, N.J. (1963).
4. L. Cooper and D. Steinberg, *Methods and Applications of Linear Programming*, W. B. Saunders Company, Philadelphia (1974).
5. S. Gass, *Linear Programming: Methods and Applications*, McGraw-Hill, New York (1969).
6. B. Gotlfried and J. Weisman, *Introduction to Optimization Theory*, Prentice-Hall, Englewood Cliffs, N.J. (1973).
7. G. Hadley, *Linear Programming*, Addison-Wesley, Reading, Mass. (1962).
8. D. Luenberger, *Introduction to Linear and Nonlinear Programming*, Addison-Wesley, Reading, Mass. (1973).
9. K. Murty, *Linear Programming*, John Wiley and Sons, New York (1983).

4

Duality and the Linear Complementarity Problem

In this chapter, it will be shown that each linear programming problem has a special correspondence to another "dual" linear programming problem. One will be a minimization while the other will be a maximization problem. A solution of either problem will yield a solution of the other. The number of variables of one problem will be equal to the number of constraints of the other problem. Since a good estimate of the number of simplex iterations required to solve a problem is twice the number of constraints of the problem, it will usually be more economical to solve one of the dual problems than the other. These relationships are developed in the first two sections. It should be noticed that the sign restriction of the i-th variable in one problem will determine the sign restriction of the i-th constraint in the other problem. Moreover, an unrestricted i-th variable in one problem will lead to equality in the i-th constraint of the other problem.

Section 3 deals briefly with the model stability of a linear programming problem. In Section 4 a new algorithm known as the dual simplex procedure, is developed.

In the final three sections, the complementary slackness theorem is developed, which leads to a reformulation of the linear programming problem as a linear complementarity problem. A method, known as Lemke's complementary pivoting algorithm, is presented for solving the latter problem.

1. DUAL LINEAR PROGRAMMING PROBLEMS

We begin by considering the linear programming problem in the form used earlier for the construction of the initial simplex tableau:

$$\text{Minimize } c^T x \text{ subject to } Ax = b \text{ and } x \geq 0 \text{ (here } A \text{ is } mxn) \tag{1}$$

This problem will be called the *primal* problem (the problem is often said to be the *unsymmetric primal* problem). The *dual* problem is defined to be

$$\text{Maximize } b^T w \text{ subject to } A^T w \leq c \tag{2}$$

The dual problem is feasible whenever there exists a w so that $A^T w \leq c$. The following result is sometimes called the *weak duality theorem*.

PROPOSITION 1   If x and w are feasible solutions of the primal and dual problems, respectively, then $b^T w \leq c^T x$.
Proof: Clearly, $w^T A \leq c^T$. Since $x \geq 0$ and $Ax = b$, then

$$w^T b = w^T A x \leq c^T x$$

Thus, when both problems are feasible the value of the objective function of the primal (minimization) problem is bounded below by any value of the dual objective function. Moreover, the value of the objective function of the dual (maximization) problem is bounded above by any value of the primal objective function. Notice also that when the primal problem has an unbounded minimum, the dual problem cannot be feasible. Likewise, when the dual problem has an unbounded maximum, then the primal problem cannot be feasible. Neither problem may be feasible (see Exercise 42). However, if both problems are feasible and $c^T \bar{x} = b^T \bar{w}$, for feasible solutions $\bar{x}$ and $\bar{w}$ of the primal and dual problems, respectively, then for any other feasible solution, x, of the primal problem, it follows that

$$c^T x \geq b^T \bar{w} = c^T \bar{x}$$

This implies that $\bar{x}$ gives an optimal solution of the primal problem. Likewise, $\bar{w}$ is an optimal solution of the dual problem. This establishes the next result.

PROPOSITION 2   If $\bar{x}$ and $\bar{w}$ are feasible solutions of the primal and dual problems, respectively, such that $c^T \bar{x} = b^T \bar{w}$, then $\bar{x}$ and $\bar{w}$ solve their respective problems. Moreover, the optimal minimum value of the primal (minimization) objective function is equal to the optimal maximum value of the dual (maximization) objective function.

We shall now state and prove the fundamental result concerning dual linear programming problems.

THEOREM 1   (Fundamental duality theorem) If either the primal or the dual problem has a finite optimal solution, then the other problem has a finite optimal solution as well. Moreover, the optimal values of the two objective functions are equal, i.e.,

$$\min \{ c^T x : Ax = b \text{ and } x \geq 0 \} = \max \{ b^T w : A^T w \leq c \}$$

Proof: Without loss of generality assume that the BFS $x_B = B^{-1}b$ and $x_D = [0 \ldots 0]^T$, where $B = [A^{(1)} \ldots A^{(m)}]$, solves (1). The final simplex tableau is then

$$\left[\begin{array}{cc|c|c} I_m & B^{-1}D & 0 & B^{-1}b \\ \hline 0 & c_B^T B^{-1}D - c_D^T & 1 & c_B^T B^{-1}b \end{array} \right] \tag{3}$$

where $c_B^T B^{-1}D - c_D^T \leq 0$ and $z_0 = c^T{}_B B^{-1}b$. Recall that $[I_m \ B^{-1}D] = B^{-1}A$. Define $w \, \varepsilon \, \mathbf{R}_{n \times 1}$ by

$$w = (c_B^T B^{-1})^T$$

Then

$$\begin{aligned} A^T w = A^T (c_B^T B^{-1})^T = (c_B^T B^{-1} A)^T = (c_B^T [I_m \ B^{-1}D])^T \\ = [c_B^T \ \ c_B^T B^{-1}D]^T \leq [c_B^T \ \ c_D^T]^T = c \end{aligned} \tag{4}$$

and w is a feasible solution of (2). Moreover,

$$b^T w = b^T (c_B^T B^{-1})^T = (c_B^T B^{-1} b)^T = z_0{}^T = z_0$$

(since z_0 is real). Thus, Proposition 2 implies that w solves the dual problem. To summarize to this point we have verified the theorem for the case when the primal problem has a finite optimal solution. We next need to show that the result holds whenever our dual problem has a finite optimal solution.

Suppose that w solves the dual problem. We shall rewrite the dual problem in a form analogous to that of the primal problem. First write $w = w_1 - w_2$, where $w_1 = [w_{11} \ldots w_{1m}]^T \geq 0$ and $w_2 = [w_{21} \ldots w_{2m}]^T \geq 0$. Next add slack variables $w_3 = [w_{31} \ldots w_{3m}]^T \geq 0$ to the constraints $A^T w \leq c$, and finally rewrite $\max b^T w$ as $-\min(-b^T w)$. Then the dual problem can be replaced with the following problem:

$$-\text{Minimize} -b^T w_1 + b^T w_2 + 0^T w_3 \text{ subject to } A^T w_1 - A^T w_2 + I w_3 = c^T,$$
$$w_1 \geq 0, \, w_2 \geq 0 \text{ and } w_3 \geq 0$$

i.e.,

$$-\text{Minimize} -b^T w_1 + b^T w_2 + 0^T w_3 \text{ subject to } [-A^T \ \ A^T \ \ -I] \begin{bmatrix} w_1 \\ \hline w_2 \\ \hline w_3 \end{bmatrix} = -c^T,$$

$$w_1 \geq 0, \, w_2 \geq 0 \text{ and } w_3 \geq 0$$

which is in a form similar to our primal problem. Moreover, this last problem as a finite optimal solution. Thus, we know from the first part of our proof that the dual of this last problem has a finite optimal solution and the two objective functions for these problems are equal at their optimal values. But the dual of this last problem is

$$-\text{Maximize} -c^T x \text{ subject to } [-A^T \quad A^T \quad -I]^T x \le \begin{bmatrix} -b \\ \hline b \\ \hline 0 \end{bmatrix}$$

which is equivalent to

$$\text{Minimize } c^T x \text{ subject to } -Ax \le -b, Ax \le b, \text{ and } -x \le 0$$

or

$$\text{Minimize } c^T x \text{ subject to } Ax = b \text{ and } x \ge 0$$

Notice that in the second half of the above proof, the column of unrestricted variables w lead to equality constraints in the dual problem.

Often the initial tableau of the simplex procedure contains all the columns of the $m \times m$ identity matrix I_m. When this happens, the inverse, B^{-1}, of the basis matrix B can be found in the final tableau. The ith column of B^{-1} is simply the column of X_B that corresponds to the column in the original tableau where $I^{(i)}$ was found. Recall that the entries in the last row of a simplex tableau are of the form $z_j - c_j = c_B^T B^{-1} A^{(j)} - c_j$. But $w = (c_B^T B^{-1})^T$ and it follows that the components of w are the suitable components of the last row of the final simplex tableau added to the corresponding components of c, namely $z_j = (z_j - c_j) + c_j$. In particular, to find the ith component of w locate the column in the origin tableau where $I^{(i)}$ was found. Then go to the corresponding column of the final simplex tableau. If that column is the jth column, then $w_i = (z_j - c_j) + c_j$, which is the jth entry in the last row added to c_j. Moreover $b^T w = c^T x = z_0$.

EXAMPLE 1   Recall Example 6 of Chapter 3,

$$\text{Minimize } x_1 + x_2 + x_3 + x_4 + x_5 + x_6 \text{ subject to}$$
$$16x_1 + 2x_2 \qquad\qquad + x_5 + 10x_6 = 4$$
$$-x_1 \qquad\qquad + x_4 \qquad + x_6 = 0$$
$$5x_1 + 4x_2 + x_3 \qquad\qquad = 8$$
$$\text{all } x_i \ge 0$$

The initial simplex tableau was

$$\begin{bmatrix} 16 & 2 & 0 & 0 & 1 & 10 & 0 & 4 \\ -1 & 0 & 0 & 1 & 0 & 1 & 0 & 0 \\ 5 & 4 & 1 & 0 & 0 & 0 & 0 & 8 \\ \hline 19 & 5 & 0 & 0 & 0 & 10 & 1 & 12 \end{bmatrix}$$

and the final tableau was

$$\begin{bmatrix} 8 & 1 & 0 & 0 & 1/2 & 5 & \vdots & 0 & \vdots & 2 \\ -1 & 0 & 0 & 1 & 0 & 1 & \vdots & 0 & \vdots & 0 \\ -27 & 0 & 1 & 0 & -2 & -20 & \vdots & 0 & \vdots & 0 \\ -21 & 0 & 0 & 0 & -5/2 & -15 & \vdots & 1 & \vdots & 2 \end{bmatrix}$$

It follows from Theorem 1 and the remarks that followed that $w = [w_1 \, w_2 \, w_3]^T$, where

$$w_1 = -5/2 + 1 = -3/2, \ w_2 = 0 + 1 \text{ and } w_3 = 0 + 1 = 1$$

solves the dual problem:

$$\text{Maximize } 4w_1 + 0w_2 + 8w_3 \text{ subject to}$$
$$16w_1 - w_2 + 5w_3 \leq 1$$
$$2w_1 \qquad + 4w_3 \leq 1$$
$$w_3 \leq 1$$
$$w_2 \leq 1$$
$$w_1 \leq 1$$
$$10w_1 + w_2 \qquad \leq 1$$

Notice that the optimal value of each cost function is 2.

EXAMPLE 2   Maximize $4w_1 + w_2$ subject to

$$2w_1 - w_2 \leq 2$$
$$2w_1 + w_2 \leq 3$$
$$w_1 \qquad \leq 1$$

Solution: The dual problem is

$$\text{Minimize } 2x_1 + 3x_2 + x_3 \text{ subject to}$$
$$2x_1 + 2x_2 + x_3 = 4$$
$$-x_1 + x_2 \qquad = 1$$
$$\text{all } x_i \geq 0$$

Adding an artificial variable to this last problem gives

$$\text{Minimize } 2x_1 + 3x_2 + x_3 + Mx_4 \text{ subject to}$$
$$2x_1 + 2x_2 + x_3 = 4$$
$$-x_1 + x_2 + x_4 = 1$$
$$\text{all } x_i \geq 0$$

Now apply the simplex procedure

$$
\begin{bmatrix}
2 & 2 & 1 & 0 & 0 & 0 & 4 \\
-1 & 1 & 0 & 1 & 0 & 0 & 1 \\
-2 & -3 & -1 & 0 & 1 & 0 & 0 \\
0 & 0 & 0 & -1 & 0 & 1 & 0
\end{bmatrix}
\begin{array}{l} R_3+R_1 \\ \xrightarrow{} \\ R_4+R_2 \end{array}
\begin{bmatrix}
2 & 2 & 1 & 0 & 0 & 0 & 4 \\
-1 & \textcircled{1} & 0 & 1 & 0 & 0 & 1 \\
0 & -1 & 0 & 0 & 1 & 0 & 4 \\
-1 & 1 & 0 & 0 & 0 & 1 & 1
\end{bmatrix}
\begin{array}{l} R_1-2R_2 \\ R_3+R_2 \\ \xrightarrow{} \\ R_4-R_2 \end{array}
$$

$$
\begin{bmatrix}
4 & 0 & 1 & -2 & 0 & 0 & 2 \\
-1 & 1 & 0 & 1 & 0 & 0 & 1 \\
-1 & 0 & 0 & 1 & 1 & 0 & 5 \\
0 & 0 & 0 & -1 & 0 & 1 & 0
\end{bmatrix}
$$

It follows that an optimal solution of the original problem occurs when $w = [w_1\ w_2]^T$, where $w_1 = 0 \cdot M + 0 + 1 = 1$ and $w_2 = -M + 1 + M = 1$.

   Now consider a linear programming problem that has equality constraints as well as inequality constraints of each sign type. Moreover, allow some of the variables to be nonrestricted, some to be nonnegative, and some to be nonpositive. Rearranging the constraints and variables leads to a problem of the following form:

$$\text{Minimize } c_1^T x_1 + c_2^T x_2 + c_3^T x_3 \text{ subject to}$$

$$A_{11}x_1 + A_{12}x_2 + A_{13}x_3 = b_1$$

$$A_{21}x_1 + A_{22}x_2 + A_{13}x_3 \le b_2$$

$$A_{31}x_1 + A_{32}x_2 + A_{33}x_3 \ge b_3$$

$$x_1 \ge 0, x_2 \le 0, x_3 \text{ unrestricted}$$

where x_1, x_2, x_3, b_1, b_2, and b_3 are now column matrices, and the coefficient matrix A of the constraints has the partitioned form

$$
A = \begin{bmatrix}
A_{11} & A_{12} & A_{13} \\
A_{21} & A_{22} & A_{23} \\
A_{31} & A_{32} & A_{33}
\end{bmatrix}
$$

Recall that the unrestricted variables of x_3 can be replaced by $x_3 = x_4 - x_5$, where x_4 and x_5 are nonnegative column matrices. Thus, after adding and subtracting slack variables x_6 and x_7 (both nonnegative column matrices), the problem becomes

Minimize $c_1^T x_1 - c_2^T(-x_2) + c_3^T x_4 - c_3^T x_5 + 0^T x_6 + 0^T x_7$ subject to

$$
\begin{aligned}
A_{11}x_1 - A_{12}(-x_2) + A_{13}x_4 - A_{13}x_5 \qquad\qquad &= b_1 \\
A_{21}x_1 - A_{22}(-x_2) + A_{23}x_4 - A_{23}x_5 + x_6 \qquad &= b_2 \\
A_{31}x_1 - A_{32}(-x_2) + A_{33}x_4 - A_{33}x_5 \quad\ - x_7 &= b_3
\end{aligned}
$$

$$x_1 \geq 0, -x_2 \geq 0, x_3 \geq 0, x_4 \geq 0, x_5 \geq 0, x_6 \geq 0, x_7 \geq 0$$

Since the problem is now in primal form, the dual can easily be expressed as

Maximize $b_1^T w_1 + b_2^T w_2 + b_3^T w_3$ subject to

$$
\begin{aligned}
A_{11}^T w_1 + A_{21}^T w_2 + A_{31}^T w_3 &\leq c_1 \\
-A_{12}^T w_1 - A_{22}^T w_2 - A_{32}^T w_3 &\leq -c_2 \\
A_{13}^T w_1 + A_{23}^T w_2 + A_{33}^T w_3 &\leq c_3 \\
-A_{13}^T w_1 - A_{23}^T w_2 - A_{33}^T w_3 &\leq -c_3 \\
w_2 &\leq 0 \\
-w_3 &\leq 0
\end{aligned}
$$

i.e.,

Maximize $b_1^T w_1 + b_2^T w_2 + b_3^T w_3$ subject to

$$
\begin{aligned}
A_{11}^T w_1 + A_{21}^T w_2 + A_{31}^T w_3 &\leq c_1 \\
A_{12}^T w_1 + A_{22}^T w_2 + A_{32}^T w_3 &\geq c_2 \\
A_{13}^T w_1 + A_{23}^T w_2 + A_{33}^T w_3 &= c_3
\end{aligned}
\qquad (6)
$$

w_1 unrestricted, $w_2 \leq 0, w_3 \geq 0$

Notice in particular that
1. One problem is a minimization and the other a maximization problem.
2. The number of constraints of one problem equals the number of variables of the other.
3. The scalar coefficients of one objective function are the components of the column matrix of the right hand side of the constraints of the other problem.
4. The coefficient matrix of the constraints of one problem is the transpose of the coefficient matrix of the constraints of the other problem.
5. An unrestricted ith variable in one problem corresponds to an ith equality constraint in the dual problem.
6. A nonnegative ith variable in the minimization problem corresponds to an ith constraint of the "$\leq$" type in the maximization problem. Also, a nonnegative jth variable in the maximization problem corresponds to a jth constraint of the type "$\geq$" in the minimization problem.
7. A nonpositive ith variable in the minimization problem corresponds to

an ith constraint of the type "$\geq$" in the maximization problem. Moreover, a nonpositive jth variable in the maximization problem corresponds to a jth constraint of the type "$\leq$" in the minimization problem.

One special case of (5) is noteworthy. The dual of

$$\text{Minimize } c^T x \text{ subject to } Ax \geq b \text{ and } x \geq 0 \tag{7}$$

is

$$\text{Maximize } b^T w \text{ subject to } A^T w \leq c \text{ and } w \geq 0 \tag{8}$$

Problem (7) is often called the *symmetric* primal problem, while (8) is called the *symmetric* dual problem. Clearly, any linear programming problem can be expressed in each of the forms (1), (2), (7), and (8) (see Exercise 4.8).

EXAMPLE 3   Write the dual of the problem

$$\text{Minimize } x_1 + 2x_2 + 5x_3 + 4x_4 \text{ subject to}$$

$$\begin{array}{rcl} 2x_1 + x_2 + x_3 & = & 7 \\ x_1 + 4x_2 & \leq & 8 \\ 6x_1 \quad + 3x_3 + x_4 & \geq & 9 \\ 2 \leq x_4 \leq 4 \end{array}$$

$$x_1 \geq 0, x_2 \leq 0, x_3 \text{ unrestricted, and } x_4 \geq 0$$

Solution: Replace the constraint $2 \leq x_4 \leq 4$ with two constraints $x_4 \leq 4$ and $x_4 \geq 2$, and treat x_4 as an unrestricted variable. Then the problem becomes

$$\text{Minimize } x_1 + 2x_2 + 5x_3 + 4x_4 \text{ subject to}$$

$$\begin{array}{rcl} 2x_1 + x_2 + x_3 & = & 7 \\ x_1 + 4x_2 & \leq & 8 \\ 6x_1 \quad + 3x_3 + x_4 & \geq & 9 \\ x_4 & \leq & 4 \\ x_4 & \geq & 2 \end{array}$$

$$x_1 \geq 0 \text{ and } x_2 \leq 0$$

The dual is

$$\text{Maximize } 7w_1 + 8w_2 + 9w_3 + 4w_4 + 2w_5 \text{ subject to}$$

$$\begin{array}{rcl} 2w_1 + w_2 + 6w_3 & \leq & 1 \\ w_1 + 4w_2 & \geq & 2 \\ w_1 \quad + 3w_3 & = & 5 \\ w_3 + w_4 + w_5 & = & 4 \end{array}$$

$$w_2 \leq 0, w_3 \geq 0, w_4 \leq 0, w_5 \geq 0$$

2. INTERPRETATIONS OF THE DUAL PROBLEM (POST-OPTIMAL ANALYSIS)

A meaningful interpretation of the linear programming problem that is dual to the original problem is not always possible. However, many such dual problems are meaningfully related to the original problem. The duality results in the last section can be useful when interpreting the meanings of the dual variables. To illustrate, consider the symmetric primal problem (7) and its dual (8) found in the last section. It follows (Exercise 4.11) that when x and w are optimal solutions of (7) and (8) respectively, then $c^T x = b^T w$. Recall that $z_0 = c^T x$. Now let $z_0 = c^T x = \sum_{i=1}^{m} b_i w_i$. Then consider w fixed, allow b_i to vary, and compute

$$\frac{\partial z_0}{\partial b_i} = w_i \geq 0 \qquad (9)$$

It follows that the dual variable represents the nonnegative rate of change of z_0 per unit rate of change of b_i. It should be pointed out that the value of $\partial z_0 / \partial b_i$ in (9) is not necessarily unique, but depends upon the optimal solution w of (8).

EXAMPLE 4   Recall Example 1 of Chapter 1. When initially formulated, Example 1 had form (7) of this chapter. Thus (8) is the dual of the original problem. From (9) it follows that the variable w_i in the dual problem must be the value of a unit of material M_i in the manufacturing process. The dual problem is to minimize the value of the materials on hand subject to the constraints which state that the total value of the materials, $\sum_{j=1}^{m} a_{ji} w_j$, in a unit of the product P_i does not exceed the cost, c_i, of a unit of the product. From (9), it follows that an additional unit of material M_i would add w_i dollars to the optimum value of z_0, if that material could be made available. Should the manufacturer decide to increase (or decrease) his supply of materials, he would consider the values of the dual variables in order to determine the best increase (or decrease) in his revenue.

The dual variables are often called *shadow prices, marginal rates of change, imputed values* (or *costs*), *sensitivity rates*, etc.

3. POST-OPTIMAL ANALYSIS

Duality has an important application concerning the model stability of a linear programming problem. In particular, suppose that the coefficient matrix A of a linear programming problem is subject to a small change. This change could be due to a change in conditions, or to unavoidable errors in measurement. Recall that a problem is model stable whenever small changes in A, b or c lead to small changes in the optimal solutions x^* and the values of $c^T x^*$. It can be shown that the symmetric problems (7) and (8) are model stable if and only if

$$Ay \geq 0 \text{ and } 0 \neq y \geq 0 \text{ imply that } c^T y > 0$$

and

$$A^T u \leq 0 \text{ and } u \geq 0 \text{ imply that } b^T u < 0$$

The first condition is called the *regularity condition for the symmetric primal problem*, while the second is called the *regularity condition for the symmetric dual problem*. For the unsymmetric primal problem (1) and its dual (2) the regularity conditions are

$$Ay = 0 \text{ and } 0 \neq y \geq 0 \text{ imply that } c^T y > 0$$

and

$$A^T u \leq 0 \text{ and } u \neq 0 \text{ imply that } b^T u < 0$$

respectively. As in the symmetric case, it can be shown that these regularity conditions are necessary and sufficient for the unsymmetric primal and dual problems to be model stable. The proofs of these statements are lengthy and tedious. The interested reader should consult the excellent discussion found in Murty [10, pp. 209 – 220].

Murty also discusses model stability for linear programming problems which possess both equality and inequality constraints, as well as, nonnegative and unrestricted variables. Unfortunately, the standard techniques for converting a problem into a standard unsymmetric or symmetric form can also convert a stable problem into an unstable one.

An example of an unsymmetric primal problem, which is due to Robinson [13], is the following:

$$\text{Minimize } 3x_1 + x_2 + x_3 + 3x_4 \text{ subject to}$$
$$x_1 + 4/3x_2 + 2x_3 \qquad = 3/2$$
$$x_2 + 3x_3 \qquad = 3/2$$
$$x_1 + x_2 + x_3 + x_4 = 1$$
$$\text{all } x_i \geq 0$$

An optimal value of the objective function is 1 which occurs at the point $x = [0 \quad 3/4 \quad 1/4 \quad 0]^T$. Moreover, $w = [1/2 \quad -1/6 \quad 1/2]^T$ solves the unsymmetric dual problem. But when a_{21} is changed from 4/3 to $4/3 + \varepsilon$, where $\varepsilon > 0$, then an optimal value of the primal objective function is 2 and occurs when $x = [1/2 \quad 0 \quad 1/2 \quad 0]^T$. The unsymmetric dual problem has an infinite number of solutions of the form $w = [0 \quad -2/3 \quad 1/3]^T + (1/\varepsilon)[4/3 \quad -4/9 \quad -4/3]^T$. The difficulty in this example is that the primal constraints do not satisfy the regularity condition.

Exercises

4.1 Write the duals of the following problems.

(a) Minimize $x_1 + 3x_2 + x_3$ subject to $7x_1 + x_2 = 8$, $x_2 + 3x_3 = 9$, $x_1 \geq 0$, $x_2 \geq 0$ and $x_3 \geq 0$.

(b) Maximize $w_1 + w_2 - w_3$ subject to $w_1 + w_2 \leq 1$, $w_2 + 2w_3 \leq 2$ and $w_1 + 4w_3 \leq 5$.

(c) Minimize $2x_1 + x_2 + 2x_3$ subject to $x_1 + 3x_2 \geq 10$, $x_2 + 3x_3 \geq 5$, $x_1 \geq 0$, $x_2 \geq 0$ and $x_3 \geq 0$.

(d) Maximize $2w_1 + w_2 + 2w_3$ subject to $w_1 + w_2 \leq 5$, $w_2 + w_3 \leq 7$, $w_1 + 4w_3 \leq 12$, $w_1 + w_2 + 8w_3 \leq 20$, $w_1 \geq 0$, $w_2 \geq 0$ and $w_3 \geq 0$.

(e) Maximize $2x_1 + 3x_2$ subject to $2x_1 + 3x_2 + 4x_3 = 20$, $12x_1 - 4x_2 - x_3 \leq 1$, $x_2 \geq 0$ and $x_3 \geq 0$.

(f) Minimize $15x_1 - 10x_2 + 3x_3$ subject to $2x_1 + 2x_2 - 4x_3 \geq 20$, $2x_1 - x_2 + 3x_3 = 30$, $2 \leq x_3 \leq 10$ and $x_1 \geq 0$.

(g) Minimize $4x_1 + x_2 - x_3$ subject to $4x_1 + 7x_2 = 5$, $x_1 + x_2 - x_3 = 1$, $x_1 \geq 0$, $x_2 \geq 0$ and $x_3 \geq 0$.

(h) Maximize $w_1 - w_2 + 2w_3$ subject to $w_1 - w_2 \leq 4$, $2w_1 + 3w_2 \leq 5$, and $4w_1 + w_3 \geq -2$.

(i) Maximize $x_1 + x_2 - 3x_3$ subject to $2x_1 + x_2 - x_3 \geq 1$, $-5x_1 + x_2 \geq 2$, $x_1 \geq 0$ and $x_2 \geq 0$.

(j) Maximize $w_1 - w_2 + w_3$ subject to $2w_1 + w_3 \leq 10$, $w_1 + w_2 \leq 21$, $-w_1 + w_3 \geq -1$, $w_1 \geq 0$, $w_2 \geq 0$ and $w_3 \geq 0$.

(k) Example 5 of Chapter 3.

4.2 Write the dual of

Maximize $5w_1 + w_2$ subject to $w_1 - w_2 \leq -4, -w_1 + w_2 \leq -2$, $w_1 \geq 0, w_2 \geq 0$

and show that neither problem is feasible.

4.3 Solve the following problem by solving the corresponding dual problem:

Maximize $4w_1 + 8w_3$ subject to $16w_1 - w_2 + 5w_3 \leq 1$, $2w_1 + 4w_3 \leq 1$, $w_1 \leq 1, w_2 \leq 1, w_3 \leq 1, 10w_1 + w_2 \leq 1, w_1 \geq 0, w_2 \geq 0$ and $w_3 \geq 0$.

4.4 Give the solution of the duals of the problems found in Exercises 3.3 of Chapter 3.

4.5 Write the dual of the problem found in Example 2 of Chapter 1.

Discuss the importance of the dual problem to a feed company that is about to manufacture a special feed containing the required amounts of nutrients. (Hint: Consider w_j to be the price of nutrient I_j in the new feed.)

4.6 Write the dual of the problem found in Example 3 of Chapter 1. Discuss the importance of the dual problem to an independent trucking company that would like to buy the chemical product at the storage facilities and then transport and sell the product to the retail outlets. (Hint: There are two prices involved – the price the trucking company pays for the chemical and the price it receives for the chemical.)

4.7 Suppose that a small manufacturing company produces three products from four raw materials. The amount of the raw material required for each product, the profit per unit of each product, and the daily amount of the raw material in stock is found in the following table:

			Products		
		P_1	P_2	P_3	Current stock
	M_1	2	1	3	100
Materials	M_2	1	2	4	150
	M_3	5	0	2	100
	M_4	1	3	1	200
	profit	2	1	2	

First find a production quota that maximizes the company's profit, then determine the supply that the company should increase in order to best increase its maximum profit.

4.8 (a) Let $A = [B \quad D] \in \mathbf{R}_{m \times n}$ where $B = [A^{(1)} \ldots A^{(m)}]$ is nonsingular. Show that the unsymmetric primal problem (1) is equivalent to the problem

$$\text{Minimize } (c_D^T - c_B^T B^{-1} D) x_D \text{ subject to } B^{-1} D x_D \leq B^{-1} b \text{ and } x_D \geq 0$$

(Hint: Consider the system of equations (8).) What is the analogous result when B is any basis matrix?

(b) Rewrite the following problem in symmetric primal form

$$\text{Minimize } x_1 + x_2 + x_3 + x_4 \text{ subject to}$$
$$2x_1 \quad + 2x_3 + x_4 = 8$$
$$- x_2 + x_3 + 3x_4 = 9$$
$$\text{all } x_i \geq 0$$

4.9   Show that when the problem

$$\text{Minimize } c^T x \text{ subject to } Ax = b$$

and its dual are both feasible, then both objective functions are constant over their corresponding sets of feasible solutions.

4.10   Prove that there exists an $x \geq 0$ such that $Ax = b$ if and only if $A^T w \geq 0$ implies that $b^T w \geq 0$. (Hint: Consider the problem minimize $0^T x$ subject to $Ax = b$ and $x \geq 0$.) This result is known as the Farkas lemma. Geometrically, the result says that either b is in the convex cone generated by the columns of A or there exists a hyperplane that separates b and the cone generated by the columns of A.

4.11   When both the symmetric primal and dual problems are feasible, then prove that there exist optimal solutions x and w to each respective problem such that $c^T x = b^T w$.

4.   THE DUAL SIMPLEX PROCEDURE

Once again consider Problem (1) of Section 1,

$$\text{Minimize } c^T x \text{ subject to } Ax = b \text{ and } x \geq 0 \qquad (1)$$

Recall that the dual of this unsymmetric primal problem is

$$\text{Maximize } b^T w \text{ subject to } A^T w \leq c \qquad (2)$$

As in Section 1, assume that $A = [B \quad D]$, where $B = [A^{(1)} \dots A^{(m)}]$ is nonsingular, and consider matrix (3) of that section

$$\left[\begin{array}{c|c|c|c} I_m & B^{-1}D & 0 & B^{-1}b \\ \hline 0 & c_B^T B^{-1} D - c_D^T & 1 & c_B^T B^{-1} b \end{array} \right] \qquad (3)$$

Again let $w = (c_B^T B^{-1})^T$ and recall from (4) that whenever $c_B^T B^{-1} D - c_D^T \leq 0$, then w is a feasible solution of (2). When this happens the set $\{A^{(1)}, \dots, A^{(m)}\}$ is called a *dual feasible* basis, and the solution $x_B = B^{-1}b$ and $x_D = [0 \dots 0]^T$ is said to be *dual feasible*. Clearly, when it also happens that $B^{-1}b \geq 0$, then $x_B = B^{-1}b$ and $x_D = [0 \dots 0]^T$ solves (1), while w solves the dual problem.

In the standard primal simplex procedure, we start with a feasible solution, and then pivot, while maintaining feasibility, until we obtain dual feasibility. Another approach, known as the *dual simplex procedure*, is to start with a dual feasible solution and then, while maintaining dual feasibility, pivot until feasibility (primal) is obtained. This amounts to solving the dual maximization problem with a primal simplex procedure.

Suppose now that we have a dual feasible solution in matrix (3). We then need to determine the correct pivot element in $B^{-1}D$. Suppose that the basic variable $x_i = (B^{-1}b)_{(i)}$ is negative. Then consider the i-th row of $B^{-1}D$. If x_{ij} is to be the new pivot element, then the last row of (3) will be transformed as

$$z_0 - ((z_j - c_j)/x_{ij})x_i \text{ and } z_l - c_l - ((z_j - c_j)/x_{ij})x_{il}$$

where $l = m + 1, \ldots, n$. Since we are now attempting to solve the dual problem, we want

$$z_0 - ((z_j - c_j)/x_{ij})x_i \geq z_0$$

or

$$((z_j - c_j)/x_{ij})x_i \leq 0$$

But $x_i < 0$ and $z_j - c_j \leq 0$. Thus, we want $x_{ij} < 0$. Since we also must maintain dual feasibility, then we must have

$$z_l - c_l - ((z_j - c_j)/x_{ij})x_{il} \leq 0 \tag{10}$$

Since $z_l - c_l \leq 0$, $x_{ij} < 0$ and $z_j - c_j \leq 0$, (10) automatically holds when $x_{il} \geq 0$. When $x_{il} < 0$, we must have

$$(z_l - c_l)/x_{il} \geq (z_j - c_j)/x_{ij}$$

Thus, dual feasibility will be maintained if we pivot at x_{ij} where $x_{ij} < 0$, $x_i < 0$, and

$$(z_j - c_j)/x_{ij} = \min \{(z_l - c_l)/x_{il} : x_{il} < 0\}$$

It is left as an exercise (Exercise 4.20) for the reader to verify that (1) is not feasible if $x_i < 0$ and each $x_{ij} \geq 0$.

EXAMPLE 5 Use the dual simplex procedure to solve the following problem

$$\text{Minimize } x_1 + x_2 + \quad x_4 + x_5 \text{ subject to}$$
$$2x_1 + x_2 - x_3 + x_4 + 2x_5 = 8$$
$$x_1 + 2x_2 \quad + 6x_4 + x_5 \leq 10$$
$$\text{all } x_i \geq 0$$

Solution: First rewrite the problem as

$$\text{Minimize } x_1 + x_2 \quad + x_4 + x_5 \text{ subject to}$$
$$-2x_1 - x_2 + x_3 - x_4 - 2x_5 \quad = -8$$
$$x_1 + 2x_2 \quad + 6x_4 + x_5 + x_6 = 10$$
$$\text{all } x_i \geq 0$$

As the problem now stands, it is clear that the basic solution in the first primal simplex tableau will not be a feasible solution. However, it is dual feasible and the dual simplex procedure may be applied as follows:

$$\min \{(-1)/(-2), (-1)/(-1)\} = 1/2$$

$$\begin{bmatrix} \boxed{-2} & -1 & 1 & -1 & -2 & 0 & 0 & -8 \\ 1 & 2 & 0 & 6 & 1 & 1 & 0 & 10 \\ -1 & -1 & 0 & -1 & -1 & 0 & 1 & 0 \end{bmatrix} \begin{array}{l} R_2 + (1/2)R_1 \\ R_3 - (1/2)R_1 \\ \hline \\ -(1/2)R_1 \end{array} \longrightarrow$$

$$\begin{bmatrix} 1 & 1/2 & -1/2 & 1/2 & 1 & 0 & 0 & 4 \\ 0 & 3/2 & 1/2 & 11/2 & 0 & 1 & 0 & 6 \\ 0 & -1/2 & -1/2 & -1/2 & 0 & 0 & 1 & 4 \end{bmatrix}$$

Thus, an optimal solution occurs when $x_1 = 4$ and $x_2 = x_3 = x_4 = x_5 = 0$. (here $B = \{A^{(1)}, A^{(6)}\}$). Notice that the use of the dual simplex procedure in this case eliminated the need for artificial variables.

EXAMPLE 6 (Post-optimal analysis—adding a constraint.) Solve the following problem:

$$\begin{aligned} \text{Minimize } x_1 + x_2 \quad &+ x_4 + x_5 \text{ subject to} \\ 2x_1 + x_2 - x_3 + x_4 + 2x_5 &= 8 \\ x_1 + 2x_2 \quad + 6x_4 + x_5 &\leq 10 \\ x_1 + x_2 + x_3 \quad - x_5 &\geq 6 \\ \text{all } x_i &\geq 0 \end{aligned}$$

Solution: Notice that this is just Example 5 with the additional constraint $x_1 + x_2 + x_3 - x_5 \geq 6$. Unfortunately, the optimal solution x^* $= [4 \quad 0 \quad 0 \quad 0 \quad 0]^T$ of Example 5 does not satisfy this constraint (see Exercise 2.43).

To take advantage of the work done in Example 5, we rewrite the new constraint as $-x_1 - x_2 - x_3 + x_5 + x_7 = -6$, where x_7 is a slack variable. Then add the row and column corresponding to this new constraint and variable to the optimal tableau of Example 5, perform the row operation $R_3 + R_1$ to restore the tableau to canonical form, and then continue with the dual simplex procedure.

$$\begin{bmatrix} 1 & 1/2 & -1/2 & 1/2 & 1 & 0 & 0 & 0 & 4 \\ 0 & 3/2 & 1/2 & 11/2 & 0 & 1 & 0 & 0 & 6 \\ -1 & -1 & -1 & 0 & 1 & 0 & 1 & 0 & -6 \\ 0 & -1/2 & -1/2 & -1/2 & 0 & 0 & 0 & 1 & 4 \end{bmatrix} \begin{array}{l} \\ \\ R_3 + R_1 \\ \hline \end{array} \longrightarrow$$

$$\begin{bmatrix} 1 & 1/2 & -1/2 & 1/2 & 1 & 0 & 0 & 0 & 4 \\ 0 & 3/2 & 1/2 & 11/2 & 0 & 1 & 0 & 0 & 6 \\ 0 & -1/2 & \boxed{3/2} & -1/2 & 2 & 0 & 1 & 0 & -2 \\ 0 & -1/2 & -1/2 & -1/2 & 0 & 0 & 0 & 1 & 4 \end{bmatrix}$$

dual feasible
$B = \{A^{(1)}, A^{(6)}, A^{(7)}\}$
$\min\{(-1/2)/(-3/2), (-1/2)/(-1/2)\}$
$= 1/3$

$$\begin{matrix} R_1 - (1/3)R_3 \\ R_2 + (1/3)R_3 \\ \xrightarrow{\hspace{2cm}} \\ R_4 - (1/3)R_3 \\ (-2/3)R_2 \end{matrix} \begin{bmatrix} 1 & 2/3 & 0 & 1/3 & 1/3 & 0 & -1/3 & 0 & 14/3 \\ 0 & 4/3 & 0 & 17/3 & 2/3 & 1 & 1/3 & 0 & 16/3 \\ 0 & 1/3 & 1 & -1/3 & -4/3 & 0 & -2/3 & 0 & 4/3 \\ 0 & -1/3 & 0 & -2/3 & -2/3 & 0 & -1/3 & 1 & 14/3 \end{bmatrix}$$

Hence, an optimal solution occurs when $x_1 = 14/3$, $x_2 = 0$, $x_3 = 4/3$, $x_4 = x_5 = 0$, $x_6 = 16/3$ and $x_7 = 0$.

Exercises

4.12 If $c_B^T B^{-1} D - c_D^T \not\leq 0$ in Problem (1), then define $e_D = [1 \ldots 1]^T$, $\bar{x} = [x_1 \ \ldots \ x_n \ x_{n+1}]^T$, $\bar{c} = [c_1 \ \ldots \ c_n \ 0]^T$, $\bar{b} = [b_1 \ \ldots \ b_m \ M]^T$, and

$$\bar{A} = \begin{bmatrix} B & D & 0 \\ 0 & e_D^T & 1 \end{bmatrix}$$

where M is a very large but unspecified positive number. Consider the following problem

$$\text{Minimize } \bar{c}^T \bar{x} \text{ subject to } \bar{A}\bar{x} = \bar{b} \text{ and } \bar{x} \geq 0 \qquad (11(M))$$

Show that Problem (1) is feasible if Problem $11(M)$ is feasible for any $M > 0$. Moreover, if (1) is feasible, then there exists a $\bar{M} > 0$ so that $11(M)$ is feasible whenever $M \geq \bar{M}$. For $11(M)$ show that matrix (3) becomes

$$\begin{bmatrix} I_m & B^{-1}D & 0 & 0 & B^{-1}b \\ 0 & e_D^T & 1 & 0 & M \\ 0 & c_B^T B^{-1}D - c_D^T & 0 & 1 & c_B^T B^{-1}b \end{bmatrix}$$

and that a pivot about the entry in row $m+1$ and column k, where $z_k - c_k = \max\{z_j - c_j\}$, produces a dual feasible solution of $11(M)$. Finally, determine sufficient conditions for (1) to be infeasible, to be unbounded, and to have a finite minimum, respectively. (Hint: Maximize the dual of $11(M)$ and consider the coefficient h of the optimal value of $z_0 = hM + k$.)

4.13 Use the dual-simplex procedure to solve the following problems.

(a) Minimize $4x_1 + x_2 + x_3$ subject to

$$x_1 + x_2 + 2x_3 \leq 6$$
$$2x_1 + x_2 - x_3 \geq 4$$
$$\text{all } x_i \geq 0$$

(b) Minimize $x_1 + x_2 + x_3 + x_4 + x_5 + x_6$ subject to

$$2x_1 + x_2 \qquad\qquad + x_6 \geq 27$$
$$x_1 + 2x_2 + x_3 \qquad\qquad \geq 34$$
$$x_2 + 2x_3 + x_4 \qquad\qquad \geq 34$$
$$x_3 + 2x_4 + x_5 \qquad\quad \geq 30$$
$$x_4 + 2x_5 + x_6 \geq 15$$
$$x_1 \qquad\qquad\qquad + x_5 + 2x_6 \geq 12$$

$$\text{all } x_i \geq 0$$

(c) Minimize $10x_1 + 20x_2 + 24x_3 + 30x_4$ subject to

$$2x_1 - 3x_2 \qquad - x_4 \geq 8$$
$$2x_1 + 3x_2 + 2x_3 + 7x_4 \geq 28$$
$$\text{all } x_i \geq 0$$

(d) Minimize $x_1 - (1/2)x_2$ subject to

$$-x_1 + 2x_2 \geq 12$$
$$x_1 + x_2 \geq 8$$
$$\text{all } x_i \geq 0$$

(Hint: Show that this problem has an unbounded minimum by showing that the related problem in Exercise 4.12 has an unbounded minimum in which $h < 0$ in the term $z_0 = hM + k$.)

4.14 (Post-optimal analysis—adding a constraint.) Join the constraint $3x_1 - 6x_2 + 7x_4 < 1$ to those of Example 6 and solve the resulting linear programming problem. Start with the optimal tableau of Example 6 and use the dual-simplex algorithm.

4.15 (Post-optimal analysis—adding a constraint.) Let matrix (3) correspond to an optimal tableau for Problem (1). If an additional constraint $d^T x \leq h$ is joined to the constraints $Ax = b$, show that the initial matrix becomes

$$\left[\begin{array}{ccc:c:c:c}
B & D & 0 & 0 & b \\
\hline
d_B^T & d_D^T & 1 & 0 & h \\
\hline
-c_B^T & -c_D^T & 0 & 1 & 0
\end{array}\right]$$

where d_B^T and d_D^T are defined in the same fashion as c_B^T and c_D^T, and matrix (3) becomes

$$\left[\begin{array}{ccc:c:c:c}
I_m & B^{-1}D & 0 & 0 & B^{-1}b \\
\hline
0 & d_D^T - d_B^T B^{-1}D & 1 & 0 & h - d_B^T B^{-1}b \\
\hline
0 & c_B^T B^{-1}D - c_D^T & 0 & 1 & c_B^T B^{-1}b
\end{array}\right]$$

4.16 (Post-optimal analysis—adding a constraint.) Join the constraint $3x_1 + 6x_2 - x_5 = 16$ to those of Example 6 and solve. (Hint: Rewrite this constraint as $3x_1 + 6x_2 - x_5 + x_8 = 20$, where x_8 is an artificial variable and solve as in Example 6. What would a positive value of x_8 in an optimal solution mean?)

4.17 (Post optimal analysis—adding a constraint.) Join the constraint $3x_1 + 6x_2 - x_5 = 20$ to those of Example 6 and solve.

4.18 (Post optimal analysis—changing b.) Change the column $b = [4 \quad 0 \quad 8]^T$ in Example 6 of Chapter 3 to the column $\bar{b} = [2 \quad 2 \quad 1]^T$ and solve the resulting problem starting with the optimal tableau of that example. (Hint: In matrix (3), $B^{-1}b$ must be replaced by $B^{-1}\bar{b}$.) But the basic matrix for the optimal solution of Example 6 is $B = [A^{(2)} A^{(4)} A^{(3)}]$. Moreover,

$$B^{-1} = \begin{bmatrix} 1/2 & 0 & 0 \\ 0 & 1 & 0 \\ -2 & 0 & 1 \end{bmatrix}$$

since $I = [A^{(5)} A^{(4)} A^{(3)}]$. (Why?)

4.19 Outline the dual simplex procedure.

4.20 Let B be a basis matrix for a dual feasible solution of problem (1). If there exists an index i for which the i-th component of $B^{-1}b$ is negative and the entire i-th row $(B^{-1}D)_{(i)}$ of $B^{-1}D$ is nonnegative, then prove that (1) is not feasible. Moreover, (2) has an unbounded maximum.

5. COMPLEMENTARY SLACKNESS

Up to this point, when it was necessary to solve a linear programming problem, we expressed the problem in the unsymmetric primal form (1) and then applied the simplex procedure. Other approaches are available which begin with a problem expressed in the symmetric primal form (7). Exercise 4.8 enables us to express any linear programming problem in form (7).

Consider Problem (7). Subtracting slack variables $u = [u_1 \ldots u_m]^T$, let us rewrite (7) in the following form:

$$\text{Minimize } c^T x \text{ subject to } Ax - u = b, x \geq 0 \text{ and } u \geq 0 \qquad (12)$$

Next rewrite Problem (8) by adding slack variables $v = [v_1 \ldots v_n]^T$ to obtain the form:

$$\text{Maximize } b^T w \text{ subject to } A^T w + v = c, w \geq 0 \text{ and } v \geq 0 \qquad (13)$$

The next result gives necessary and sufficient conditions for problems (12) and (13) to have optimal solutions.

THEOREM 2 (Complementary Slackness) Supposte that $[x^T \ u^T]^T$ and $[w^T \ v^T]^T$ are feasible solutions of (12) and (13), respectively. Then these solutions are optimal if and only if $v^T x + u^T w = 0$.
Proof: The solutions are optimal if and only if $c^T x = b^T w$, i.e.,

$$0 = c^T x - b^T w = (A^T w + v)^T x - (Ax - u)^T w = (w^T A + v^T)x - (x^T A^T - u^T)w$$
$$= w^T Ax + v^T x - x^T A^T w + u^T w = v^T x + u^T w$$

since $w^T Ax = (w^T Ax)^T = x^T A^T w$.

Now let

$$M = \begin{bmatrix} 0 & -A^T \\ A & 0 \end{bmatrix}, r = \begin{bmatrix} v \\ u \end{bmatrix}, s = \begin{bmatrix} x \\ w \end{bmatrix}, \text{ and } q = \begin{bmatrix} c \\ -b \end{bmatrix} \qquad (14)$$

It now follows from the complementary slackness theorem that solving Problems (7) and (8) is equivalent to finding column vectors r and s such that

$$Ir - Ms = q, r \geq 0, s \geq 0, \text{ and } r^T s = 0 \qquad (15)$$

The variables r_i and s_i are said to be *complementary*.(Here r_i denotes the i-th component of r, etc.)

6.   THE LINEAR COMPLEMENTARITY PROBLEM

Let $M = [M_{ij}] \in \mathbf{R}_{n \times n}$ and $q = [q_1 \ldots q_n]^T \in \mathbf{R}_{n \times 1}$. The *linear complementarity problem*, denoted by $LCP(q,M)$, is to find column vectors $r = [r_1 \ldots r_n]^T$ and $s = [s_1 \ldots s_n]^T$ such that

$$Ir - Ms = q, r \geq 0, s \geq 0 \text{ and } r^T s = 0 \qquad (16)$$

Using (14) and (15), any linear programming problem in symmetric primal form can be expressed as a linear complentarity problem. In a later chapter, it will be shown that a quadratic programming problem can also be expressed as a $LCP(q,M)$. Hence, any algorithm that solves an $LCP(q,M)$ can be used to solve linear programming as well as some nonlinear programming problems. Many other applied problems can be expressed as linear complementarity problems. The list includes bimatrix game problems.

For each $i = 1, \ldots, n$, select either the i-th column $I^{(i)}$ of the identity matrix I or the i-th column $-M^{(i)}$ of the matrix $-M$. Denote the column selected by $B^{(i)}$. Then the convex cone

$$\text{coni}\{B^{(1)}, \ldots, B^{(n)}\} = \{\sum_{i=1}^{n} \alpha_i B^{(i)} : \alpha_i \geq 0\}$$

is called a *complementary cone* of M. Geometrically problem (16) has a solution if and only if q belongs to one of the 2^n complementary cones of M.

Any n of the variables r_i and s_i are *basic* whenever their corresponding coefficient matrix is nonsingular. A *basic solution* is a solution that is obtained by solving for a set of basic variables whenever the corresponding nonbasic variables are set equal to zero. A basic solution is *degenerate* whenever it has more than n of the variables r_i and s_i equal to zero.

7. LEMKE'S COMPLEMENTARY PIVOTING ALGORITHM

Several iterative algorithms have been developed for solving linear complementarity problems. The best known are the principal pivoting algorithm of Cottle and Dantzig [4] and the complementary pivoting algorithm by Lemke [7]. Additional discussions of the principal pivoting algorithm can be found in the references, [3] and[10],listed at the end of this chapter. In this section we shall outline Lemke's method. Notice that the $LCP(q,M)$, Problem (16), has a trival solution when $q \geq 0$, namely $r = q$ and $s = 0$. Therefore, an algorithm is needed only when $q \not\geq 0$.

Let $e = [1 \ldots 1]^T \in \mathbf{R}_{n+1}$, let s_0 be a variable, and consider the problem of finding $r = [r_1 \ldots r_n]^T$, $s = [s_1 \ldots s_n]^T$, and s_0 such that

$$Ir - Ms - es_0 = q, r \geq 0, s \geq 0, s_0 \geq 0, \text{ and } r_i s_i = 0 \qquad (17)$$

for at least $n-1$ of the i.

A basic feasible solution of Problem (16) in which there is exactly one basic variable from each complementary pair, r_i and s_i, of variables is known as a *complementary basic feasible solution*. An *almost complementary basic feasible* solution of problem (16) is a basic feasible solution of Problem (17) in which s_0 is a basic variable and there is exactly one basic variable from each of exactly $n-1$ complementary pairs of variables.

In Lemke's algorithm to be described, we shall move from one almost complementary basic feasible solution of (16) to another. If at any stage we have a complementary basic feasible solution of (16), then the condition, $r^T s = 0$, will automatically be satisfied and the problem will be solved. We begin with a tableau of the form:

$$
\begin{array}{ccccccc}
r_1 \ldots r_n & s_1 & & s_n & s_0 & q \\
\hline
1 \ \ldots \ 0 & -m_{11} & \ldots & -m_{nn} & -1 & q_1 \\
\vdots \quad \vdots & \vdots & & \vdots & \vdots & \vdots \\
0 \ \ldots \ 1 & -m_{n1} & \ldots & -m_{nn} & -1 & q_n
\end{array}
\tag{18}
$$

The following steps summarize the algorithm:

Step 1 Pick the smallest q_i. Since $q \not\geq 0$, the smallest q_i is negative. In the column under s_0, pivot at the entry corresponding to the smallest q_i. The variable s_0 is now basic.

Step 2 Find the pair of nonbasic complementary variables, r_j and s_j. In the column under s_j select and pivot in the usual simplex way. This will introduce column $-M^{(j)}$ into the admissible basis. If s_0 becomes nonbasic, then go to Step 4.

Step 3 Note the variable that just dropped from the basic vector. Select and pivot in the column of the variable that is complementary to the dropping variable. (Use the usual selection and pivoting techniques in this column.)

Step 4 If you have a complementary basic feasible solution, then stop—the problem is solved. If not, return to Step 3.

The algorithm fails if at any step the pivoting column is nonpositive. In this case the problem may or may not have a solution. When Problem (17) is degenerate, the complementary pivoting algorithm may or may not solve the problem. In this case it is possible to have the algorithm cycle.

EXAMPLE 7 Formulate the problem.

Minimize $2x_1 + 3x_2$ subject to $x_1 + 2x_2 \geq 4$, $3x_1 + x_2 \geq 6$ and all $x_i \geq 0$

as a $LCP(q,M)$ and solve with the complementary pivoting algorithm.
Solution: For this problem

$$
M = \begin{bmatrix} 0 & 0 & -1 & -3 \\ 0 & 0 & -2 & -1 \\ 1 & 2 & 0 & 0 \\ 3 & 1 & 0 & 0 \end{bmatrix}, \quad q = \begin{bmatrix} 2 \\ 3 \\ -4 \\ -6 \end{bmatrix}, \quad r = \begin{bmatrix} v_1 \\ v_2 \\ u_1 \\ u_2 \end{bmatrix} \quad \text{and} \quad s = \begin{bmatrix} x_1 \\ x_2 \\ w_1 \\ w_2 \end{bmatrix}
$$

The solution of the $LCP(q,M)$ is now generated using tableau (18)

Tableau 1

r_1	r_2	r_3	r_4	s_1	s_2	s_3	s_4	s_0	q
1	0	0	0	0	0	1	3	−1	2
0	1	0	0	0	0	2	1	−1	3
0	0	1	0	−1	−2	0	0	−1	−4
0	0	0	1	−3	−1	0	0	(−1)	−6

$R_1 - R_4$
$R_2 - R_4$
$R_3 - R_4$ →
$- R_4$

r_4 goes out, s_4 comes in
min $\{8/3, 9/1\} = 8/3$

Tableau 2

r_1	r_2	r_3	r_4	s_1	s_2	s_3	s_4	s_0	q
1	0	0	−1	3	1	1	(3)	0	8
0	1	0	−1	3	1	2	1	0	9
0	0	1	−1	2	−1	0	0	0	2
0	0	0	−1	3	1	0	0	1	6

$R_2 - (1/3)R_1$
$(1/3)R_1$

r_1 goes out, s_1 comes in
min $\{(8/3)/1, (19/3)/2, 2/2, 6/3\} = 1$

Tableau 3

r_1	r_2	r_3	r_4	s_1	s_2	s_3	s_4	s_0	q
1/3	0	0	−1/3	1	1/3	1/3	1	0	8/3
−1/3	1	0	−2/3	2	2/3	5/3	0	0	19/3
0	0	1	−1	(2)	−1	0	0	0	2
0	0	0	−1	3	1	0	0	1	6

$R_1 - (1/2)R_3$
$R_2 - R_3$
$R_4 - (3/2)R_2$ →
$(1/2)R_2$

r_3 goes out s_3 comes in
min $\{(5/3)/(1/3), (13/3)/(5/3)\} = 13/5$

Tableau 4

r_1	r_2	r_3	r_4	s_1	s_2	s_3	s_4	s_0	q
1/3	0	−1/2	1/6	0	5/6	1/3	1	0	5/3
−1/3	1	−1	1/3	0	5/3	(5/3)	0	0	13/3
0	0	1/2	−1/2	1	−1/2	0	0	0	1
0	0	−3/2	1/2	0	5/2	0	0	1	3

$R_1 - (1/5)R_2$
$(3/5)R_2$ →

r_2 goes out, s_2 comes in
min $\{(4/5)/(1/2), (13/5)/1, 3/(5/2)\} = 6/5$

Tableau 5

r_1	r_2	r_3	r_4	s_1	s_2	s_3	s_4	s_0	q
2/5	−1/5	−3/10	1/10	0	1/2	0	1	0	4/5
−1/5	3/5	−3/5	1/5	0	1	1	0	0	13/5
0	0	1/2	−1/2	1	−1/2	0	0	0	1
0	0	−3/2	1/2	0	(5/2)	0	0	1	3

$R_1 - (1/5)R_4$
$R_2 - (2/5)R_4$
$R_3 + (1/5)R_4$
$(2/5)R_4$

Tableau 6

r_1	r_2	r_3	r_4	s_1	s_2	s_3	s_4	s_0	q
2/5	−1/5	0	0	0	0	0	1	−1/5	1/5
−1/5	3/5	0	0	0	0	1	0	−2/5	7/5
0	0	1/5	−2/5	1	0	0	0	1/5	8/5
0	0	−3/5	1/5	0	1	0	0	2/5	6/5

Here, a solution to the $LCP(q,M)$ is $r_1 = r_2 = r_3 = r_4 = 0$, $s_1 = 8/5$, $s_2 = 6/5$, $s_3 = 7/5$ and $s_4 = 1/5$. A solution to the original linear programming problem occurs at $x_1 = 8/5$ and $x_2 = 6/5$.

The pivot operations have been performed as in Chapter 3. The solution is easily checked by showing that $Ir - Ms = q$ and $r^T s = 0$.

EXAMPLE 8   Solve the $LCP(q,M)$, where $q = [1 \quad 2 \; -5]^T$ and

$$M = \begin{bmatrix} 1 & -3 & -3 \\ -2 & 2 & -4 \\ 1 & -1 & -5 \end{bmatrix}$$

Solution: Since the column corresponding to s_3 in the second iteration

r_1	r_2	r_3	s_1	s_2	s_3	s_0	q	
1	0	0	-1	3	3	-1	1	$R_1 - R_3$
0	1	0	2	-2	4	-1	2	$R_2 - R_3$
0	0	1	-1	1	5	(-1)	-5	$-R_3$
1	0	-1	0	2	-2	0	6	
0	1	-1	3	-3	-1	0	7	
0	0	-1	1	-1	-5	1	5	

is nonpositive, the complementary pivoting algorithm fails. In this example the problem does not have a solution.

Matrix M is *copositive-plus* whenever $x \geq 0$ implies that $x^T M x \geq 0$, and $x \geq 0$ together with $x^T M x = 0$ imply that $(M + M^T)x = 0$. Any positive semidefinite matrix (i.e., a matrix M for which $x^T M x \geq 0$ for all x) is copositive-plus. When problem (17) is nondegenerate and M is copositive-plus, it can be shown that Lemke's algorithm solves the $LCP(q,M)$ in a finite number of steps ([4] and [10]). Moreover, any $LCP(q,M)$ can be transformed into a nondegenerate problem [3]. However, this is not always necessary for computation purposes.

Exercises

4.21   When possible, solve the $LCP(q,M)$ for the following cases:

(a) $q = [-2 \; -8 \quad 1]^T$ and $M = \begin{bmatrix} -1/2 & 0 & 1 \\ 0 & -1/2 & 1 \\ -1 & -1 & 1 \end{bmatrix}$

(b) $q = [-5 \quad 3]^T$ and $M = \begin{bmatrix} 1 & 1 \\ 2 & 3 \end{bmatrix}$

(c) $q = [-2 \quad 1]^T$ and $M = \begin{bmatrix} -1 & -1 \\ 2 & -3 \end{bmatrix}$

4.22   Write a Fortran program that uses the complementary pivoting algorithm to solve an $LCP(q,M)$.

4.23   Write the linear programming problem found in Example 5 of Chapter 3 as a $LCP(q,M)$ and solve using the complementary pivoting algorithm.

4.24   Write the linear programming problem found in Example 6 of Chapter 3 as a $LCP(q,M)$ and solve using the complementary pivoting algorithm. (Hint: Use Exercise 4.8.)

4.25   (Complementary slackness for unsymmetric dual problems) Consider Problems (1) and (2). If $Ax = b, x \geq 0, A^T w + v = c, w \geq 0$ and $v \geq 0$, then prove that x and w are optimal solutions for problems (1) and (2), respectively, if and only if $v^T x = 0$.

4.26   Show that whenever the complementary pivoting algorithm fails because of a nonpositive pivot column, then a ray, $\{[r^T s^T s_0]^T + \lambda d^T : \lambda \geq 0\}$, of almost complementary basic feasible solutions is determined. When the algorithm fails for this reason, the failure is referred to as *ray termination*.

4.27   Find an example of a $LCP(q,M)$ that ends in ray termination and has no solution.

4.28   Let $M \in \mathbf{R}_{2 \times 2}$ be nonsingular and nonnegative. Show that the $LCP(q,M)$ has a solution for all $q \in \mathbf{R}_{2 \times 1}$ if and only if $I^{(k)} = Mz, z > 0$ has a solution whenever $M^{(k)} \not\geq 0$ [6]. When a matrix $M \in \mathbf{R}_{n \times n}$ has the property that the $LCP(q,M)$ has a solution for all $q \in \mathbf{R}_{n \times 1}$, then M is said to be a *Q-matrix*.

4.29   When M is a Q-matrix, then the principle submatrix obtained from M by deleting the i-th row and column of M is a Q-matrix or $I^{(i)}$ belongs to a complementary cone of M that has $-M^{(i)}$ as a generator [6] (Note: When the same rows and columns are deleted from a square matrix, the resulting matrix is said to be a principal submatrix.)

4.30   Let $M \in \mathbf{R}_{n \times n}$. If $M \geq 0$ and each $m_{ii} > 0$, then the $LCP(q,M)$ has a solution for each q [11].

4.31   (a) Let $M \in \mathbf{R}_{n \times n}$ be a Q-matrix. Let $B = [B^{(1)} \dots B^{(n)}]$ be nonsingular, where the $B^{(i)}$ are the generators, either $I^{(i)}$ or $-M^{(i)}$, of a complementary cone of M. When both sides of $Ir - Ms = q$, in Problem (16), are multiplied on the right by B^{-1}, show that the result is a $LCP(\bar{q}, \bar{M})$ where $\bar{M}$ is a Q-matrix. The matrix $\bar{M}$ is called a *principal pivot transform* of M.
(b) Let M be a Q-matrix. Then show that some submatrix B as described in (a) must exist. (Hint: If not, then every principal submatrix of M must be singular. An inductive argument can then be used to show that every row and column of M contains exactly one nonzero entry, which is a constradiction, [12].).

4.32   The matrix $M \in \mathbf{R}_{n \times n}$ is a Q-matrix if and only if the union of all the complementary cones of M is $\mathbf{R}_{n \times 1}$.

4.33   (a) Show that the dual of the problem

$$\text{Minimize } c^T x \text{ subject to } Ax = b \text{ and } 0 \le x \le d \tag{19}$$

is

$$\text{Maximize } b^T w - d^T \bar{w} \text{ subject to } A^T w - \bar{w} \le c \text{ and } \bar{w} \ge 0$$

(b) Prove that any x that satisfies $Ax = b$ and $0 \le x \le d$ is a solution of (19) if and only if there exists $\bar{w} \ge 0$ and $v \ge 0$ such that $A^T w - \bar{w} + v = c$ and $v^T x + \bar{w}^T (d - x) = 0$.

REFERENCES

1.   M. Bazaraa and J. Jarvis, *Linear Programming and Network Flows,* John Wiley and Sons, New York (1977).
2.   M. Bazaraa and C. M. Shetty, *Nonlinear Programming,* John Wiley and Sons, New York (1979).
3.   A. Berman and R. J. Plemmons, *Nonnegative Matrices in the Mathematical Sciences,* Academic Press, New York (1979).
4.   R. W. Cottle and G. B. Dantzig, Complementary Pivot Theory of Mathematical Programming, *Linear Algebra and Its Applications 1,* (1968), pp. 103 – 125.
5.   S. Gass, *Linear Programming,* McGraw-Hill, New York (1969).
6.   M. W. Jeter and W. C. Pye, Some Properties of Q-matrices, *Linear Algebra and Its Applications 57,* (1984), pp. 169 – 180.
7.   C. E. Lemke, Bimatrix Equilibrium Points and Mathematical Programming, *Management Science 11,* (1965), pp. 681 – 689.
8.   C. E. Lemke, The Dual Method of Solving the Linear Programming Problem, *Naval Research Logistics Quarterly 1,* (1954), pp. 36 – 47.
9.   G. Mitra, *Theory and Application of Mathematical Programming,* Academic Press, New York (1976).

10. K. Murty, *Linear Programming*, John Wiley and Sons, New York (1983).
11. K. Murty, On the Number of Solutions of the Linear Complementarity Problem and Spanning Properties of Complementary Cones, *Linear Algebra and Its Applications 5*, (1972), pp. 65 – 108.
12. T. D. Parsons, Applications of Principal Pivoting, *Proceedings of the Princeton Symposium on Mathematical Programming*, Princeton University Press, Princeton, N.J. (1970), pp. 567 – 581.
13. S. M. Robinson, A Characterization of Stability in Linear Programming, *Operations Research 25* (1977), pp. 435 – 447.
14. G. Zoutendijk, *Mathematical Programming Methods*, North-Holland, New York (1976).

5

Other Simplex Procedures

This chapter deals with several additional computational procedures that are commonly used in linear programming. The simplex method found in Chapter 3 is often called the primal simplex procedure. A closely related procedure, known as the revised simplex procedure, may be computationally more efficient when the matrix A of the constraints is sparse (A is *sparse* when it has a small percentage of nonzero entries), or when A has many more columns than it has rows. The revised simplex algorithm can also be efficiently applied to the maximal flow problem of the next chapter. Several variations of the revised simplex procedure are discussed in this chapter. Another procedure known as the primal-dual algorithm (which also has important applications in the network flow problems of Chapter 6) is also presented. These algorithms are motivated, but not developed in the same degree of detail as was the primal simplex procedure in Chapter 3. Additional development can be found in the references at the end of the chapter. Post-optimal analysis is also dealt with briefly in this chapter.

1. THE PRIMAL SIMPLEX TABLEAU REVISITED

For $A \in \mathbf{R}_{m \times n}$ consider the unsymmetric primal problem

$$\text{Minimize } c^T x \text{ subject to } Ax = b \text{ and } x \geq 0 \tag{1}$$

Let $z_0 = c^T x$. Then problem (1) is equivalent to finding a solution $x \in \mathbf{R}_{n \times 1}$ and $z_0 \in \mathbf{R}$ of

$$Ax = b$$
$$-c^T x + z_0 = 0 \tag{2}$$

for which z_0 is minimal and $x \geq 0$. The constraints (2) can be represented as the augmented matrix

$$\left[\begin{array}{c:c:c} A & 0 & b \\ \hline -c^T & 1 & 0 \end{array}\right] \tag{3}$$

where 0 is a column of m zeros.

For simplicity of notation and without loss of generality, we shall assume that $A = [B \quad D]$, where $B = [A^{(1)} \ldots A^{(m)}]$ is nonsingular. The matrix in (3) can now be expressed as

$$\left[\begin{array}{c:c:c:c} B & D & 0 & b \\ \hline -c_B^T & -c_D^T & 1 & 0 \end{array}\right] \tag{4}$$

where $c_B = [c_1 \ldots c_m]^T$ and $c_D = [c_{m+1} \ldots c_n]^T$. Multiplying (4) on the left by

$$\left[\begin{array}{c:c} B & 0 \\ \hline -c_B^T & 1 \end{array}\right]^{-1} = \left[\begin{array}{c:c} B^{-1} & 0 \\ \hline c_B^T B^{-1} & 1 \end{array}\right]$$

gives

$$\left[\begin{array}{c:c:c:c} I_m & B^{-1}D & 0 & B^{-1}b \\ \hline 0 & c_B^T B^{-1}D - c_D^T & 1 & c_B^T B^{-1}b \end{array}\right] \tag{5}$$

Matrix (5) can also be obtained from matrix (4) by the usual pivoting method using elementary row operations. Matrix (5) represents the system of equations

$$\begin{aligned} x_B + \quad B^{-1}Dx_D \quad &= \quad B^{-1}b \\ (c_B^T B^{-1}D - c_D^T)x_D + z_0 &= c_B^T B^{-1}b \end{aligned} \tag{6}$$

Since $z_0 = c_B^T B^{-1}b - (c_B^T B^{-1}D - c_D^T)x_D$, it is clear that $x_B = c_B^T B^{-1}b$ and $x_D = [0 \ldots 0]^T$ is an optimal solution of (1) whenever $x_B \geq 0$ and $c_B^T B^{-1}D - c_D^T \leq 0$. Finally, notice that (5) can be constructed whenever B^{-1} is known.

Now suppose that the jth entry of $c_B^T B^{-1}D - c_D^T$ is positive and we wish to obtain a new basic feasible solution by pivoting in the corresponding column of $B^{-1}D$. For simplicity let the column be

$$(B^{-1}D)^{(j)} = [x_{ij} \ldots x_{mj}]^T$$

and suppose that the pivot element is x_{ij}. Let

$$\bar{B} = [A^{(1)} \ldots A^{(i-1)} \ A^{(j)} A^{(i+1)} \ldots A^{(m)}]$$

(Notice that column $A^{(j)}$ has replaced column $A^{(i)}$ in $\bar{B}$.) We would like to compute $\bar{B}^{-1}$ in order to generate the next simplex tableau. This can easily be done since we already have B^{-1}. To see this, consider the following sequence of elementary row operations

$$[\bar{B}|I_m] \to \cdots \to [I^{(1)} \ldots I^{(i-)} (B^{-1}D)^{(j)} I^{(i+1)} \ldots I^{(n)} | B^{-1}]$$

which is a sequence of operations used to compute B^{-1}. The column

$(B^{-1}D)^{(j)}$ must still be transformed into $I^{(i)}$. The row operations required for this purpose are the following

$$[I^{(1)} \ldots I^{(i-1)}(B^{-1}D)^{(j)} I^{(i+1)} \ldots I^{(n)} \,\vdots\, B^{-1}] \xrightarrow[(1/x_{ij})R_i]{R_\ell - (x_{\ell j}/x_{ij})R_i} [I_m \,\vdots\, \bar{B}^{-1}]$$

where $\ell = 1, \ldots, i-1, i+1, \ldots, m$. This implies that $\bar{B}^{-1}$ can be computed by transforming the augmented matrix $[\bar{B}^{-}D^{(j)} \,\vdots\, B^{-1}]$ as indicated in (8).

$$[B^{-1}D^{(j)} \,\vdots\, B^{-1}] \xrightarrow[(1/x_{ij})R_i]{R_\ell - (x_{\ell j}/x_{ij})R_i} [I_{(i)} \,\vdots\, \bar{B}^{-1}], \ell \neq i \qquad (8)$$

The new simplex tableau (5) corresponding to $\bar{B}$ can then be generated by computing the necessary components $\bar{B}^{-1}\bar{D}$, $\bar{B}^{-1}b$, etc. Finally, notice that $\bar{B}^{-1}b$ can be obtained by performing the same sequence of row operations,

$$[B^{-1}D^{(j)} \,\vdots\, B^{-1} \,\vdots\, B^{-1}b] \xrightarrow[(1/x_{ij})R_i]{R_\ell - (x_{\ell j}/x_{ij})R_i} [I_m \,\vdots\, \bar{B}^{-1} \,\vdots\, \bar{B}^{-1}b], \ell \neq i \qquad (9)$$

2. THE REVISED SIMPLEX PROCEDURE

In practice the primal simplex procedure usually produces a solution to Problem (1) in approximately m to $2m$ iterations. Thus, when the number of columns of A far exceeds the number of rows of A, it is apparent that a lot of computational effort is wasted on those columns which never contain pivot elements. Moreover, storage space is wasted when these columns are retained. The *revised simplex* procedure is a modification of the primal simplex procedure that attempts to minimize this waste. The method is summarized in the steps below.

Step 1 Determine an initial admissible basis and compute the inverse, B^{-1}, of the corresponding matrix. Also compute $x_B = B^{-1}b$.

Step 2 Compute the elements in $c_B^T B^{-1}D - c_D^T$. If at any point of this computation a positive entry is encountered, then go to Step 3. Otherwise, go to Step 6.

Step 3 Suppose that the jth component of $c_B^T B^{-1}D - c_D^T$ is positive, then compute $B^{-1}D^{(j)}$. If $B^{-1}D^{(j)} \leq 0$, then go to Step 7. Otherwise, go to Step 4.

Step 4 Determine the pivot element of $B^{-1}D^{(j)}$ using the usual ratio test.

Step 5 Carry out the transformation in (9) to get $\bar{B}^{-1}$ and $\bar{B}^{-1}b$. Denote the new matrix $\bar{B}$ by B and return to Step 2.

Step 6 Stop. Since $c_B^T B^{-1}D - c_D^T \leq 0$ an optimal solution is $x_B = B^{-1}b$ and $x_D = [0 \ldots 0]^T$.

Step 7 Stop. The problem has an unbounded minimum.

The method outlined above is also known as the *inverse matrix* form of the revised simplex procedure.

EXAMPLE 1   Use the revised simplex procedure to solve the problem

$$\text{Minimize } x_1 + 2x_2 + 3x_3 - 4x_4 + 20x_5 + 6x_6 \text{ subject to}$$

$$
\begin{aligned}
x_1 \quad\quad + x_3 + x_4 \quad\quad\quad\quad &= 8 \\
x_2 + x_3 \quad\quad + 3x_5 \quad &= 7 \\
x_4 + 2x_5 + x_6 &= 9
\end{aligned}
$$

$$\text{all } x_i \geq 0$$

Solution: Clearly the columns $A^{(1)}, A^{(2)}$ and $A^{(6)}$ form a basis for a basic feasible solution. If $B = [A^{(1)} \quad A^{(2)} \quad A^{(6)}]$, then clearly $B^{-1} = B = I_m$ and $B^{-1}b = [8 \quad 7 \quad 9]^T$. Moreover for this B, $c_B^T = [1 \quad 2 \quad 6]$ and $c_D^T = [3 \ -4 \quad 20]$. Also, $D = [A^{(3)} \quad A^{(4)} \quad A^{(5)}]$. Thus,

$$
c_B^T B^{-1} D - c_D^T = [1 \quad 2 \quad 6] \begin{bmatrix} 1 & 0 & 0 \\ 0 & 1 & 0 \\ 0 & 0 & 1 \end{bmatrix} \begin{bmatrix} 1 & 1 & 0 \\ 1 & 0 & 3 \\ 0 & 1 & 2 \end{bmatrix} - [3 \ -4 \quad 20]
$$

$$= [0 \quad 11 \quad \cdot \ ]$$

where the computation was halted when a positive entry was encountered. Next compute $B^{-1}D^{(2)} = I_m D^{(2)} = D^{(2)} = [1 \quad 0 \quad 1]^T$. Then form the matrix in (9) and pivot in the usual manner. The result is summarized by

$$
\begin{bmatrix} ① & 1 & 0 & 0 & 8 \\ 0 & 0 & 1 & 0 & 7 \\ 1 & 0 & 0 & 1 & 9 \end{bmatrix} \xrightarrow{R_3 - R_1} \begin{bmatrix} 1 & 1 & 0 & 0 & 8 \\ 0 & 0 & 1 & 0 & 7 \\ 0 & -1 & 0 & 1 & 1 \end{bmatrix}
$$

Now rename $B = [A^{(4)} \quad A^{(2)} \quad A^{(6)}]$. Then $D = [A^{(1)} \quad A^{(3)} \quad A^{(5)}]$,

$$
B^{-1} = \begin{bmatrix} 1 & 0 & 0 \\ 0 & 1 & 0 \\ -1 & 0 & 1 \end{bmatrix}, \text{ and } B^{-1}b = \begin{bmatrix} 8 \\ 7 \\ 1 \end{bmatrix}
$$

Moreover, $c_B^T = [-4 \quad 2 \quad 6]$ and $c_D^T = [1 \quad 3 \quad 20]$. Next compute

$$
c_B^T B^{-1} D - c_D^T = [-4 \quad 2 \quad 6] \begin{bmatrix} 1 & 0 & 0 \\ 0 & 1 & 0 \\ -1 & 0 & 1 \end{bmatrix} \begin{bmatrix} 1 & 1 & 0 \\ 0 & 1 & 3 \\ 0 & 0 & 2 \end{bmatrix} - [1 \quad 3 \quad 20]
$$

$$= [-11 \quad -11 \quad -2]$$

Thus, an optimal solution to the problem occurs when $x_4 = 8$, $x_2 = 7$, $x_6 = 1$, and $x_1 = x_3 = x_5 = 0$.

Clearly, the revised simplex method is similar to the primal simplex method. As suggested by the last example, the revised simplex procedure often requires more computations in a single iteration than does the primal simplex method. However, in many practical problems the matrix A has far more columns than rows. When this is the case, the total number of pivots (and hence computations) in the revised simplex procedure may be substantially less than in the primal simplex procedure. Many practical problems have hundreds of rows and thousands of columns.

When the matrix A is sparse, then the computation of $c_B^T B^{-1} D - c_D^T$ in Step 2 of the revised simplex procedure may be greatly simplified provided that D is also sparse. Unfortunately, the matrices X_B do not normally remain sparse as the iterations continue.

The primary reason for the use of the revised simplex procedure is the reduction in the amount of information that must be retained (stored) during the iterations. It is a method that is often used to work a large problem on a high speed computer. The next two sections will discuss improvements in the revised simplex procedure for such applications.

The revised simplex procedure starts with an admissible basis and a basic feasible solution. These, if not apparently available, can be obtained using the techniques of Chapter 3. Modifications of the two-phase method have been developed specifically for this purpose, and can be found in some of the references listed at the end of this chapter.

3.   THE PRODUCT FORM OF THE INVERSE

We shall now discuss a variation of the revised simplex procedure known as the *product form of the inverse*. The method is used primarily because it requires less computer storage than does the standard revised simplex procedure. We shall shortly see that it is not necessary to store the matrix B^{-1} at each iteration.

Without loss of generality, suppose that $\{A^{(1)}, \ldots, A^{(m)}\}$ is an admissible basis for a basic feasible solution to problem (1). Moreover, suppose that B^{-1}, where $B = [A^{(1)} \ldots A^{(m)}]$, is known. If it has been determined that $A^{(j)}$ should enter the admissible basis in the place of $A^{(i)}$, then the next iteration of the revised simplex procedure can be carried out once the inverse of $\bar{B} = [A^{(1)} \cdots A^{(i-1)} A^{(j)} A^{(i+1)} \cdots A^{(m)}]$ has been computed. Notice that

$$\begin{aligned} B^{-1} \bar{B} &= [B^{-1} A^{(1)} \cdots B^{-1} A^{(i-1)} \ \ B^{-1} A^{(j)} \ \ B^{-1} A^{(i+1)} \ \cdots \ B^{-1} A^{(m)}] \\ &= [I^{(m)} \cdots I^{(m-1)} \ X_B^{(j)} \ I^{(i+1)} \cdots I^{(m)}] \end{aligned} \tag{10}$$

The matrix on the far right-hand side of equation (10) can be transformed into the identity matrix I_m by the following elementary row operations.

$$[I^{(m)} \dots I^{(m-1)} X_B^{(j)} I^{(i+1)} \dots I^{(m)}] \quad \xrightarrow[\quad (1/x_{ij})R_i \quad]{R_k - (x_{kj}/x_{ij})R_i} \quad I_m \tag{11}$$

where $X_B^{(j)} = [x_{1j} \cdots x_{kj} \cdots x_{ij} \cdots x_{mj}]^T$ and $k \neq i$. It follows that $\bar{B}^{-1}$ can be computed by performing the elementary row operations (11) on B^{-1}. An alternate way of doing this is to first perform the row operations (11) on I_m to get

$$I_m \quad \xrightarrow[\quad (1/x_{ij})R_i \quad]{\substack{R_k - (x_{kj}/x_{ij})R_i \\ (i \neq k)}} \quad \begin{bmatrix} 1 \dots & -x_{1j}/x_{ij} & \dots 0 \\ \vdots & \vdots & \vdots \\ 0 \dots & 1/x_{ij} & \dots 0 \\ \vdots & \vdots & \vdots \\ 0 \dots & -x_{mj}/x_{ij} & \dots 1 \end{bmatrix} = F \tag{12}$$

which is a pivot matrix. Then multiply B^{-1} on the left by F to get $\bar{B}^{-1} = FB^{-1}$. Since F differs from I_m in only one column, it is not necessary to store (or retain) the entire matrix. Only the column $[-x_{1j}/x_{ij} \dots -x_{mj}/x_{ij}]^T$ and the row index i of the pivot element must be retained in order to reconstruct F. After k iterations, the inverse of the basic matrix has the form $F_k F_{k-1} \dots F_1 B^{-1}$, where each F_i is a pivot matrix. This matrix can be constructed from the $k(m+1)$-tuples $(i, [-x_{1j}/x_{ij} \dots -x_{mj}/x_{ij}])$ and the $m \times m$ matrix B. When the initial basis matrix is I_m, the basic matrix after k iterations has the simpler form $F_k F_{k-1} \dots F_1$.

Suppose now that the inverse of the basis matrix after k iterations is $F_k F_{k-1} \cdots F_1$. Let $B^{-1} = F_k F_{k-1} \cdots F_1$. Notice that the calculations

$$x_B = B^{-1}b = F_k(F_{k-1}(\cdots (F_1 b) \cdots))$$

and

$$B^{-1} D^{(j)} = F_k(F_{k-1}(\cdots (F_1 D^{(j)}) \cdots))$$

from the revised simplex method are equivalent to performing the same sequence of row operations on b and $D^{(j)}$ that were required to produce B^{-1}.

This version of the revised simplex procedure is clearly not the best method available for hand calculations. It does offer some advantages when using a high speed computer. A more detailed discussion of this method and its variations can be found in [18] of the references at the end of the chapter. One variation is briefly discussed in the next section.

4. THE ELIMINATION FORM OF THE INVERSE

Up to this point we have always modified the Gauss-Jordan method of elimination to obtain a simplex procedure for solving linear programming problems. However, the numerically superior method of Gaussian elimination can also be adapted to the revised simplex procedure. For large problems in particular, it is often desirable to adapt the use of LU-decompositions to the revised simplex procedure. The result is a technique known as the elimination form of the inverse. The use of the triangular matrices in this process tends to preserve matrix sparsity, which is not normally the case when inverse matrices are used.

Without loss of generality, suppose that

$$B = [A^{(1)} \ldots A^{(i-1)} A^{(i)} A^{(i+1)} \ldots A^{(m)}]$$

is a basis matrix for a basic feasible solution of problem (1). Observe that B^{-1} was used in Steps 1 and 4 of the revised simplex procedure to solve the matrix equations $Bx_B = b$ and $BX_B^{(j)} = D^{(j)}$ for x_B and $X_B^{(j)}$, respectively. If $B = LU$ is an LU-decomposition of B, then the solution of these two equations could be obtained without computing B^{-1}. Moreover, if we let $q^T = c_B^T B^{-1}$ in Step 2, then q is the solution of $U^T L^T q = B^T q = c_B$. Hence, all of the operations in the revised simplex procedure that involve B^{-1} can be adapted to use an LU-decomposition of B. All that is needed is a systematic method of moving from an LU-decomposition of B in one pivot to an LU-decomposition of $\bar{B}$ in the next pivot.

Suppose that it has been determined that the column $A^{(i)}$ will leave and the column $A^{(j)}$, where $j = m+k$, will enter the admissible basis during the next pivot. Rather than replacing $A^{(i)}$ by $A^{(j)}$ in the new admissible basis and pivoting at a_{ij} (to get $I^{(i)}$), we shall write the new admissible basis as $\{A^{(1)}, \ldots, A^{(i-1)}, A^{(i+1)}, \ldots, A^{(m)} A^{(j)}\}$ and let

$$\bar{B} = [A^{(1)} \ldots A^{(i-1)} A^{(i+1)} \ldots A^{(m)} A^{(j)}]$$

The LU-decomposition of B can then be easily upgraded to obtain an LU-decomposition for $\bar{B}$

To see this, first consider the problem of solving $BX_B^{(j)} = D^{(j)}$ for $X_B^{(j)}$, where in this case $D^{(j)} = A^{(m+k)}$. Recall from Chapter 2 that when the LU-decomposition, $B = LU$, was used to solve this matrix equation for $X_B^{(j)}$, it was necessary to first solve $Ly = A^{(m+k)}$ for y, where $y = UX_B^{(j)}$. In particular, $y = L^{-1}A^{(m+k)}$ where $L^{-1} = F_k P_k \cdots F_1 P_1$ is the sequence of row permutations and pivots that must be performed on B to get U. If this sequence of operations was performed on the entire matrix $[B|D]$, then $y = [y_1 \ldots y_m]^T$ is already known. Thus, consider

$$L^{-1}\bar{B} = [L^{-1}A^{(1)} \ldots L^{-1}A^{(i-1)} \; L^{-1}A^{(i+1)} \ldots L^{-1}A^{(m)} \; L^{-1}A^{(m+k)}]$$
$$= [U^{(1)} \ldots U^{(i-1)} \; U^{(i+1)} \ldots U^{(m)} \; y]$$

$$= \begin{bmatrix}
u_{11} \cdots & u_{1\,i-1} & u_{1\,i+1} & \cdots u_{1m} & y_1 \\
\vdots & \vdots & \vdots & \vdots & \vdots \\
0 \cdots & u_{i-1\,i-1} & u_{i-1\,i+1} \cdots & u_{i-1\,m} & y_{i-1} \\
0 \cdots & 0 & u_{i\,i+1} & \cdots u_{im} & y_i \\
0 \cdots & 0 & u_{i+\,i+1} \cdots & u_{i+1\,m} & y_{i+1} \\
0 \cdots & 0 & 0 & & \cdot \\
\vdots & \vdots & \vdots & \vdots & \vdots \\
0 \cdots & 0 & 0 & \cdots u_{mm} & y_m
\end{bmatrix} \qquad (13)$$

Matrix (13) can be transformed into an upper triangular, matrix $\bar{U}$, by performing a sequence of $m - i$ row permutations and $m - i$ row pivots on $L^{-1}\bar{B}$, i.e.,

$$\bar{U} = F_m P_m \cdots F_{i+1} P_{i+1} L^{-1} \bar{B}$$

where each $F_{i+\ell}$ has the term

$$F_{i+\ell} = \begin{bmatrix}
1 & \cdots & 0 & 0 & \cdots & 0 \\
\vdots & & \vdots & \vdots & & \vdots \\
\vdots & & \vdots & \vdots & & \vdots \\
0 & \cdots & 1 & 0 & \cdots & 0 \\
0 & \cdots & f_{i+\ell} & 1 & \cdots & 0 \\
\vdots & & \vdots & \vdots & & \vdots \\
\vdots & & \vdots & \vdots & & \vdots \\
0 & \cdots & 0 & 0 & \cdots & 1
\end{bmatrix}$$

and either $f_{i+\ell} = -(u_{i+\ell\,i+\ell}/u_{i+\ell-1\,i+\ell})$ or its reciprocal. Thus,

$$\bar{B} = L P_{i+1}^{-1} \, F_{i+1}^{-1} \; \cdots \; P_m^{-1} \, F_m^{-1} \, \bar{U}$$

is an LU-decomposition of $\bar{B}$. To summarize, the upper triangular factor, $\bar{U}$, of $\bar{B}$ is found by transforming (13) by a sequence of row permutations and pivots. The factor $\bar{L}$ is simply $\bar{L} = L P_{i+1}^{-1} \, F_{i+1}^{-1} \; \cdots \; P_m^{-1} \, F_m^{-1}$.

EXAMPLE 2   Rework Example 1 using the elimination form of the inverse.

Solution   Again let $B = [A^{(1)} A^{(2)} A^{(6)}] = I$. Then $U = I$ and $L = I$ are the upper and lower triangular factors of B. Clearly, $x_B = [8 \quad 7 \quad 9]^T$, $c_B = [1 \quad 2 \quad 6]^T$ and $c_D = [3 \quad -4 \quad 20]^T$. Observe that the solution of $B^T q = c_B$ is $q = [1 \quad 2 \quad 6]^T$. Thus, $c_B^T B^{-1} D - c_D^T = [0 \quad 11 \quad \cdot]$, where the computation was halted when a positive entry was encountered. It follows that $A^{(4)}$ will enter the admissible basis in the next pivot. Clearly, $B^{-1} A^{(4)} = A^{(4)}$ and the usual ratio tests indicates that $A^{(1)}$ will leave the admissible basis. Thus, the new basis matrix will be

$$\bar{B}^{-1} = [A^{(2)} \, A^{(6)} \, A^{(4)}] = \begin{bmatrix} 0 & 0 & 1 \\ 1 & 0 & 0 \\ 0 & 1 & 1 \end{bmatrix}$$

Since $L = I$, then $L^{-1}\bar{B} = \bar{B}$. If P is the permutation matrix

$$P = \begin{bmatrix} 0 & 1 & 0 \\ 0 & 0 & 1 \\ 1 & 0 & 0 \end{bmatrix}$$

then the LU-decomposition of $\bar{B}$ is $\bar{B} = \bar{L}\bar{U}$, where

$$PL^{-1}\bar{B} = \begin{bmatrix} 1 & 0 & 0 \\ 0 & 1 & 1 \\ 0 & 0 & 1 \end{bmatrix} = \bar{U}$$

and $\bar{L} = (PL^{-1})^{-1} = LP^{-1} = LP^T = P^T$.

Now rename $B = \bar{B}$, $U = \bar{U}$ and $L = \bar{L}$. Notice that $c_B = [2 \quad 6 \quad -4]^T$ and $c_D = [1 \quad 3 \quad 20]^T$. Next consider $B^T q = c_B$. Let $p = L^T q$. Solving $U^T p = c_B$ for p gives $p = [2 \quad 6 \quad -10]^T$. Next solve $L^T q = p$ for q to get $q = [-10 \quad 2 \quad 6]^T$. Then

$$c_B^T B^{-1} D - c_D^T = q^T D - c_D^T = [-10 \quad 2 \quad 6] \begin{bmatrix} 1 & 1 & 0 \\ 0 & 1 & 3 \\ 0 & 0 & 2 \end{bmatrix} - [1 \quad 3 \quad 20]^T$$

$$= [-11 \quad -11 \quad -2]$$

Again we have the conclusion that an optimal solution to the problem is given by $x = [0 \quad 7 \quad 0 \quad 8 \quad 0 \quad 1]^T$.

EXAMPLE 3   Use the elimination form of the inverse to solve the following problem:

$$\text{Minimize } x_1 + 2x_2 + 11x_3 + 3x_4 \text{ subject to}$$

$$x_1 + x_2 + x_3 + 2x_4 = 3$$
$$x_2 + 2x_3 + 4x_4 = 3$$
$$4x_3 + 2x_4 = 4$$
$$\text{all } x_i \geq 0$$

Solution: Let $B = [A^{(1)} \ A^{(2)} \ A^{(3)}]$. Clearly, a LU-decomposition for B is $B = IB$. Also, the matrix equation $Bx_B = [3 \ 3 \ 4]^T$, where $x_B = [x_1 \ x_2 \ x_3]^T$, has the solution $x_B = [1 \ 1 \ 1]^T$. This completes step 1 of the revised simplex procedure. To begin Step 2, we must solve

$$\begin{bmatrix} 1 & 0 & 0 \\ 1 & 1 & 0 \\ 1 & 2 & 4 \end{bmatrix} q = B^T q = c_B = \begin{bmatrix} 1 \\ 2 \\ 11 \end{bmatrix}$$

for q. Easily, $q = [1 \ 1 \ 2]^T$. Thus,

$$c_B^T B^{-1} D - c_D^T = [1 \ 1 \ 2] \begin{bmatrix} 2 \\ 4 \\ 2 \end{bmatrix} - 3 = 10 - 3 = 7$$

and $A^{(4)}$ will enter the admissible basics on the next pivot.

In Step 3, we need to solve

$$\begin{bmatrix} 1 & 1 & 1 \\ 0 & 1 & 2 \\ 0 & 0 & 4 \end{bmatrix} y = By = A^{(4)} = \begin{bmatrix} 2 \\ 4 \\ 2 \end{bmatrix}$$

for y. The solution is clearly $y = [-3/2 \ 3 \ 1/2]^T$. By the standard ratio test min $\{1/3, 1/(1/2)\}$, the column $A^{(2)}$ will leave the admissible basis on the next pivot. The new basic matrix will be $\bar{B} = [A^{(1)} \ A^{(3)} \ A^{(4)}]$. (Notice that $A^{(4)}$ was entered as the far right column rather than as the middle column.) This completes Step 4.

We next need to obtain a LU-decomposition for $\bar{B}$. Consider

$$L^{-1}\bar{B} = [A^{(1)} A^{(3)} A^{(4)}] = \begin{bmatrix} 1 & 1 & 2 \\ 0 & 2 & 4 \\ 0 & 4 & 2 \end{bmatrix}$$

Let

$$P = \begin{bmatrix} 1 & 0 & 0 \\ 0 & 0 & 1 \\ 0 & 1 & 0 \end{bmatrix} \text{ and } F = \begin{bmatrix} 1 & 0 & 0 \\ 0 & 1 & 0 \\ 0 & -1/2 & 1 \end{bmatrix}$$

Then

$$\bar{U} = FPL^{-1}\bar{B} = \begin{bmatrix} 1 & 0 & 0 \\ 0 & 1 & 0 \\ 0 & -1/2 & 1 \end{bmatrix}\begin{bmatrix} 1 & 0 & 0 \\ 0 & 0 & 1 \\ 0 & 1 & 0 \end{bmatrix}\begin{bmatrix} 1 & 1 & 2 \\ 0 & 2 & 4 \\ 0 & 4 & 2 \end{bmatrix} = \begin{bmatrix} 1 & 1 & 2 \\ 0 & 4 & 2 \\ 0 & 0 & 3 \end{bmatrix}$$

and $\bar{L}^{-1} = FPL^{-1} = FP$.

Next rename $B = \bar{B}$, $U = \bar{U}$ and $L = \bar{L}$. To complete Step 5, we need to solve $Bx_B = b = \begin{bmatrix} 3 & 3 & 4 \end{bmatrix}^T$, where $x_B = \begin{bmatrix} x_1 & x_3 & x_4 \end{bmatrix}^T$. Since $LUx_B = Bx_B = b$, then

$$\begin{bmatrix} 1 & 1 & 2 \\ 0 & 4 & 2 \\ 0 & 0 & 3 \end{bmatrix}\begin{bmatrix} x_1 \\ x_3 \\ x_4 \end{bmatrix} = Ux_B = L^{-1}b = FPb = \begin{bmatrix} 3 \\ 4 \\ 1 \end{bmatrix}$$

Using back substitution, $x_B = \begin{bmatrix} x_1 & x_3 & x_4 \end{bmatrix}^T = \begin{bmatrix} 3/2 & 5/6 & 1/3 \end{bmatrix}^T$. This completes Step 5.

Returning to Step 2 we need to solve

$$U^T L^T q = B^T q = c_B = \begin{bmatrix} 1 & 11 & 3 \end{bmatrix}^T$$

for q. Let $p = L^T q$. First solve

$$U^T p = \begin{bmatrix} 1 & 0 & 0 \\ 1 & 4 & 0 \\ 2 & 2 & 3 \end{bmatrix}p = c_B = \begin{bmatrix} 1 \\ 11 \\ 3 \end{bmatrix}$$

for p. The solution is $p = \begin{bmatrix} 1 & 5/2 & -4/3 \end{bmatrix}^T$. To solve $L^T q = p$ for q, notice that $q^T = p^T L^{-1} = p^T FP$. Hence,

$$q^T = p^T L^{-1} = p^T FP = \begin{bmatrix} 1 & 5/2 & -4/3 \end{bmatrix}\begin{bmatrix} 1 & 0 & 0 \\ 0 & 1 & 0 \\ 0 & -1/2 & 1 \end{bmatrix}\begin{bmatrix} 1 & 0 & 0 \\ 0 & 0 & 1 \\ 0 & 1 & 0 \end{bmatrix}$$

$$= \begin{bmatrix} 1 & 19/6 & -4/3 \end{bmatrix}\begin{bmatrix} 1 & 0 & 0 \\ 0 & 0 & 1 \\ 0 & 1 & 0 \end{bmatrix} = \begin{bmatrix} 1 & -4/3 & 19/6 \end{bmatrix}$$

Thus,

$$c_B^T B^{-1} D - c_D^T = \begin{bmatrix} 1 & -4/3 & 19/6 \end{bmatrix}\begin{bmatrix} 1 \\ 1 \\ 0 \end{bmatrix} - 2 = -7/3$$

and the problem has an optimal solution when $x = \begin{bmatrix} 3/2 & 0 & 5/6 & 1/3 \end{bmatrix}^T$.

It is clear from this last example that the primal simplex procedure is better for hand calculations. But for the reasons mentioned earlier, the elimination form of the inverse is often used in large computer systems.

Exercises

5.1   Rework Example 5 of Chapter 3 using the following revised simplex methods:
(a)   The inverse matrix form of the revised simplex procedure.
(b)   The elimination form of the inverse.

5.2   Rework Example 6 of Chapter 3 using the following revised simplex methods:
(a)   The inverse matrix form.
(b)   The elimination form of the inverse.

5.3   Rework the problems in Exercise 3.3 using the inverse matrix version of the revised simplex procedure.

5.4   Rework the problems in Exercise 3.3 using the elimination form of the inverse.

5. THE PRIMAL-DUAL ALGORITHM

In this section we shall describe another method, known as the *primal-dual* algorithm, for solving linear programming problems. Many of the special techniques for solving network flow problems (see Chapter 6) are specializations of this technique.

Again consider the unsymmetric primal problem

$$\text{Minimize } c^T x \text{ subject to } Ax = b \text{ and } x \geq 0 \tag{1}$$

and its dual

$$\text{Maximize } b^T w \text{ subject to } A^T w \leq c \tag{2}$$

By multiplying the individual constraints in (1) by -1 (when necessary), we can assume that $b \geq 0$. Recall (Exercise 4.25) that the complementary slackness conditions for optimal solutions x and w of these respective problems are

$$(c - A^T w)^T x = 0, A^T w \leq c, Ax = b, \text{ and } x \geq 0$$

The primal-dual algorithm begins with a feasible solution of w of (2), i.e., a column matrix w so that $A^T w \leq c$. When $c \geq 0$, such a solution is $w = 0$. When $c \not\geq 0$, a feasible solution w can still be constructed using the technique outlined in Exercise 4.12, or by observation. Once such a w is determined, let

$$\Omega = \{i : (A^T)_{(i)} w - c_i = 0\}$$

where $(A^T)_{(i)}$ denotes row i of A^T. We shall try to construct a solution to

$Ax = b$ and $x \geq 0$, where $x_i = 0$ whenever $i \notin \Omega$. Thus, setting all $x_i = 0$, where $i \notin \Omega$, we seek a solution of

$$\sum_{i \in \Omega} x_i A^{(i)} = b \text{ and } x_i \geq 0 \, (i \in \Omega) \tag{14}$$

Clearly if this can be done, then the complementary slackness conditions for optimality will be satisfied and x solves (1) while w solves (2).

Rather than tackle Problem (14) directly, we shall attempt to determine a solution to the following problem which is known as *restricted primal* problem:

$$\text{Minimize} \sum_{j=1}^{m} y_j \quad \text{subject to}$$

$$\sum_{i \in \Omega} x_i A^{(i)} + y = b \tag{15}$$

$$y \geq 0 \text{ and } x_i \geq 0 \text{ for each } i \in \Omega$$

where y is a column of artificial variables y_j. If in an optimal solution of problem (15), we have that $y = [0 \ldots 0]^T$, i.e., $\sum_{j=1}^{m} y_j = 0$, then clearly we have a solution of (14) and hence (1).

When $y \neq 0$, i.e., $\sum_{j=1}^{m} y_j \neq 0$, consider the dual of (15),

$$\text{Maximize } b^T r \quad \text{subject to}$$

$$(A^T)_{(i)} r \leq 0, i \in \Omega \tag{16}$$

$$r \leq 1$$

Problem (16) is called a *restricted dual* problem. Recall from Chapter 4, that an optimal solution for (16) can be found in the final simplex tableau for Problem (15). Let r be such a solution. We shall try to generate a new feasible solution for (2) using column r.

Consider $\bar{w} = w + \theta r$, where $\theta > 0$. In order for $\bar{w}$ to be a feasible solution for (2) we must have that $A^T \bar{w} \leq c$. Clearly, when $(A^T)_{(i)} r \leq 0$ it follows that

$$(A^T)_{(i)} \bar{w} = (A^T)_{(i)} w + \theta (A^T)_{(i)} r \leq (A^T)_{(i)} w \leq c_i \tag{17}$$

(This includes the case where $i \in \Omega$.) When $(A^T)_{(i)} r > 0$, (17) will again hold provided

$$(A^T)_{(i)} w + \theta (A^T)_{(i)} r \leq c_i$$

i.e.,

$$\theta \leq (c_i - (A^T)_{(i)} w)/(A^T)_{(i)} r$$

Thus, $A^T w \leq c$ provided

$$\theta = \min \{(c_i - (A^T)_{(i)} w)/(A^T)_{(i)} r : (A^T)_{(i)} r > 0\} \qquad (18)$$

Notice that $\theta > 0$. Also if x_j is positive in (15), then complementary slackness implies that $(A^T)_{(j)} r = 0$ in (16). Thus, if the entire process is repeated using $\bar{w}$ as the feasible solution of (2), then $j \in \bar{\Omega} = \{i : (A^T)_{(i)} \bar{w} - c_i = 0\}$ and, hence, x_j will appear again in the next restricted primal problem (15). Thus, the potential exists for a smaller optimal value of the objective function $\sum_{j+1}^{m} y_j$ in the next iteration. The process continues until a restricted primal problem is found that has $\sum_{j=1}^{m} y_j = 0$ for its optimal objective value, or until $\sum_{j=1}^{m} y_j > 0$ and each $(A_T)_{(i)} r \leq 0$, where r is the corresponding optimal solution of (16). Clearly in the first case, problems (1) and (2) are solved. It is left as an exercise (see Exercise 5.5) for the reader to show that in the second case the dual problem (2) is unbounded and, hence, the primal problem (1) is not feasible. The algorithm will terminate in a finite number of steps whenever the restricted primal problems are not degenerate (see Exercise 5.8).

The primal-dual algorithm is summarized in the following steps:

Step 1 Determine a feasible solution for Problem (2), i.e., determine a w so that $A^T w \leq c$.

Step 2 Let $\Omega = \{i : (A^T)_{(i)} w - c_i = 0\}$.

Step 3 Solve the restricted primal problem.

$$\text{Minimize} \sum_{j=1}^{m} y_j \text{ subject to}$$

$$\sum_{i \in \Omega} x_i A^{(i)} + y = b$$

$$y \geq 0 \text{ and } x_i \geq 0 \, (i \in \Omega)$$

If the optimal solution gives a value of 0 for the objective function $\sum_{j=1}^{m} y_j$, i.e., if $y = [0 \ldots 0]^T$, then the complementary slackness conditions hold for both problems (1) and (2), and we can stop. Problem (1) has been solved. Otherwise, go to Step 4.

Step 4 When $y \neq 0$ in Step 3, then determine the corresponding optimal solution r for the restricted dual problem.

$$\text{Maximize } b^T r \text{ subject to}$$

$$(A^T)_{(i)} r \leq 0, i \in \Omega$$

$$r \leq 1$$

The solution can be read from the final simplex tableau in Step 3.

Step 5 Next compute each $(A^T)_{(i)}r$, for $i \notin \Omega$. If some $(A^T)_{(i)}r > 0$, then go to Step 7.

Step 6 If $(A^T)_{(i)}r \leq 0$ for all $i \notin \Omega$, then stop. Problem (2) is unbounded and Problem (1) is not feasible.

Step 7 Compute θ, where

$$\theta = \min \{(c_i - (A^T)_{(i)}w)/(A^T)_{(i)}r : (A^T)_{(i)}r > 0\}$$

Notice that $\theta > 0$.

Step 8 Replace w by $w + \theta r$. (Notice that $w + \theta r$ is also a dual feasible solution of (2).)

Step 9 Return to Step 2.

Notice that in (15) the cost coefficients are all zeros and ones, while in (16) the right-hand side of the constraints are likewise zeros and ones. The simplex problems (15) and (16) are dependent on the original cost coefficients in only an indirect sense, and the solutions of these problems are independent of the original cost coefficients. Simply stated the primal-dual algorithm starts with a feasible solution of (2) then continues to generate feasible solutions of (2) while maintaining complementary slackness until (15) produces a feasible solution of (1).

EXAMPLE 4   Use the primal-dual algorithm to solve the following problem:

$$\text{Minimize } 2x_1 + 8x_2 + 2x_3 \text{ subject to}$$

$$\begin{aligned} -x_1 + 4x_2 - x_3 &= 1 \\ x_1 + 2x_2 - 2x_3 &= 1 \\ \text{all } x_i &\geq 0 \end{aligned} \tag{19}$$

Solution: The dual problem is

$$\text{Minimize } w_1 + w_2 \text{ subject to}$$

$$\begin{aligned} -w_1 + w_2 &\leq 2 \\ 4w_1 + 2w_2 &\leq 8 \\ -w_1 - 2w_2 &\leq 2 \end{aligned} \tag{20}$$

If we let $w_2 = 0$ and then examine the three constraints, we find that $w = [-2 \quad 0]^T$ is a feasible solution (Note: Since $c = [2 \quad 8 \quad 2]^T > 0$, we could have started with the feasible solution $w = [0 \quad 0]^T$ without examining the dual constraints.) For $w = [-2 \quad 0]^T$ we have equality in the first and third dual constraints and, hence, $\Omega = \{1,3\}$.

To begin Step 3 we must solve the restricted primal problem

$$\text{Minimize } y_1 + y_2 \text{ subject to}$$

$$
\begin{aligned}
-x - x_3 + y_1 \quad &= 1 \\
x_1 - 2x_3 \quad + y_2 &= 1 \\
\text{all } x_i \geq 0 \text{ and } y_i &\geq 0
\end{aligned}
$$

As seen in the following tableau an optimal solution for

$$
\begin{bmatrix}
-1 & -1 & 1 & 0 & 0 & 1 \\
1 & -2 & 0 & 1 & 0 & 1 \\
\hline
0 & -3 & 0 & 0 & 1 & 2
\end{bmatrix}
$$

this problem is $x_1 = x_3 = 0$ and $y_1 = y_2 = 1$ (Note: This is only the final tableau of the solution.) Continuing to Step 4, we determine from this tableau the corresponding optimal solution $r = [1 \quad 1]^T$ restricted dual problem (16).

Next compute

$$
(A^T)_{(2)}r = [4 \quad 2]\begin{bmatrix} 1 \\ 1 \end{bmatrix} = 6 \text{ and } \theta = 16/6 = 8/3
$$

Then let $w = [-2 \quad 0]^T + (8/3)[1 \quad 1]^T = [2/3 \quad 8/3]^T$. Substituting these values into the original dual (20), we find that we have equality in only the first and second constraints. Thus for this iteration, $\Omega = \{1,2\}$.

We next must solve the restricted primal problem

Minimize $y_1 + y_2$ subject to

$$
\begin{aligned}
-x_1 + 4x_2 + y_1 \quad &= 1 \\
x_1 + 2x_2 \quad + y_2 &= 1 \\
\text{all } x_i \geq 0 \text{ and all } y_i &\geq 0
\end{aligned}
$$

The solution is generated in the following sequence of tableau.

$$
\begin{bmatrix}
-1 & 4 & 1 & 0 & 0 & 1 \\
1 & 2 & 0 & 1 & 0 & 1 \\
\hline
0 & 0 & -1 & -1 & 1 & 0
\end{bmatrix}
\xrightarrow{R_3 + (R_1 + R_2)}
\begin{bmatrix}
-1 & \boxed{4} & 1 & 0 & 0 & 1 \\
1 & 2 & 0 & 2 & 0 & 1 \\
\hline
0 & 6 & 0 & 0 & 1 & 2
\end{bmatrix}
$$

$$
\xrightarrow[\substack{R_2 - (1/2)R_1 \\ R_3 - (3/2)R_1 \\ (1/4)R_1}]{}
\begin{bmatrix}
-1/4 & 1 & 1/4 & 0 & 0 & 1/4 \\
\boxed{3/2} & 0 & -1/2 & 1 & 0 & 1/2 \\
\hline
3/2 & 0 & -3/2 & 0 & 1 & 1/2
\end{bmatrix}
$$

$$
\xrightarrow[\substack{R_1 + (1/6)R_2 \\ R_3 - R_2 \\ (2/3)R_2}]{}
\begin{bmatrix}
0 & 1 & 1/6 & 1/6 & 0 & 1/3 \\
1 & 0 & -1/3 & 2/3 & 0 & 1/3 \\
\hline
0 & 0 & -1 & -1 & 1 & 0
\end{bmatrix}
$$

An optimal solution is $x_1 = x_2 = 1/3$ and $y_1 = y_2 = 0$. It follows that $x = [1/3 \quad 1/3 \quad 0]^T$ and $w = [2/3 \quad 8/3]^T$ are optimal solutions of Problems (19) and (20), respectively.

Exercises

5.5   Prove that when in the primal-dual algorithm a restricted primal problem is found that has a positive optimal objective value, i.e., $\sum_{j=1}^{m} y_j > 0$, and each $(A^T)_{(i)} r \leq 0$, where r is the corresponding optimal solution of the restricted dual, then the dual problem (2) is unbounded and the primal Problem (1) is not feasible (Hint: Consider $w + \theta r$ as $\theta \to \infty$.)

5.6   Use the primal-dual algorithm to solve the following problems.

(a)   Minimize $x_1 + x_2 + x_3 + x_4 + x_5 + x_6$ subject to

$$
\begin{array}{rcl}
16x_1 + 2x_2 \quad\quad\quad + x_5 + 10x_6 &=& 4 \\
- x_1 \quad\quad\quad + x_4 \quad + \quad x_6 &=& 0 \\
5x_1 + 4x_2 + x_3 \quad\quad\quad &=& 8 \\
\text{all } x_i \geq 0 &&
\end{array}
$$

(b) Minimize $2x_1 + x_2 + 2x_3$ subject to

$$
\begin{array}{rcl}
x_1 \quad - 2x_3 &=& 2 \\
x_1 \quad + x_3 &=& 1 \\
x_2 + x_3 &=& 1 \\
\text{all } x_i \geq 0 &&
\end{array}
$$

(c) Minimize $2x_1 - x_2 + x_3$ subject to

$$
\begin{array}{rcl}
x_1 + x_2 - x_3 &=& 1 \\
x_1 - x_2 + x_3 &=& 1 \\
- x_1 + x_2 + x_3 &=& 1 \\
\text{all } x_i \geq 0 &&
\end{array}
$$

5.7   For a given iteration of the primal-dual algorithm, prove that the columns of an admissible basis of an optimal BFS of (15) along with an additional column of A are admissible in the next iteration.

5.8   If each of the restricted primal problems (15) is nondegenerate, then prove that no admissible basis of a BFS of a restricted primal problem is repeated, and that (1) is solved in a finite number of steps.

6.   PARAMETRIC LINEAR PROGRAMMING

In this section we shall examine the post-optimal analysis of the linear programming problem

$$\text{Minimize } c^T x \text{ subject to } Ax = b \text{ and } x \ge 0 \, (A \in \mathbf{R}_{m \times n})$$

by allowing either the entire column c of cost coefficients to be replaced with $c + \lambda h$, where $h \in \mathbf{R}_{n \times 1}$ and $\lambda \ge 0$, or the entire right hand column b of the constraint to be replaced by $b + \lambda h$, where $h \in \mathbf{R}_{m \times 1}$ and $\lambda \ge 0$. We shall use the matrix notation of the first section of this chapter to describe the process and to carry out the iterations.

As in the first section suppose without loss of generality that $B = [A^{(1)} \ldots A^{(m)}]$ is the basic matrix for an optimal solution. Recall that matrix (5),

$$\left[\begin{array}{ccccc} I_m & B^{-1}D & 0 & B^{-1}b \\ \hline o & c_B^T B^{-1}D - c_D^T & 1 & c_B^T B^{-1}b \end{array}\right] \tag{5}$$

which corresponds to the optimal simplex tableau for the optimal solution, can be obtained from

$$\left[\begin{array}{cccc} B & 0 & 0 & b \\ \hline -c_B^T & -c_D^T & 1 & o \end{array}\right] \tag{4}$$

by multiplying (4) on the left by

$$\left[\begin{array}{cc} B & 0 \\ -c_B^T & 1 \end{array}\right]^{-1} = \left[\begin{array}{cc} B^{-1} & 0 \\ c_B^T B^{-1} & 1 \end{array}\right]$$

or by the usual pivoting method using elementary row operations. Also, matrix (5) represents the system of equations

$$\begin{aligned} x_B + B^{-1}Dx_D &= B^{-1}b \\ (c_B^T B^{-1}D - c_D^T)x_D + z_0 &= c_B^T B^{-1}b \end{aligned} \tag{6}$$

We begin with a study of the case where c is replaced with the parametric cost coefficients $c + \lambda h$.

The parametric cost problem: In this problem we want to find solutions of

$$\text{Minimize } (c^T + \lambda h^T)x \text{ subject to } Ax = b \text{ and } x \ge 0 \tag{21}$$

where $\lambda \ge 0$. We shall assume that an optimal solution is available and that matrix (5) represents the corresponding "optimal" tableau. When changing the cost coefficients from c to $c + \lambda h$, it is clear that only the last row of (5) must be modified. Notice that $c_B^T B^{-1} D - c_D^T$ and $c_B^T B^{-1} b$ must both be replaced by

$$(c_B^T + \lambda h_B^T) B^{-1} D - (c_D^T + \lambda h_D^T) = c_B^T B^{-1} D - c_D^T + \lambda (h_B^T B^{-1} D - h_D^T)$$

and
$$\text{(22)}$$

$$(c_B^T + \lambda h_B^T) B^{-1} b = c_B^T B^{-1} b + \lambda h_B^T B^{-1} b$$

respectively, where h_B and h_D are defined in the same fashion as c_B, c_D, etc., of the first section. To simplify computations and record keeping, we shall follow the similar technique used for artificial variables of using two rows rather than one to store the components of (22). The next to the last row will contain the components $c_B^T B^{-1} D - c_D^T$ and the term $c_B^T B^{-1} b$, while the last row will store the components of $h_B^T B^{-1} D - h_D^T$ and the term $h_B^T B^{-1} b$ (the coefficients of λ). Thus, (4) will be replaced by

$$\begin{bmatrix} B & D & 0 & 0 & b \\ -c_B^T & -c_D^T & 1 & 0 & 0 \\ -h_B^T & -h_D^T & 0 & 1 & 0 \end{bmatrix} \tag{23}$$

Carrying out the row operations required to transform (4) into (5) is equivalent to multiplying (23) on the left by

$$\begin{bmatrix} B & 0 & 0 \\ -c_B^T & 1 & 0 \\ -h_B^T & 0 & 1 \end{bmatrix}^{-1} = \begin{bmatrix} B^{-1} & 0 & 0 \\ c_B^T B^{-1} & 1 & 0 \\ h_B^T B^{-1} & 0 & 1 \end{bmatrix} \tag{24}$$

to get

$$\begin{bmatrix} I_m & B^{-1} D & 0 & 0 & B^{-1} b \\ 0 & c_B^T B^{-1} D - c_D^T & 1 & 0 & c_B^T B^{-1} b \\ 0 & h_B^T B^{-1} D - h_D^T & 0 & 1 & h_B^T B^{-1} b \end{bmatrix} \tag{25}$$

Matrix (25) represents the equations

$$x_B + B^{-1} D x_D = B^{-1} b$$
$$(c_B^T B^{-1} D - c_D^T) x_D + z_0 = c_B^T B^{-1} b$$
$$(h_B^T B^{-1} D - h_D^T) x_D + \bar{z}_0 = h_B^T B^{-1} b$$

where

$$z = z_0 + \bar{z}_0 \lambda = c_B^T B^{-1} b + \lambda h_B^T B^{-1} b - ((c_B^T B^{-1} D - c_D^T) + \lambda (h_B^T B^{-1} D - h_D^T)) x_D$$

But $c_B^T B^{-1} D - c_D^T \leq 0$, since (5) corresponds to an optimal solution of (1), and $\lambda \geq 0$. Thus if

$$h_B^T B^{-1} D - h_D^T \leq 0 \tag{26}$$

then an optimal solution to (21) is $c_B^T B^{-1} b + \lambda h_B^T B^{-1} b$, which is given by $x = [x_B^T \ x_D^T]^T$, where $x_B = B^{-1} b$ and $x_D = [0 \dots 0]^T$. Moreover, this holds for any $\lambda \geq 0$.

Suppose that some of the components of $h_B^T B^{-1} D - h_D^T$ are positive. Then $x = [x_B^T \ x_D^T]^T$, where $x_B = B^{-1} b$ and $x_D = [0 \dots 0]^T$ will still be an optimal solution provided that $0 \leq \lambda \leq \lambda_1$, where

$$\lambda_1 = \min \left\{ -\frac{z_i - c_i}{\bar{z}_i - h_i} : \bar{z}_i - h_i > 0 \right\} = -\frac{z_k - c_k}{\bar{z}_k - h_k}$$

and $\bar{z}_i$ denotes the ith component of $h_B^T B^{-1} D$. When $\lambda = \lambda_1$, then the kth component of $c_B^T B^{-1} D - c_D^T + \lambda(h_B^T B^{-1} D - h_D^T)$ is zero and as in Example 7 of Chapter 3, another optimal solution can be obtained by pivoting, if possible, in the kth column. Once this is done, the process can be repeated with the new tableau, the new optimal solution, and $\lambda \geq \lambda_1$. The iterations continue until (26) holds for some optimal solution of (21), or until a λ_j is generated for which

$$\lambda_j = \min \left\{ -\frac{z_i - c_i}{\bar{z} - h_i} : \bar{z}_i - h_i > 0 \right\} = -\frac{z_\ell - c_\ell}{\bar{z}_\ell - h_\ell}$$

and $x_{i\ell} \leq 0$, for all i. As in Chapter 3, the latter situation indicates that Problem (21) would have an unbounded minimum, whenever $\lambda \varepsilon (\lambda_j, \infty)$. The process is summarized in the following steps:

Step 1 Set $k = 1$ and $\lambda_{k-1} = 0$.

Step 2 Solve problem (21) for $\lambda = 0$ and determine B.

Step 3 Determine h_B and h_D.

Step 4 Compute $h_B^T B^{-1} D - h_D^T$ and $h_B^T B^{-1} b$.

Step 5 If $h_B^T B^{-1} D - h_D^T \leq 0$, then stop. The current basic feasible solution is optimal for all $\lambda \varepsilon [\lambda_{k-1}, \infty)$.

Step 6 If $h_B^T B^{-1} D - h_D^T \nleq 0$, then compute

$$\lambda_k = \min \left\{ -\frac{z_i - c_i}{\bar{z}_i - h_i} : \bar{z}_i - h_i > 0 \right\} = -\frac{z_\ell - c_\ell}{\bar{z}_\ell - h_\ell}$$

The current basis feasible solution is optimal over $[\lambda_{k-1}, \lambda_k]$. This could be a single point.

Step 7 If $h_B^T B^{-1} D - h_D^T \nleq 0$, $\lambda_k = -(z_\ell - c_\ell)/(\bar{z}_\ell - h_\ell)$, and each $x_{i\ell} \leq 0$, then stop. Problem (21) has an unbounded minimum over (λ_k, ∞).

Step 8 If some $x_{i\ell} > 0$, then pivot in the usual fashion in column ℓ to get another basic feasible solution (here $\lambda = \lambda_k$) and corresponding matrix B.

Step 9 Increase k by 1 and return to Step 5.

EXAMPLE 5   Solve the cost parametric problem

Minimize $(-2-\lambda)x_1 + (-1+\lambda)x_2 - 3x_3$ subject to

$$x_1 + 3x_2 + 2x_3 \leq 6$$

$$-x_1 + x_2 - 3x_3 \geq -9$$

$$\text{all } x_i \geq 0$$

Solution: This is a cost parametrization of Example 5 of Chapter 3, where $h = [-1 \quad 1 \quad 0 \quad 0 \quad 0]^T$. (Notice that two slack variables were used.) An optimal solution for $\lambda = 0$ is $x = [6 \quad 0 \quad 0 \quad 0 \quad 3]^T$ and the corresponding tableau (from Chapter 3) is

$$\begin{bmatrix} 1 & 3 & 2 & 1 & 0 & 0 & 6 \\ 0 & -4 & 1 & -1 & 1 & 0 & 3 \\ \hline 0 & -5 & -1 & -2 & 0 & 1 & -12 \end{bmatrix} \tag{27}$$

Thus $B = [A^{(1)} A^{(5)}]$, $h_B = [-1 \quad 0]^T$, and $h_D = [1 \quad 0 \quad 0]^T$.
Hence,

$$h_B^T B^{-1} D - h_D^T = [-1 \quad 0] \begin{bmatrix} 3 & 2 & 1 \\ -4 & 1 & -1 \end{bmatrix} - [1 \quad 0 \quad 0] = [-4 \quad -2 \quad -1]$$

and

$$h_B^T B^{-1} b = [-1 \quad 0] \begin{bmatrix} 6 \\ 3 \end{bmatrix} = -6$$

So matrix (25) becomes

$$\begin{bmatrix} 1 & 3 & 2 & 1 & 0 & 0 & 0 & 6 \\ 0 & -4 & 1 & -1 & 1 & 0 & 0 & 3 \\ \hline 0 & 0 & -5 & -1 & -2 & 1 & 0 & -12 \\ 0 & 0 & -4 & -2 & -1 & 0 & 1 & -6 \end{bmatrix}$$

Since $h_B^T B^{-1} D - h_D^T = [-4 \quad -2 \quad -1] \leq 0$, then follows that $x = [6 \quad 0 \quad 0 \quad 0 \quad 3]^T$ is an optimal solution for all $\lambda \geq 0$.

EXAMPLE 6 Redo the last example using $h = [1 \quad 1 \quad 1 \quad 0 \quad 0]^T$, i.e., using a cost function of $(-2+\lambda)x_1 + (-1+\lambda)x_2 + (-3+\lambda)x_3$.
Solution: Again when $\lambda = 0$, an optimal solution is $x = [6 \quad 0 \quad 0 \quad 0 \quad 3]^T$, (27) is the corresponding tableau, $h_B = [1 \quad 0]^T$, and $h_D = [1 \quad 1 \quad 0]^T$. Thus,

$$h_B^T B^{-1} D - h_D^T = [1 \quad 0] \begin{bmatrix} 3 & 2 & 1 \\ -4 & 1 & -1 \end{bmatrix} - [1 \quad 1 \quad 0] = [2 \quad 1 \quad 1]$$

and

$$h_B^T B^{-1} b = [1 \quad 0] \begin{bmatrix} 6 \\ 3 \end{bmatrix} = 6$$

Thus matrix (25) becomes

$$\begin{bmatrix} 1 & 3 & 2 & 1 & 0 & 0 & 0 & 6 \\ 0 & -4 & ① & -1 & 1 & 0 & 0 & 3 \\ 0 & -5 & -1 & -2 & 0 & 1 & 0 & -12 \\ 0 & 2 & 1 & 1 & 0 & 0 & 1 & 6 \end{bmatrix} \quad \begin{matrix} R_1 - 2R_2 \\ \xrightarrow{R_3 + R_2} \\ R_4 - R_2 \end{matrix}$$

Thus,

$$\lambda_1 = \min\{-(-5)/2, -(-1)/1, -(-2)/1\} = 1$$

and $x = [6 \ \ 0 \ \ 0 \ \ 0 \ \ 3]^T$ solves (21) when $\lambda \in [0,1]$. Next pivot as indicated to get the following matrix:

$$\begin{bmatrix} 1 & 11 & 0 & ③ & -2 & 0 & 0 & 0 \\ 0 & -4 & 1 & -1 & 1 & 0 & 0 & 3 \\ 0 & -9 & 0 & -3 & 1 & 1 & 0 & -9 \\ 0 & 6 & 0 & 2 & -1 & 0 & 1 & 3 \end{bmatrix} \quad \begin{matrix} R_2 + 1/3R_1 \\ R_3 + R_1 \\ \xrightarrow{R_4 - 2/3R_1} \\ 1/3R_1 \end{matrix}$$

Then compute

$$\lambda_2 = \min\{-(-9)/6, -(-3)/2\} = 3/2 \text{ (a tie)}$$

It follows that $x = [0 \ \ 0 \ \ 3 \ \ 0 \ \ 0]^T$ solves Problem (21) whenever $\lambda \in [1, 3/2]$. Again pivot as indicated above to obtain

$$\begin{bmatrix} 1/3 & 11/3 & 0 & 1 & -2/3 & 0 & 0 & 0 \\ 1/3 & -1/3 & 1 & 0 & ①/3 & 0 & 0 & 3 \\ 1 & 2 & 0 & 0 & -1 & 1 & 0 & -9 \\ -2/3 & -4/3 & 0 & 0 & 1/3 & 0 & 1 & 3 \end{bmatrix} \quad \begin{matrix} R_1 + 2R_2 \\ R_3 + 3R_2 \\ \xrightarrow{R_4 - R_2} \\ 3R_2 \end{matrix}$$

Next compute $\lambda_3 = \min\{-(-1)/(1/3)\} = 3$. Thus, $x = [0 \ \ 0 \ \ 3 \ \ 0 \ \ 0]^T$ solves (21) when $\lambda \in [\lambda_2, \lambda_3] = [3/2, 3]$. Pivoting as indicated, we obtain

$$\begin{bmatrix} 1 & 3 & 2 & 1 & 0 & 0 & 0 & 6 \\ 1 & -1 & 3 & 0 & 1 & 0 & 0 & 9 \\ 2 & 1 & 3 & 0 & 0 & 1 & 0 & 0 \\ -1 & -1 & -1 & 0 & 0 & 0 & 1 & 0 \end{bmatrix}$$

Clearly in this case $h_B^T B^{-1} D - h_D^T \le 0$ and, hence, $x = [0 \ \ 0 \ \ 0 \ \ 6 \ \ 9]^T$ solves (21) whenever $\lambda \in [3, \infty]$.

EXAMPLE 7   Solve the following problem

Minimize $-(4 + 10\lambda)x_1 + (3 + \lambda)x_2 - x_3 + (2 + \lambda)x_4 + (5 + 6\lambda)x_5$ subject to

$$-5x_1 \qquad\qquad + x_3 \qquad\qquad\quad + x_5 = 2$$
$$-x_1 + x_2 \qquad\qquad\qquad\qquad + x_5 = 1$$
$$-8x_1 \qquad\qquad\qquad\qquad + x_4 + x_5 = 1$$

$$\text{all } x_i \ge 0$$

Solution: First notice that $h = [-10 \quad 1 \quad 0 \quad 1 \quad 6]^T$. We shall first solve the problem for the special case where $\lambda = 0$. The first primal simplex tableau is found by

$$
\left[\begin{array}{ccccc|c|c}
-5 & 0 & 1 & 0 & 1 & 0 & 2 \\
-1 & 1 & 0 & 0 & 1 & 0 & 1 \\
-8 & 0 & 0 & 1 & 1 & 0 & 1 \\ \hline
4 & -3 & 1 & -2 & -5 & 1 & 0
\end{array}\right]
\quad \xrightarrow{R_4 - R_1 + 3R_2 + 2R_3} \quad
\left[\begin{array}{ccccc|c|c}
-5 & 0 & 1 & 0 & 1 & 0 & 2 \\
-1 & 1 & 0 & 0 & 1 & 0 & 1 \\
-8 & 0 & 0 & 1 & 1 & 0 & 1 \\ \hline
-10 & 0 & 0 & 0 & -1 & 1 & 3
\end{array}\right]
$$

Hence, $x = [0 \quad 1 \quad 2 \quad 1 \quad 0]^T$ solves the problem when $\lambda = 0$. Since $B = [A^{(3)} \quad A^{(2)} \quad A^{(4)}]$, it follows that $h_B = [0 \quad 1 \quad 1]^T$ and $h_D = [-10 \quad 6]^T$. Thus,

$$
h_B^T B^{-1} D - h_D^T = [0 \quad 1 \quad 1]\begin{bmatrix} -5 & 1 \\ -1 & 1 \\ -8 & 1 \end{bmatrix} - [-10 \quad 6] = [1 \quad -4]
$$

and

$$
h_D^T B^{-1} b = [0 \quad 1 \quad 1]\begin{bmatrix} 2 \\ 1 \\ 1 \end{bmatrix} = 2
$$

Thus, (25) becomes

$$
\left[\begin{array}{ccccc|cc|c}
-5 & 0 & 1 & 0 & 1 & 0 & 0 & 2 \\
-1 & 1 & 0 & 0 & 1 & 0 & 0 & 1 \\
-8 & 0 & 0 & 1 & 1 & 0 & 0 & 1 \\ \hline
-10 & 0 & 0 & 0 & -1 & 1 & 0 & 3 \\
1 & 0 & 0 & 0 & -4 & 0 & 1 & 2
\end{array}\right]
$$

Since $\lambda_1 = \min\{-(-10)/1\} = 10$, it follows that $x = [0 \quad 1 \quad 2 \quad 1 \quad 0]^T$ solves the problem whenever $\lambda \in [0, 10]$. Moreover, since $x_{i1} \leq 0$ for all i, it follows that the problem has an unbounded minimum whenever $\lambda > 10$.

The parametrization of b: Let $\lambda \geq 0$ and consider the problem

$$
\text{Minimize } c^T x \text{ subject to } Ax = b + \lambda h \text{ and } x \geq 0 \tag{28}
$$

where $h \in \mathbf{R}_{m \times 1}$. Using an approach analogous to that found in the first section, we shall represent this problem with the following matrix

$$
\left[\begin{array}{c|c|ccc}
B & D & 0 & b & h \\ \hline
-c_B^T & -c_D^T & 1 & 0 & 0
\end{array}\right] \tag{29}
$$

where B is a basis matrix. Multiplication of (29) on the left by

$$
\begin{bmatrix} B^{-1} & 0 \\ c_B^T B^{-1} & 1 \end{bmatrix}
$$

gives

$$
\left[\begin{array}{ccc|c|cc}
I_m & B^{-1}D & \vdots & 0 & \vdots & B^{-1}b & B^{-1}h \\
\hline
0 & c_B^T B^{-1}D - c_D^T & \vdots & 1 & \vdots & c_B^T B^{-1}b & c_B^T B^{-1}h
\end{array}\right] \tag{30}
$$

Matrix (30) can also be obtained from (29) with the usual sequence of row operations. From (30) we have that

$$
\begin{aligned}
x_B + \quad B^{-1}Dx_D &= B^{-1}b + \lambda B^{-1}h \\
(c_B^T B^{-1}D - c_D^T)x_D + z_0 &= c_B^T B^{-1}b + \lambda c_B^T B^{-1}h
\end{aligned} \tag{31}
$$

When $\lambda = 0$ and when B is a basis matrix for an optimal solution of the corresponding problem, then

$$
c_B^T B^{-1}D - c_D^T \leq 0 \text{ and } B^{-1}b \geq 0 \tag{32}
$$

Assuming (32) holds, consider what happens when λ is allowed to become positive.

If $B^{-1}h \geq 0$, then $B^{-1}b + \lambda B^{-1}h \geq 0$ for all $\lambda \geq 0$. It follows that an optimal solution for (28) is $x = [x_B^T \ x_D^T]^T$, where $x_D = [0 \ldots 0]^T$ and $x_B = B^{-1}b + \lambda B^{-1}h$. Moreover, this solution holds for $\lambda \geq 0$.

If $B^{-1}h \not\geq 0$, then it is a simple exercise to show that $B^{-1}b + \lambda B^{-1}h \geq 0$ provided that

$$
\lambda_1 = \min \{ -(B^{-1}b)_i/(B^{-1}h)_i : (B^{-1}h)_i < 0\} = -(B^{-1}b)_\ell/(B^{-1}h)_\ell \tag{33}
$$

where as usual $(B^{-1}b)_i$ denotes the i-th component in the column matrix $B^{-1}b$. In this case $x = [x_B^T \ x_D^T]^T$, where $x_D = [0 \ \ldots \ 0]^T$ and $x_B = B^{-1}b + \lambda B^{-1}h$, solves (28) for any $0 \leq \lambda \leq \lambda_1$. when $\lambda = \lambda_1$, then $(B^{-1}b)_\ell + \lambda_1 (B^{-1}h)_\ell = 0$ and another optimal solution can possibly be obtained by pivoting in row ℓ using the dual simplex procedure. (Notice, in this situation, that $(B^{-1}h)_\ell < 0$). This is possible when row ℓ of $B^{-1}D$ contains a negative entry. In this case pivot using the dual simplex procedure and then repeat the foregoing analysis with the new basis matrix.

If row ℓ of $B^{-1}D$ does not contain a negative entry, the Problem (28) is not feasible when $\lambda > \lambda_1$ (see Exercise 4.20).

EXAMPLE 8   Solve the following problem:

$$
\begin{aligned}
\text{Minimize } & x_1 + 3x_2 + x_3 + 2x_4 + x_5 \text{ subject to} \\
& 3x_2 + x_3 \quad\quad + x_5 = 1 + \lambda \\
& x_1 \quad\quad\quad\quad\quad - x_5 = 3 \\
& \quad -3x_2 \quad + x_4 \quad\quad = 2 - \lambda \\
& \text{all } x_i \geq 0
\end{aligned}
$$

Solution: An initial basis matrix is $B = [A^{(3)} A^{(1)} A^{(4)}] = I_3$. Hence, $c_B = [1 \quad 1 \quad 2]^T$, $c_B^T B^{-1} D - c_D^T = [-6 \quad -1]$, $c_B^T B^{-1} b = 8$, and

$$c_B^T B^{-1} h = [1 \quad 1 \quad 2] I_3 \begin{bmatrix} 1 \\ 0 \\ -1 \end{bmatrix} = -1$$

Matrix (30) then becomes

$$\begin{bmatrix} 0 & 3 & 1 & 0 & 1 & 0 & 1 & 1 \\ 1 & 0 & 0 & 0 & -1 & 0 & 3 & 0 \\ 0 & -3 & 0 & 1 & 0 & 0 & 2 & -1 \\ 0 & -6 & 0 & 0 & -1 & 1 & 8 & -1 \end{bmatrix} \quad \begin{matrix} R_1 + R_3 \\ R_4 - 2R_3 \\ \hline -(1/3)\,R_3 \end{matrix}$$

Clearly when $\lambda = 0$, it follows that $x = [3 \quad 0 \quad 1 \quad 2 \quad 0]^T$ solves (28). Moreover, $\lambda_1 = \min \{-2/(-1)\} = 2$. Hence, $x = [3 \quad 0 \quad 1+\lambda \quad 2-\lambda \quad 0]^T$ solves (28) when $\lambda \in [0,2]$.

    Using the dual simplex procedure and pivoting at the indicated entry of row 2, we have

$$\begin{bmatrix} 0 & 0 & 1 & 1 & 1 & 0 & 3 & 0 \\ 1 & 0 & 0 & 0 & -1 & 0 & 3 & 0 \\ 0 & 1 & 0 & -1/3 & 0 & 0 & -2/3 & 1/3 \\ 0 & 0 & 0 & -2 & -1 & 1 & 4 & 1 \end{bmatrix}$$

It follows that $x = [3 \quad (-2+\lambda)/3 \quad 3 \quad 0 \quad 0]^T$ solves (28) when $\lambda \geq 2$. The optimal value of z_0 is $4 + \lambda$.

EXAMPLE 9   Solve the following problem:

$$\text{Minimize } x_1 + 3x_2 + 6x_3 + 8x_4 + x_5 \text{ subject to}$$
$$x_1 + 3x_2 + 5x_3 + 6x_4 \qquad = 4 - 2\lambda$$
$$- x_2 \qquad - 2x_4 + x_5 = 1 - \lambda$$
$$\text{all } x_i \geq 0$$

Solution: An initial basis matrix for this problem is $B = [A^{(1)} A^{(5)}] = I_2$. Thus, $c_B = [1 \quad 1]^T$, $c_B^T B^{-1} D - c_D^T = [-1 \quad -1 \quad -4]$, $c_B^T b^{-1} = 5$, and

$$c_B^T B^{-1} h = [1 \quad 1] I_2 \begin{bmatrix} -2 \\ -1 \end{bmatrix} = -3$$

Easily, matrix (30) becomes

$$\begin{bmatrix} 1 & 3 & 5 & 6 & 0 & 0 & 4 & -2 \\ 0 & -1 & 0 & -2 & 1 & 0 & 1 & -1 \\ 0 & -1 & -1 & -4 & 0 & 1 & 5 & -3 \end{bmatrix} \quad \begin{matrix} R_1 + 3R_2 \\ R_3 - R_2 \\ \hline -R_2 \end{matrix}$$

Again we are beginning with an optimal solution, $x = [4 \quad 0 \quad 0 \quad 0 \quad 1]^T$, to

Problem (28) when $\lambda = 0$. Computing λ_1, we have

$$\lambda_1 = \min\{-4/(-2), -1/(-1)\} = 1$$

Thus, $x = [4 \quad -2\lambda \ 0 \ 0 \ 0 \ 1-\lambda]^T$ is an optimal solution of Problem (28) when $\lambda \in [0,1]$.

Since $\min\{-1/(-1), -4(-2)\} = 1$, we pivot at the indicated position to obtain

$$\begin{bmatrix} 1 & 0 & 5 & 0 & 3 & \vdots & 0 & \vdots & 7 & -5 \\ 0 & 1 & 0 & 2 & -1 & \vdots & 0 & \vdots & -1 & 1 \\ \hline 0 & 0 & -1 & -2 & -1 & \vdots & 1 & \vdots & 4 & -2 \end{bmatrix}$$

Next compute $\lambda_2 = \min\{-7/(-5)\} = 7/5$. Then $x = [7-5\lambda \quad -1+\lambda \quad 0 \quad 0 \quad 0]^T$ solves Problem (28) when $\lambda \in [1,7/5]$. Since $(c_B^T B^{-1} h)_1 = -5$ and $B^{-1}D = [5 \ 0 \ 3] \geqslant 0$ for this basis matrix, it follows that Problem (28) is not feasible when $\lambda > 7/5$.

7. DEGENERACY AND CYCLING

We shall now briefly examine degeneracy and cycling in the primal simplex procedure. Similar discussions for the other simplex procedures mentioned in this chapter can be found in the references at the end of the chapter.

Recall from Chapter 3 that a degenerate basic feasible solution of Problem (1),

$$\text{Minimize } c^T x \text{ subject to } Ax = b \text{ and } x \geq 0 \tag{1}$$

is a solution $x = [x_B^T \quad x_D^T]$, where $x_D = [0 \ldots 0]^T$ and some component of $x_B = B^{-1}b$ is also zero (here B is a basis matrix). Geometrically, this means that b is in the convex hull of at most $m-1$ of the column matrices $B^{(i)}$ (see Example 6 of Chapter 3). Algebraically, a degenerate solution will result when there is a tie in computing θ in the primal simplex procedure (see Example 5 of Chapter 3). When degenerate solutions are present, the possibility exists that $\theta = 0$ in the next iteration (see Example 6 of Chapter 3). This would lead to a different basis matrix but would not decrease the value of the objective function. It is then possible, but not likely in a practical problem, that the primal simplex procedure could continue indefinitely in a repeating sequence of basis matrices without ever improving the value of the objective function or satisfying the optimally criteria $c_B^T B^{-1} D - c_D^T \leq 0$.

Several techniques have been devised for the purpose of avoiding degeneracy in the primal simplex procedure. The method presented here is Charnes! Additional details and additional techniques can be found in the references. The techniques by Wolfe is well suited for use in computer programs.

Charnes' method is to parametrize, or perturb, b in the following fashion

$$\text{Minimize } c^T x \text{ subject to } Ax = b + \sum_{i=1}^{n} \varepsilon^i A^{(i)} \text{ and } x \geq 0 \qquad (34)$$

where ε is a small, but arbitrary, positive number. It is left as an exercise (Exercise 5.22) for the reader to show that Problem (34) is nondegenerate. Moreover, if B is a basis matrix for an optimal solution $x_B(\varepsilon) = B^{-1}(b + \sum_{i=1}^{n} \varepsilon^i A^i)$ and $x_D(\varepsilon) = [0 \ldots 0]^T$ of (34), then B is also a basis matrix of an optimal solution $x_B = B^{-1}b$ and $x_D = [0 \ldots 0]^T$ of (1) (Exercise 5.23). Assuming that $z_k - c_k > 0$, pivot at x_{lk} where (here $\bar{b} = B^{-1}b$)

$$\theta = \frac{\bar{b}_l + \sum_{j=1}^{n} x_{lj}\varepsilon^j}{x_{lk}} = \min_{1 \leq i \leq m} \left\{ \frac{\bar{b}_i + \sum_{j=1}^{n} x_{ij}\varepsilon^j}{x_{ik}} : x_{ik} > 0 \right\}$$

In each iteration, the value of θ will be positive (Exercise 5.24). Moreover, θ will be uniquely determined in each iteration (Exercise 5.25).

Geometrically we are perturbing b slightly so that it will be in the interior of the cone generated by any basis matrix. The ε is never specified.

Most problems that arise from real world applications do not cycle even if they are degenerate. Hence, techniques for avoiding degeneracy are not in common use. The above method does not readily adapt to the revised simplex procedures discussed in the beginning sections of this chapter. Dantzig, Orden, and Wolfe have developed degeneracy techniques for those procedures. For degeneracy in the dual simplex procedure, the interested reader should see Beale.

8. DECOMPOSITION

Techniques exist for decomposing very large linear programming problems into a sequence of smaller problems. Usually these techniques are applied to those linear programming problems in which A has a special structure. Some of the more common examples are those for which

$$A = \begin{bmatrix} A_1 & 0 & \ldots & 0 \\ 0 & A_2 & \ldots & 0 \\ \vdots & & \vdots & \vdots \\ 0 & 0 & \ldots & A_k \end{bmatrix}$$

These techniques are discussed in many of the references at the end of this chapter.

9. REINVERSION

In large problems round-off error can accumulate and render the expressions B^{-1} and $B^{-1}b$ inaccurate. Hence, in large problems the accuracy of B^{-1} should be checked. This can be done in several ways including
1. Compute $B^{-1}B$
2. Compute $\|b - B^{-1}x_B\|$
3. Compute $\|c_B - w^T B\|$

If the error reaches an unacceptable level, then B^{-1} should be recomputed directly from B before continuing.

Other reinversion techniques exist for other procedures and for other reasons (such as preserving sparsity). Additional information concerning this topic can be found in many of the chapter's references.

Exercises

5.9   Outline the procedure for solving Problem (28).

5.10   Solve the following parametric linear programming problems

$$\text{Minimize } (c^T + \lambda h^T)x \text{ subject to}$$

$$x_1 + 2x_2 - x_3 \qquad = 8$$
$$x_2 + 2x_3 + x_4 \le 6$$
$$\text{all } x_i \ge 0$$

where $c^T = [4 \ 1 \ 2]$ and
(a) $h^T = [1 \ -1 \ 1]^T$
(b) $h^T = [-1 \ -1 \ -1]^T$

5.11   Solve the following problems
(a)   Minimize $(x_1 + 6\lambda) - x_2 + 2(x_3 + \lambda) - 3x_4$ subject to

$$3x_1 \qquad + \quad x_3 \qquad - \quad x_4 = 1$$
$$7x_1 \qquad + x_2 \qquad\qquad - \quad x_4 = 0$$
$$-8x_1 \qquad\qquad\qquad + \quad x_4 \le 1$$
$$\text{all } x_i \ge 0$$

(b)   Maximize $(1 + 2\lambda)x_1 + (1 + 4\lambda)x_2 + (1 + \lambda)x_3$ subject to

$$x_1 \qquad + x_2 \qquad\qquad = 2$$
$$x_2 \qquad + x_3 \ge 1$$
$$x_1 \qquad\qquad + x_3 \le 5$$
$$\text{all } x_i \ge 0$$

5.12   For the specified values of $h = [h_1 \quad h_2]^T$ solve the following:

$$\text{Minimize } 4x_1 + x_2 + 2x_3 \qquad \text{subject to}$$
$$x_1 + 2x_2 - x_3 \qquad = 8 + \lambda h_1$$
$$x_2 + 2x_3 + x_4 \le 6 + \lambda h_2$$
$$\text{all } x_i \ge 0$$

(a) $h = [1 \quad 1]^T$
(b) $h = [-2 \quad -3]^T$

5.13   Solve the following problems
(a)   Minimize $x_1 + x_2 \quad + x_4 + x_5$ subject to

$$2x_1 + x_2 - x_3 + x_4 + 2x_5 = 8 + \lambda$$
$$x_1 + 2x_2 \quad + 6x_4 + x_5 \le 10 - \lambda$$
$$\text{all } x_i \ge 0$$

(b)   Minimize $x_1 + x_2 \quad + x_4 + x_5$ subject to

$$2x_1 + x_2 - x_3 + x_4 + 2x_5 = 8$$
$$x_1 + 2x_2 \quad + 6x_4 + x_5 \le 10 + \lambda$$
$$x_1 + x_2 + x_3 \quad - x_5 \ge 6 + 3\lambda$$
$$\text{all } x_i \ge 0$$

(c)   Minimize $x_1 + x_2 \quad + x_4 + x_5$ subject to

$$2x_1 + x_2 - x_3 + x_4 + 2x_5 = 8$$
$$x_1 + 2x_2 \quad + 6x_4 + x_5 \le 10 - 4\lambda$$
$$x_1 + x_2 + x_3 \quad - x_5 \ge 6 - 5\lambda$$
$$\text{all } x_i \ge 0$$

5.14   Outline a procedure for solving the problem : Minimize $(c^T + \lambda d^T)x$ subject to $Ax = b + \beta h$ and $x \ge 0$ ($\lambda \ge 0$ and $\beta \ge 0$).

5.15   Let λ and β be parameters and solve the following problem

$$\text{Minimize } (4 + 2\lambda)x_1 + (1 - 2\lambda)x_2 + (2 + \lambda)x_3 \text{ subject to}$$
$$x_1 \qquad + 2x_2 \qquad - x_3 \qquad = 8 - 4\beta$$
$$x_2 \qquad + 2x_3 + x_4 \le 6 - 6\beta$$
$$\text{all } x_i \ge 0$$

5.16   Graph the optimal values of the objective function of Example 6.

5.17   Suppose that matrix (5) of Problem (1) has the property that $B^{-1}b \not\ge 0$

and $c_B^T B^{-1} D - c_D^T \not\geq 0$. Try to devise a direct procedure for solving the problem.

5.18 Use the primal simplex procedure to solve the problem

$$\text{Minimize} -(3/4)x_1 + 150x_2 - (1/50)x_3 + 6x_4 \text{ subject to}$$

$$(1/4)x_1 - \quad 60x_2 - (1/25)x_3 + 9x_4 + x_5 \qquad\qquad = 0$$

$$(1/2)x_1 - \quad 90x_2 - (1/50)x_3 + 3x_4 \quad + x_6 \qquad = 0$$

$$x_3 \qquad\qquad\qquad\qquad\quad + x_7 = 0$$

$$\text{all } x_i \geq 0$$

Always select the pivot so that it lies in the column with the smallest index j, where $z_j - c_j > 0$. When the pivots are selected in this manner the basis matrices will cycle and will not lead to a solution. This example of a cycling problem was devised by Beale (for an additional reference, see Gass).
(b) Solve the above problem using the four step procedure outlined in Section 7.

5.19 Use the primal simplex procedure to solve the following problem

$$\text{Minimize} \quad 2x_1 + 3x_2 - \quad x_3 - \quad x_4 \text{ subject to}$$

$$-2x_1 - 9x_2 + \quad x_3 + 9x_4 = 0$$

$$(1/3)x_1 + \quad x_2 - (1/3)x_3 - 2x_4 = 0$$

$$x_1 + \quad x_2 + \quad x_3 + \quad x_4 = 1$$

$$\text{all } x_i \geq 0$$

This problem, which is due to Tucker, can cycle.

5.20 A solution, r and s, of the linear complementarity problem $LCP(q,M)$ is *degenerate* whenever $r_i = s_i = 0$ for some i. In this case, the complementary pivoting may not solve the problem in a finite number of steps. Why?

5.21 Consider the linear complementarity problem $LCP(q,M)$ of Chapter 4. Let $R,S,$ and $Q = [q \mathrel{\vdots} I]$ all belong to $\mathbf{R}_{n \times (n+1)}$. Define the row of a matrix to be *lexico-positive* if and only if its first nonzero entry is positive. Then consider the following matrix version of the $LCP(q,M)$:
Find $R,S \in \mathbf{R}_{n \times (n+1)}$ such that

$$Q = R - MS, \quad R^T S = 0$$

and both R and S are nonnegative in the lexicographic sense. Show that all basic solutions of this last problem are nondegenerate in the lexicographic sense. Moreover, a solution of the latter problem yields a solution of $LCP(q,M)$.

5.22   Prove that there exists an $\delta > 0$ such that problem (34) is nondegenerate whenever $0 < \varepsilon < \delta$. (Hint: Show that the components of $x_B(\varepsilon)$ have the form $\bar{b}_i + \varepsilon^i + \Sigma_{j=m+1}^n x_{ij} \varepsilon^j$.)

5.23   Prove that any basis matrix B for Problem (34) is also a basis matrix for Problem (1).

5.24   Verify that $\theta > 0$ in each iteration of the primal simplex procedure applied to Problem (34).

5.25   In Section 7, verify that θ is uniquely determined in each iteration. (Hint: Notice that

$$\theta = \min_{1 \le i \le n} \left\{ \frac{\bar{b}_i + \varepsilon^i + \sum_{j=m+1}^{n} x_{ij} \varepsilon^j}{x_{ik}} : x_{ik} > 0 \right\}$$

REFERENCES

1.   R. H. Bartels and G. H. Golub, The Simplex Method of Linear Programming Using LU-Decomposition, *Comm. ACM 12*, 5, (1969), pp. 266 – 268.
2.   M. Bazaraa and J. J. Jarvis, *Linear Programming and Network Flows*, John Wiley and Sons, New York (1977).
3.   E. M. L. Beale, *Mathematical Programming in Practice*, John Wiley and Sons, New York (1968).
4.   E. M. L. Beale, Cycling in the Dual Simplex Algorithm, Naval Research Logistics Quarterly 2, no. 4 (1955), pp. 169 – 170.
5.   A. Charnes, Optimality and Degeneracy in Linear Programming, *Econometrica 20* (1952), pp. 160 – 275.
6.   G. B. Dantzig, Computional Algorithm of the Revised Simplex Method, Rand Report RM-1266, The Rand Corporation, Santa Monica, Ca. (1953).
7.   G. B. Dantzig, L. R. Ford, and D. R. Fulkerson, A Primal-Dual Algorithm for Linear Programs, *Linear Inequalities and Related Systems*, ed. H. W. Kuhn and A. W. Tucker, Princeton University Press, Princeton, N.J., (1955), pp. 171 – 181.
8.   G. B. Dantzig and W. Orchard-Hayes, The Product Form for the Inverse in the Simplex Method, *Mathematical Tables and Aids to Computation 8*, 46, (1954), pp. 64 – 67.
9.   G. B. Dantzig, A. Orden, and P. Wolfe, The Generalized Simplex Method for Minimizing a Linear Form Under Linear Inequality Restraints, *Pacific Journal of Mathematics 5*, (1955), pp. 183 – 195.
10. G. B. Dantzig and P. Wolfe, The Decomposition Algorithm for Linear Programs, *Econometrica*, 29(4) (1961), pp. 767 – 778.

11. S. Gass, *Linear Programming: Methods and Applications*, McGraw-Hill, New York (1969).

12. D. Luemberger, *Introduction to Linear and Nonlinear Programming*, Addison-Wesley, Reading, Mass. (1973).

13. K. Murty, *Linear and Combinatorial Programming*, John Wiley and Sons, New York (1976).

14. W. Orchard-Hayes, Background Development and Extensions of the Revised Simplex Method, Rand Report RM 1433, The Rand Corporation, Santa Monica, Ca. (1954).

15. C. H. Papadimitriou and K. Steiglitz, *Combinatorial Optimization*, Prentice-Hall, Englewood Cliffs, N.J. (1982).

16. P. Wolfe, A Technique for Resolving Degeneracy in Linear Programming, Rand Report RM 2995 – PR, The Rand Corporation, Santa Monica, Ca. (1962).

17. G. Zoutendijk, *Mathematical Programming Methods*, North-Holland, New York (1976).

6

Network Programming

Many of the more practical linear programming problems can be expressed as network flow problems. Section 1 discusses several such problems. As we shall see shortly, the constraints of such problems possess an especially simple structure which allows very efficient algorithms to be designed for these problems. In this chapter, we shall develop these algorithms using three approaches. First in the third section, we shall modify the primal simiplex procedure for the uncapacitated minimal cost network flow problem (see Exercises 3.7 and 6.13 for the more general minimal cost network flow problem). Then in Section 4, we shall develop the maximal flow algorithm independently from graph theory. Last, several specializations of the primal-dual algorithm are presented for various types of network flow problems. The interested reader can find several comprehensive developments of these algorithms in the references. We begin in the first section by describing several types of network flow problems. The second section presents some material from graph theory that is used in the remaining sections.

1. LINEAR NETWORK FLOW PROBLEMS

A *network* (or *graph*) is a collection of *nodes* and *arcs*. The nodes frequently represent physical locations of plants, warehouses, distribution centers, retail outlets, pumping stations, relay centers, etc. The nodes are connected by the arcs, which may represent highways, pipelines, power lines, phone lines, rail lines, etc. In a *network flow problem*, a commodity (for example, oil, electricity, messages, etc.) is to flow (or to be shipped, etc.) through the network in such a way that certain supply and demand requirements are satisfied. In this book, the arcs are directional in the sense that the commodity can flow in only one direction along a given arc. For example, if arc $a(1,2)$ connects node 1 to node 2 in the following manner

a(1,2)

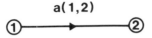

then the commodity can only flow from node 1 to node 2 along $a(1,2)$. When the commodity can flow from either node to the other, then two arcs are required as illustrated below

a(1,2)

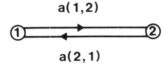

a(2,1)

Suppose now that the network has m nodes and n arcs. Arrange the arcs in a row so that $a(i,j)$ comes before $a(k,\ell)$ if and only if $i \le k$, and $j \le \ell$ whenever $i = k$. Then for each arc $a(i,j)$ form a $m \times 1$ column matrix having a one in the ith position, a negative one in the jth position and zeros elsewhere. Let A be the matrix of all such columns, where the ordering of the columns corresponds to the ordering of the arcs. The matrix A is called the *node-arc incidence* matrix of the network. For example, the node-arc matrix for network (1) below

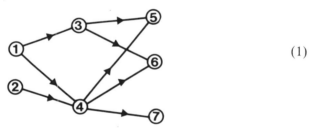

(1)

is

$$a(1,3) \quad a(1,4) \ a(2,4) \ a(3,5) \ a(3,6) \ a(4,5) \ a(4,6) \ a(4,7) \quad \text{arcs/nodes}$$

$$A = \begin{bmatrix} 1 & 1 & 0 & 0 & 0 & 0 & 0 & 0 \\ 0 & 0 & 1 & 0 & 0 & 0 & 0 & 0 \\ -1 & 0 & 0 & 1 & 1 & 0 & 0 & 0 \\ 0 & -1 & -1 & 0 & 0 & 1 & 1 & 1 \\ 0 & 0 & 0 & -1 & 0 & -1 & 0 & 0 \\ 0 & 0 & 0 & 0 & -1 & 0 & -1 & 0 \\ 0 & 0 & 0 & 0 & 0 & 0 & 0 & -1 \end{bmatrix} \begin{matrix} 1 \\ 2 \\ 3 \\ 4 \\ 5 \\ 6 \\ 7 \end{matrix}$$

Let x_{ij} denote the amount of the commodity that flows along $a(i,j)$. Then $x_{ij} \ge 0$. Frequently there is an upper bound u_{ij} on the capacity of $a(i,j)$.

When this happens, $0 \le x_{ij} \le u_{ij}$ and the arc is said to be *capacitated*. Let b_j denote the difference between the amount flowing out of a node and the amount flowing into the node. Then $Ax = b$. When $b_i > 0$, then node i is called a *source* (or *supply*) node. If $b_i < 0$, then node i is called a *sink* (or *demand*) node and $|b_i|$ is called the *demand* at node i. Node i is called an *intermediate* (or *transshipment*) node whenever $b_i = 0$. The cost of shipping one unit of the commodity along arc $a(i,j)$ will be denoted by c_{ij}. The linear network flow problem that seeks to minimize the shipping cost $\Sigma c_{ij} x_{ij}$ of a commodity being shipped along such a network is called the *minimal cost network flow problem*. Moreover, it can be formulated as a linear programming problem as follows

$$\text{Minimize } c^T x \text{ subject to } Ax = b \text{ and } 0 \le x \le u \qquad (3)$$

where $c = [c_{11} \dots c_{1n} \dots c_{m1} \dots c_{mn}]^T$, $x = [x_{11} \dots x_{1n} \dots x_{m1} \dots x_{mn}]^T$,

$u = [u_{11} \dots u_{1n} \dots u_{m1} \dots u_{mn}]^T$ and $0 = [0 \cdots 0 \cdots 0 \cdots 0]^T$

For example, in network (1) let nodes 1 and 2 denote power generating plants which generate $b_1 > 0$ and $b_2 > 0$ megawatts of electricity, respectively. Let nodes 3 and 4 denote relay stations, where $b_3 = b_4 = 0$. Let nodes 5, 6 and 7 denote cities requiring $|b_5|$, $|b_6|$ and $|b_7|$ megawatts of electricity, respectively. Finally, let u_{ij} denote the capacity in megawatts of the power line $a(i,j)$. Then nodes 1 and 2 are source notes, while 3 and 4 are intermediate nodes, and nodes 5, 6, and 7 are sinks. Moreover $b_5 < 0$, $b_6 < 0$, and $b_7 < 0$. By solving (3) we can determine the most economical way to route the electricity through (1).

Notice that the column of the node-arc incidence matrix that corresponds to $a(i,j)$ has the form $I^{(i)} - I^{(j)}$. Also, the sum of the rows of A is a row matrix having all zero entries. Hence, the rows of A are linearly dependent and A has rank of at most $m - 1$. Moreover, when we add the equality constriants of problem (3), we want $\Sigma_{i=1}^m b_i = 0$, i.e., we want the supply at the source nodes to equal the demand at the sink nodes. More often than not we would expect the supply to exceed the demand, i.e., $\Sigma_{i=1}^m b_i > 0$. when this is the case we simply create a fictitious sink node $m+1$ having demand b_{m+1}, where $b_{m+1} + \Sigma_{i=1}^m b_i = 0$, and m fictitious uncapacitated arcs $a(i,m+1)$, where $i = 1, \dots, m$, having shipping cost $c_{i\,m+1} = 0$.

The matrix A in (3) is generally very large and as a result, the standard simplex procedures (i.e., the primal, dual, revised, and primal-dual simplex procedures) are usually not practical to directly apply. However, the special structure of A allows these procedures to be modified in such a way that solutions can be generated quickly and economically. The resulting algorithms, and others, can also be developed independently from graph theory. The present literature contains many volumes devoted entirely to algorithms for solving network flow problems. In this chapter, we shall examine a few of the more common algorithms for solving these problems. Some will be presented for the minimal cost network flow problem (3), and some will be presented for special cases of (3). We shall present algorithms that use a variety of approaches which include the specialization of the primal simplex procedure, the specialization of the primal-dual procedure, and the independent development of the procedure using graph theory. The interested reader can find several comprehensive development of these algorithms in the references.

Some special cases of the minimal cost network follow problem to be considered in this chapter are the following:

1. The transshipment problem: This is an uncapacitated minimal cost network flow problem, i.e., a problem (3) in which each $u_{ij} = \infty$.

2. The transportation problem: In Problem (3) let each b_i be positive or negative, and let each $u_{ij} = \infty$. This means that each node is a source or a sink, and that the arcs all have unlimited capacity. A typical example is network (4):

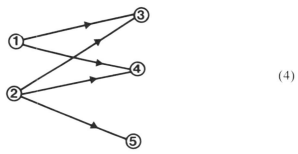

$$(4)$$

where nodes 1 and 2 are sources, and the remaining nodes are sinks. This problem is also known as the *Hitchcock problem.* If some of the u_i are allowed to be finite, the problem is called the *capacitated transportation problem.*

3. The assignment problem: This is a special case of the capacitated transportation problem in which each $|b_i| = 1$ and each $0 \le x_{ij} \le 1$. For example, suppose that three individuals, denoted by nodes 1,2, and 3, are to be assigned to three jobs, denoted by nodes 4,5, and 6. Each individual is a candidate for each job, and each job must be filled. Then network (5),

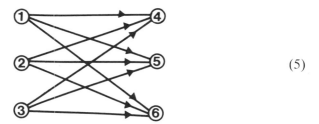

$$(5)$$

represents the problem. The node-arc incidence matrix for (5) is

$$A = \begin{bmatrix} 1 & 1 & 1 & 0 & 0 & 0 & 0 & 0 & 0 \\ 0 & 0 & 0 & 1 & 1 & 1 & 0 & 0 & 0 \\ 0 & 0 & 0 & 0 & 0 & 0 & 1 & 1 & 1 \\ -1 & 0 & 0 & -1 & 0 & 0 & -1 & 0 & 0 \\ 0 & -1 & 0 & 0 & -1 & 0 & 0 & -1 & 0 \\ 0 & 0 & -1 & 0 & 0 & -1 & 0 & 0 & -1 \end{bmatrix}$$

Since nodes 1,2, and 3 are sources and each individual must fill exactly one job, it follows that $b_1 = b_2 = b_3 = 1$. Likewise, nodes 4,5, and 6 are sinks and each must be filled by exactly one person. Thus, $b_4 = b_5 = b_6 = -1$. Let $x_{ij} = 1$ if the ith person fills the jth job, and let $x_{ij} = 0$ otherwise. Let c_{ij} be an estimate of the value of the ith person in the jth job. (For example, c_{ij} might be the time required for person i to complete job j.). The assignment problem is to maximize the total value of the job assignments. The problem can be restated as

$$\text{Maximize } c^T x \text{ subject to } Ax = b \text{ and } 0 \le x \le [1 \dots 1]^T \qquad (6)$$

where $b = [1\ 1\ 1\ -1\ -1\ -1]^T$. Of course (6) is equivalent to

$$\text{Minimize } -c^T x \text{ subject to } Ax = b \text{ and } 0 \le x \le [1 \dots 1]^T$$

4. *The maximal flow problem:* In this problem we wish to determine the maximum amount of flow that can be sent from the sources to the sinks in a given network. For simplicity suppose that node 1 is the only source, node m is the only sink, and nodes 2 through $m-1$ are intermediate notes. Then $b_2 = \cdots = b_{m-1} = 0$. Moreover, the amount of the flow entering the network at the sink must equal the amount of the flow that leaves the network at the sink. To express the maximal flow problem as a minimum cost flow problem simply add an uncapacitated fictitious arc $a(m,1)$ to the network, and assign a shipping cost coefficient of $c_{ij} = 0$ to all but the fictitious arc. Let $u_{m1} = \infty$ and $c_{m1} = 1$. All nodes are then intermediate nodes. The problem can then be restated as

$$\text{Maximize } x_{m1} \text{ (or minimize } -x_{m1}) \text{ subject to}$$

$$\sum_j (x_{ij} - x_{ji}) = 0$$

for each $i = 1, \dots, m$, and each $0 \le x_{ij} \le u_{ij}$.

5. *The shortest path problem:* Let each arc $a(i,j)$ of a network have length c_{ij}, respectively. The shortest path problem is the find a path of minimal length between two given nodes. For example, to find the shortest

path between nodes 1 and m, we can

$$\text{Minimize } \Sigma c_{ij} x_{ij} \text{ subject to}$$

$$\sum_j (x_{1j} - x_{j1}) = 1$$

$$\sum_j (x_{ij} - x_{j1}) = 0, \text{ when } i \text{ is not 1 or } m$$

$$\sum_j (x_{mj} - x_{jm}) = -1$$

and each

$$0 \le x_{ij} \le \infty$$

where one unit of an imaginary commodity flows from node 1 through the network to node m, and x_{ij} denotes the amount of the commodity that flows along arc $a(i,j)$. When $x_{ij} > 0$, then arc $a(i,j)$ is part of the path (of flow) from node 1 to node m. Clearly, the problem will have an optimal solution x for which each $0 \le x_{ij} \le 1$. If some $0 < x_{ij} < 1$, then there exit more than one path of minimal length between nodes 1 and m. It is often useful to omit the last constraint, $\Sigma_j (x_{mj} - x_{jm}) = -1$, which is the negative of the sum of the other constraints and, hence is redundant.

2. SOME BASIC GRAPH THEORY

As mentioned in Section 1, a network $\mathcal{G}$ is a collection of nodes $\mathcal{N}$ and arcs $\mathcal{A}$. A *proper* network has at least two nodes and one arc. Unless otherwise stated, all networks in this book shall be assumed to be proper. A network $\bar{\mathcal{G}}$ having nodes $\bar{\mathcal{N}}$ and arcs $\bar{\mathcal{A}}$ is a *subnetwork* of $\mathcal{G}$ whenever $\bar{\mathcal{N}} \subseteq \mathcal{N}$ and $\bar{\mathcal{A}} \subseteq \mathcal{A}$. When $\bar{\mathcal{N}} = \mathcal{N}$, the subnetwork $\bar{\mathcal{G}}$ *spans* $\mathcal{G}$. A *chain* is a finite sequence $c = \{n_0, a_0, n_1, a_1, \ldots, n_k\}$ of distinct nodes n_i and arcs a_i, where $k \ge 1$ and each a_i is either $a(n_i, n_{i+1})$ or $a(n_{i+1}, n_i)$. The chain is said to *link* nodes n_0 and n_k. Within the chain, each arc of the form $a(n_i, n_{i+1})$ is said to be a *forward* arc, and to have *orientation* of $+1$. Likewise, each arc $a(n_{i+1}, n_i)$ has *orientation* of -1 and is called a *reverse* arc. When each arc in the chain c has orientation of $+1$, then the chain is called a *path* from node n_0 to node n_k. A *cycle* is a finite sequence of nodes and arcs $\{n_0, a_0, \ldots, n_k, a_k, n_0\}$, where $k \ge 1$, $\{n_0, a_0, \ldots, n_k\}$ is a chain, and a_k is either $a(n_k, n_0)$ or $a(n_0, n_k)$. A *circuit* is a cycle in which all arcs have the same orientation. A network is *connected* whenever any two nodes in the network are linked by at least one chain. A network is *acyclic* if it does not possess any cycles. A *tree* is an acyclic network that is connected. A tree that spans the network is called a *spanning* tree. The *degree* of a node is the number of arcs in the network that are connected to the node. A node of degree one is called an *endpoint*.

To illustrate these concepts consider the following network

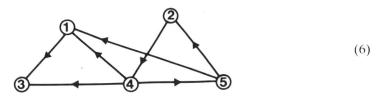

(6)

Consider the following subnetworks of (6):

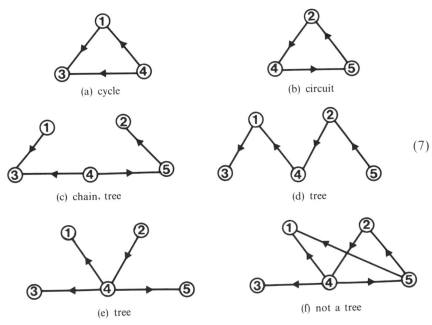

(7)

Networks 7(c), 7(d), 7(e), and 7(f) span network (6). In fact 7(c), 7(d) and 7(e) are spanning trees of (6). Network 7(f) is not a tree since it contains the circuit 7(b). Notice that the trees all have at least two endpoints. Network 7(f), which is not a tree has one endpoint, while (6) has none. All of the networks in 7 are connected. Network 7(a) is a cycle. Network 7(c) is a chain that links nodes 1 and 2, while network 7(d) is a path from node 5 to node 3. The sequence of orientation numbers for 7(d), is $+1, +1, +1, +1$, where the path is considered to be $\{5, a(5,2), 2, a(2,4), 4, a(4,1), 1, a(1,3), 3\}$. If the chain in 7(c) is considered to be $\{1, a(1,3), 3, a(4,3), 4, a(4,5), 5, a(5,2), 2\}$, then the sequence of orientation numbers is $+1, -1, +1, +1$. If the cycle 7(a) is considered to be $\{1, a(1,3), 3, a(4,3), 4, a(4,1), 1\}$, then the corresponding sequence of orientation numbers is $+1, -1, +1$.

The following theorem gives several equivalent definitions of a tree. The proof is lengthy and is not presented in this book. Proofs can be found in some of the references at the end of the chapter.

THEOREM 1   The following statements are equivalent:
(a)   A tree is an acyclic network that is connected.
(b)   A tree is an acyclic network in which the number of nodes exceeds the number of arcs by one.
(c)   A tree is a connected network in which the number of nodes exceeds the number of arcs by one.
(d)   A tree is a network in which a unique chain links each pair of nodes.

As an immediate corollary we have the following result.

COROLLARY 1   Let $\mathcal{N}$ be a network with m nodes. Lt $\mathcal{T}$ be a spanning subnetwork with $m - 1$ arcs. If the columns of the node-arc incidence matrix A that correspond to the arcs of $\mathcal{T}$ are linearly independent, then $\mathcal{T}$ is a tree.
Proof: The result follows immediately from (b) of Theorem 1 and Exercise 6.5.

The relationship between the arcs of a spanning tree of a network and the columns of the node–arc incidence matrix is strengthened further by the next fundamental theorem.

THEOREM 2   Every connected network is spanned by a tree. Moreover, the columns of the node–arc incidence matrix that correspond to the arcs of a tree are linearly independent.
Proof: Since we are only dealing with proper networks, then the network in question must contain at least two nodes and one arc. Let k be the largest positive integer such that the network contains a tree with k nodes. Clearly, $k \geq 2$ since any arc and its two endpoints form a tree. Consider any tree with k nodes. If the tree does not span the network, then there are nodes in the original network that are not in the tree. Since the network is connected, each of these nodes can be linked to any node of the tree. Select any arc that is not an arc of the tree but that links one of these nodes to a node of the tree. Then the tree along with this new arc and node form another tree having $k+1$ nodes. This contradicts the definition of k. Hence, the first tree must span the original network.

To finish the proof, we must show that the columns of the node–arc incidence matrix A that correspond to the arcs of a tree are linearly independent. The result clearly holds for any tree containing exactly two nodes and one arc. Now let p the largest integer for which all trees in the network with p nodes have corresponding linearly independent columns in A. Suppose that $p < m$, where m is the number of nodes in the network. Then there exists a tree having $p+1$ nodes and corresponding linearly dependent columns in A. Select and delete an endpoint and its joining arc from this tree (see Exercise 6.1). The result is a new tree with p nodes, and hence corresponding linearly independent columns in A. Without loss of generality, suppose that columns $A^{(1)}, A^{(2)}, \cdots, A^{(p)}, A^{(p+1)}$ of A correspond to the first tree, while columns $A^{(1)}, A^{(2)}, \cdots, A^{(p)}$ of A correspond to the second tree. Since the first set of columns is linearly dependent, then there exist scalars $\beta_1, \cdots, \beta_{p+1}$, not all zero, such that $0 = \beta_{p+1}A^{(p+1)} + \sum_{i=1}^{p} \beta_i A^{(i)}$, where $0 = [0 \cdots 0]^T$. Notice that the arc that was deleted corresponds to $A^{(p+1)}$. Since this arc corresponds to an endpoint (to simplify the notation, call this node $p+2$) of the first tree, then without loss of generality each $a_{p+2\ i} = 0$ for $i = 1, \cdots, p$, and $|a_{p+2\ p+1}| = 1$. Thus $0 = \beta_{p+1}a_{p+2\ p+1} + \sum_{i=1}^{p} \beta_i a_{p+2\ i} = \beta_{p+1}a_{p+2\ p+1}$ and this implies that $\beta_{p+1} = 0$. But this means that $0 = \sum_{i=1}^{p} \beta_i A^{(i)}$, where not all of the $\beta_i = 0$, and this contradicts the definition of p. It follows that the columns corresponding to the arcs of a tree are linearly independent.

Recall that the rows of the node-arc incidence matrix A are linearly dependent and that A has rank of at most $m-1$. Since each spanning tree has $m-1$ arcs (see Theorem 1), then the next result follows immediately from Theorem 2.

COROLLARY 2   The rank of the node-arc incidence matrix of a connected network possessing m nodes is $m-1$.

The primal and revised simplex procedures in previous chapters were applied to linear programming problems in which the coefficient matrix of the constriants had full row rank. As Corollary 2 points out, this is not the case for the node-arc incidence matrix of Problem (3). We shall follow the procedure of Chapter 4 and add an artificial, x_a, to Problem (3) as follows:

$$\text{Minimize } c^T x \text{ subject to } Ax + x_a I^{(k)} = b$$

$$\text{where } 0 \le x \le u \text{ and } 0 \le x_a \le 0$$

where k is any of the integers $1, \ldots, m$. Notice that x_a is forced to be 0. Hence, problems (3) and (7) are equivalent. The kth node in problem (7) will be called the *root* node. The network associated with (7) will be called a *rooted* network. A subnetwork containing the rooted node will also be called *rooted*. The rooted node is indicated graphically on the network by a *rooted* arc that joins only the rooted node. For example, node 4 is the rooted node in both 8(a) and 8(b).

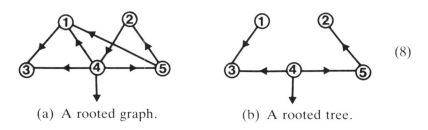

(8)

(a) A rooted graph.   (b) A rooted tree.

3. THE NETWORK SIMPLEX PROCEDURE FOR THE TRANSSHIPMENT PROBLEM

The next result lays the foundation for a special adaption of the simplex procedure to network flow problems.

THEOREM 3   Let A be the node-arc incidence matrix for a connected network having m nodes. Assume that the kth node is rooted. Then B is a basis matrix of $[A|I^{(k)}]$ if and only if $I^{(k)}$ is a column of B and the remaining columns of B correspond to a tree that spans the network.
Proof: Suppose that $I^{(k)}$ is a column of B and that the remaining columns of B correspond the arcs of a spanning tree. Consider node i of the spanning tree. There exists a chain in the tree that links node i to node k. It follows from Exercise 6.6 that the ith column, $I^{(i)}$ of I_m, is in the space spanned by the columns of B. Since this holds for all m nodes, it follows that B spans $\mathbf{R}_{m \times 1}$. Thus, rank $B = m$ and B is a basis matrix for $[A|I^{(k)}]$.

Next assume that B is a basis matrix for $[A|I^{(k)}]$. As in the first part of the proof, it follows from Exercise 6.6 that rank $[A|I^{(k)}] = m$. Since rank $A = m - 1$, then any basis matrix B must contain $I^{(k)}$. It follows from Corollary 1 that the remaining columns of B correspond to the arcs of a spanning tree (see also Theorem 1(b) and Exercise 6.5).

COROLLARY 3   The root arc is in every spanning tree of the rooted graph. Moreover, any node can be selected to be the rooted node.

It follows from the next proposition that the solutions of a system $Bx = b$, where B is a basis matrix of $[A|I^{(k)}]$, can be efficiently computed by properly selecting the equations of $Bx = b$ and applying back substitution.

PROPOSITION 1   Any basis matrix B of $[A|I^{(k)}]$, where A is the node–arc matrix of a connected network having m nodes and a rooted kth node, can be transformed into a lower triangular matrix by suitable permutations of the rows and columns of B.

Proof: Let $\mathcal{T}$ be the tree corresponding to B. Select an endpoint of $\mathcal{T}$ that is not the root node of $\mathcal{T}$.(see Exercise 6.1). The row of B corresponding to this endpoint has only one nonzero entry, the column of B that corresponds to this endpoint has only two nonzero entries. Permute this row with the first row of B, and permute this column with the first column of B. The result is a matrix of the form

$$\begin{bmatrix} \pm 1 & 0 \\ \pm I^{(j)} & B' \end{bmatrix} \tag{9}$$

Deleting the first row and column of matrix (9) leaves a collect of linearly independent columns which correspond to the tree, $\mathcal{T}'$, obtained by deleting the selected endpoint and joining arc from $\mathcal{T}$. This tree is also rooted and must contain an endpoint different from the root. Thus, the rows and columns of (9) can be permuted to obtain

$$\begin{bmatrix} \pm 1 & 0 & 0 \\ \hline \pm I^{(p)} & \pm 1 & 0 \\ & \pm I^{(\ell)} & B'' \end{bmatrix}$$

If this procedure is repeated $m - 1$ times, the result will be a lower triangular matrix.

EXAMPLE 1   If B represents the rooted tree 8(b), then

$$
B = \begin{array}{c}
\begin{array}{ccccc}
a(1,3) & a(4,3) & a(4,5) & a(5,2) & a(r)
\end{array} \\
\begin{bmatrix}
1 & 0 & 0 & 0 & 0 \\
0 & 0 & 0 & -1 & 0 \\
-1 & -1 & 0 & 0 & 0 \\
0 & 1 & 1 & 0 & 1 \\
0 & 0 & -1 & 1 & 0
\end{bmatrix}
\end{array}
\begin{array}{c}
\text{arcs/nodes} \\
1 \\
2 \\
3 \\
4 \\
5
\end{array}
$$

where $a(r)$ denotes the rooted arc. Notice in tree 8(b) that node 1 is an endpoint (different from the rooted node). The procedure outlined in Proposition 1 for transforming B into a lower triangular matrix can be carried out as follows:

$$\begin{bmatrix} 1 & 0 & 0 & 0 & 0 \\ 0 & 0 & 0 & -1 & 0 \\ -1 & -1 & 0 & 0 & 0 \\ 0 & 1 & 1 & 0 & 1 \\ 0 & 0 & -1 & 1 & 0 \end{bmatrix} \quad \text{permute}$$

$$\begin{bmatrix} 1 & 0 & 0 & 0 & 0 \\ -1 & -1 & 0 & 0 & 0 \\ 0 & 0 & 0 & -1 & 0 \\ 0 & 1 & 1 & 0 & 1 \\ 0 & 0 & -1 & 1 & 0 \end{bmatrix}$$

permute

$$\begin{bmatrix} 1 & 0 & 0 & 0 & 0 \\ -1 & -1 & 0 & 0 & 0 \\ 0 & 0 & -1 & 0 & 0 \\ 0 & 1 & 0 & 1 & 1 \\ 0 & 0 & 1 & -1 & 0 \end{bmatrix}$$

permute

$$\begin{bmatrix} 1 & 0 & 0 & 0 & 0 \\ -1 & -1 & 0 & 0 & 0 \\ 0 & 0 & -1 & 0 & 0 \\ 0 & 0 & 1 & -1 & 0 \\ 0 & 1 & 0 & 1 & 1 \end{bmatrix}$$

To solve $Bx = b$, we can use

$$\begin{bmatrix} 1 & 0 & 0 & 0 & 0 \\ -1 & -1 & 0 & 0 & 0 \\ 0 & 0 & -1 & 0 & 0 \\ 0 & 0 & 1 & -1 & 0 \\ 0 & 1 & 0 & 1 & 1 \end{bmatrix} \begin{bmatrix} x_{13} \\ x_{43} \\ x_{52} \\ x_{45} \\ x_a \end{bmatrix} = \begin{bmatrix} b_1 \\ b_3 \\ b_2 \\ b_5 \\ b_4 \end{bmatrix}$$

to obtain

$$\begin{aligned} x_{13} &= b_1 \\ -x_{13} - x_{43} &= b_3 \\ -x_{52} &= b_2 \\ x_{52} - x_{45} &= b_5 \\ x_{43} + x_{45} + x_a &= b_4 \end{aligned} \qquad (10)$$

If $b = [12 \; 18 \; -20 \; -10 \; 0]^T$, then the system of equations (10) gives the basic solution $x_{13} = 12$, $x_{43} = 8$, $x_{52} = -18$, $x_{45} = -18$, and all other $x_{ij} = 0$. Notice that this is not a feasible solution. This solution can also be obtained directly from 8(b) using the fact that $A_{(k)}x = b_k$ at the kth node, i.e., the flow out of node k minus the flow into the node is the demand at the node. In the rooted tree 8(b) start at node 1. All 12 units of the supply at node 1 must flow to node 3. Thus, $x_{13} = 12$. Node 3 has a demand of $|-20|$. Thus, node 3 needs all of the flow from node 1 plust 8 units from node 4. Hence, $x_{43} = 8$. Next go to node 2. In tree 8(b) nothing flows from node 2 which is a supply node with 18 units. Thus the amount flowing from node 2 minus the amount flowing into node 2, x_{52}, from node 5 must equal 18. Thus, $x_{52} = -18$. Finally, $x_{45} - x_{52} = 0$ which implies that $x_{45} = -18$. All other $x_{ij} = 0$. This is a basic solution that is not feasible. In summary, 8(b) has the following appearance:

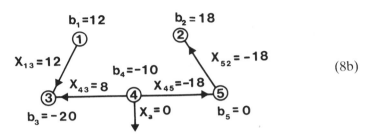

$$(8b)$$

EXAMPLE 2   For network (1), let $b = [25 \ 15 \ 0 \ 0 \ -10 \ -20 \ -10]^T$ and find a BFS.

Solution: Consider network (11) which spans network (1).

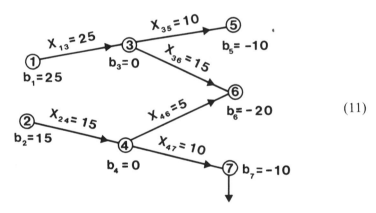

$$(11)$$

Since network (11) is connected and has seven nodes and 6 arcs, it is a tree by Theorem 1. The supply node 1 must ship all of its units to the intermediate node 3, i.e., $x_{13} = 25$. Since node 3 is the only node in this tree to ship to sink node 5, then it must follow that $x_{35} = 10$. Moreover, node 3 ships all of the remaining units to node 6. Thus, $x_{36} = 15$ and all but 5 units of the demand at sink 6 has been satisfied. Next consider node 2. Ship $x_{24} = 15$ units to node 4. Node 4 must ship $x_{46} = 5$ units to sink 6 and $x_{47} = 10$ units to sink 7. Let $x_{ij} = 0$ for any arc not present in tree (11). We then have a BFS of $Ax = b$ for network (1).

Notice that we have always worked from the endpoints of the tree to the rooted node when computing the basic solution in each of these examples.

Now let $A^{(p)}$ be the column of a basis B that corresponds to the arc $a(i,j)$ of a spanning tree of a rooted network that is connected (Presently $a(i,j)$ is not the rooted arc). Then

$$z_{ij} - c_{ij} = c_B^T B^{-1} A^{(p)} - c_{ij} = w^T A^{(p)} - c_{ij} \tag{12}$$

where $w = [w_1 \ldots w_m]^T = c_B^T B^{-1}$ is the column of the dual variables. Since the only nonzero entries of $A^{(p)}$ are 1 in the i-th position and -1 in the jth position, it follows that equation (12) can be simplified to be

$$z_{ij} - c_{ij} = w_i - w_j - c_{ij} \tag{13}$$

Recall that for the basic columns of A, equation (13) must have the form

$$0 = z_{ij} - c_{ij} = w_i - w_j - c_{ij} \tag{14}$$

An optimal solution is obtained when all

$$w_i - w_j - c_{ij} = z_{ij} - c_{ij} \leq 0 \tag{15}$$

Now consider the basic column $A^{(r)}$ that corresponds to the rooted arc. Since $A^{(r)} = I^{(k)}$, equations (12) and (14) imply that

$$w_k = 0 \tag{16}$$

EXAMPLE 3   Assign the following cost coefficients to network (1)

$$
\begin{array}{lll}
c_{13} = 2 & c_{35} = 3 & c_{46} = 4 \\
c_{14} = 1 & c_{36} = 2 & c_{47} = 1 \\
c_{24} = 2 & c_{45} = 4 &
\end{array}
$$

and compute each $z_{ij} - c_{ij}$ for the BFS of Example 2.
Solution: To keep tract of all the pertinent information relating to such a problem, we shall attach a label (b_i, w_i) to each node, and a label $(c_{ij}, x_{ij}, z_{ij} - c_{ij})$ to each arc of a spanning tree corresponding to a BFS of the uncapacitated minimal cost network flow problem associated with network (1). Thus, for tree (11) we currently have

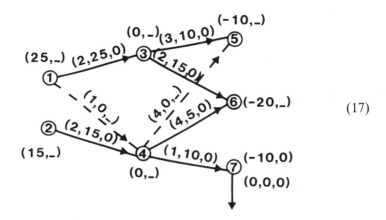

$$(17)$$

Starting at the rooted node with $w_7 = 0$, we have that $0 = w_4 - w_7 - c_{47} = w_4 - 0 - 1$. Thus, $w_4 = 1$. Then $0 = w_2 - w_4 - c_{24} = w_2 - 1 - 2$. Hence $w_2 = 3$. Also, $0 = w_4 - w_6 - c_{46} = 1 - w_6 - 4$, which implies that $w_6 = -3$. Continuing in this fashion, we have that $w_3 = -1$, $w_5 = -4$, and $w_1 = 1$. Clearly, $z_{ij} - c_{ij} = 0$ for all the basic arcs.

Next compute $z_{ij} - c_{ij}$ for the nonbasic arcs. First, $z_{14} - c_{14} = w_1 - w_4 - c_{14} = 1 - 1 - 1 = -1$. Second, $z_{45} - c_{45} = w_4 - w_5 - c_{45} = 1 - (-4) - 4 = 1$. Thus the labels in (17) can be completed to give

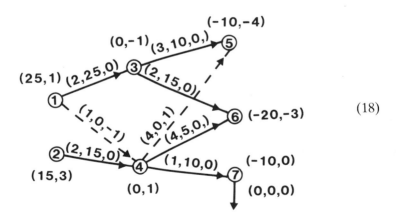

$$(18)$$

Since $z_{44} - c_{45} = 1$ we shall attempt to find a basic feasible solution containing x_{45} as a basic variable. Let $x_{45} = \theta$, where $\theta \geq 0$. Consider network (19). Since $x_{45} = \theta$, then $x_{35} = 10 - \theta$. Also,

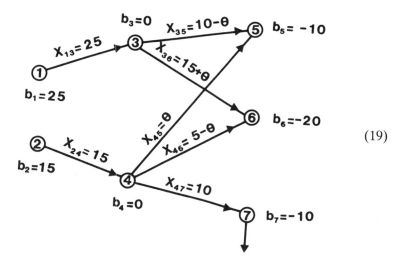

(19)

$x_{13} = 25$ and $x_{24} = 15$. Since $x_{13} = x_{35} + x_{36}$, it follows that $x_{36} = 15 + \theta$. Clearly node 4 must satisfy the entire demand at sink 7. Hence, $x_{47} = 10$. Finally, $x_{46} = 5 - \theta$, since $x_{24} = x_{45} + x_{46} + x_{47}$.

Recall that in the primal simplex procedure, θ is increased until some variable in (19) becomes zero. Here, this happens whenever $x_{45} = \theta = 5$ and $x_{46} = 0$. Notice that the corresponding network (20) is a tree by Theorem 1.

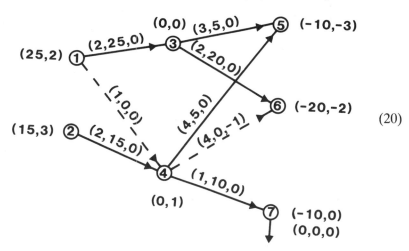

(20)

The labels in (20) were computed in the same manner as were the labels of (18). Since each $z_{ij} - c_{ij} \leq 0$ in (20), then the BFS $x_{13} = 25$, $x_{14} = 0$, $x_{24} = 15$, $x_{35} = 5$, $x_{36} = 20$, $x_{45} = 5$, $x_{46} = 0$, and $x_{47} = 10$ solves the transshipment problem corresponding to network (1) and the given values for c_{ij} and b_i.

The network simplex procedure can be adapted to the minimal cost network flow problem (3). However, space does not permit that presentation here (see Exercise 6.13). The interested reader should consult the references at the end of the chapter.

When the components of b are all integers, then the special composition of the node-arc incidence matrix gives rise to integral valued basic feasible solutions to the transshipment problem. This is also true for the minimal cost network flow problem provided each u_i is an integer.

Exercises

6.1   Prove that every tree having two or more nodes has two or more endpoints.

6.2   Prove that the sum of the degrees of the nodes of a network is twice the number of arcs of the network.

6.3   Prove that in a tree there is a unique chain linking any two nodes.

6.4   Prove that when an endpoint of a tree having at least two endpoints and the arc joining the endpoint are removed, the resulting network is still a tree.

6.5   Show that the columns of the node-arc incidence matrix A which correspond to the arcs of a cycle of a network are linearly dependent. (Hint, consider a linear combination of such columns, where the scalars are the orientation numbers of the corresponding arcs).

6.6   Let $c = \{n_0, a_0, n_1, a_1, \ldots, a_{k-1}, n_k\}$ be a chain having $\alpha_0, \alpha_1, \ldots, \alpha_{k-1}$ as its sequence of orientation numbers. Show that $I^{(n_k)} - I^{(n_0)} = -\sum_{i=0}^{k-1} \alpha_i B^{(i)}$, where each $B^{(i)}$ is the column of the node-arc incidence matrix A that corresponds to arc a_i.

6.7   Outline the network simplex procedure for the transshipment problem.

6.8   Solve the transshipment problem when

$$b = [40 \ 30 \ 0 \ 0 \ 0 \ -50 \ -20]^T, \ c = [1 \ 3 \ 2 \ 1 \ 1 \ 2 \ 1 \ 1 \ 1 \ 2 \ 1]^T \text{ and}$$

$$A = \begin{bmatrix} 1 & 1 & 0 & 0 & 0 & 0 & 0 & 0 & 0 & 0 & 0 \\ 0 & 0 & 1 & 1 & 0 & 0 & 0 & 0 & 0 & 0 & 0 \\ -1 & 0 & 0 & 0 & 1 & 1 & 0 & 0 & 0 & 0 & 0 \\ 0 & -1 & 0 & 0 & -1 & 0 & 1 & 1 & 1 & -1 & 0 \\ 0 & 0 & -1 & 0 & 0 & 0 & -1 & 0 & 0 & 1 & 1 \\ 0 & 0 & 0 & -1 & 0 & -1 & 0 & -1 & 0 & 0 & 0 \\ 0 & 0 & 0 & 0 & 0 & 0 & 0 & 0 & -1 & 0 & -1 \end{bmatrix}$$

6.9   Solve the following transshipment problems
(a) $b = [50\ 0\ 0\ 0\ 0\ 0\ -50]^T$, each $c_{ij} = 1$, and

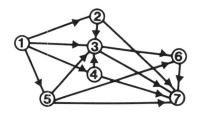

(b) $b = [10\ 20\ 30\ -15\ -15\ -15\ -5]^T$, $c_{14} = 2$, $c_{15} = 3$, $c_{25} = 1$, $c_{24} = 2$, $c_{26} = 3$, $c_{27} = 4$, $c_{36} = 1$, $c_{37} = 1$, and

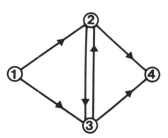

(c) $b = [40\ 0\ 0\ -30]^T$, $c_{12} = 3$, $c_{13} = 4$, $c_{23} = -1$, $c_{24} = 1$, $c_{32} = 2$, $c_{34} = 1$, and

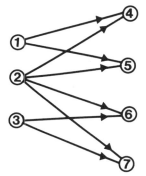

(d) $b = [10\ 50\ 40\ -30\ -40\ -30]^T$, $c_{14} = c_{25} = c_{36} = 1$, all other $c_{ij} = 2$.

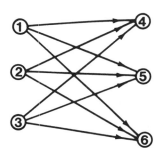

6.10 (a) Show that the transportation problem is never unbounded.
(b) Show that the transportation problem is always feasible. (Hint: For each source i and sink j define $x_{ij} = -(b_ib_j)/d$, where d is the total demand.)

6.11 (The *Northwest Corner Rule* for generating a BFS.)   Suppose that an arc joins each source node of a transportation problem to each sink node. Arrange the data for the problem in the following format

		Sinks		Supply
	$m+1$	$m+2$ $\cdots$	$m+n$	
1	$c_{1\,m+1}$	$c_{1\,m+2}$	$c_{1\,m+n}$	b_1
Sources 2	$c_{2\,m+1}$	$c_{2\,m+2}$	$c_{2\,m+n}$	b_2
$\vdots$				$\vdots$
m	$c_{m\,m+1}$	$c_{m\,m+2}$	$c_{m\,m+n}$	b_m
	$|b_{m+1}|$	$|b_{m+2}|$ $\cdots$	$|b_{m+n}|$	

Demand

(i) Select the arc corresponding to the northwest corner of this tableau, and ship the maximum amount, x_{ij}, by this arc. This either exhausts the supply at the corresponding source node or it completely satisfies the demand at the corresponding sink.

(ii) Reduce the supply at the source node used. Reduce the demand at the sink node. Delete the row or the column that corresponds to the exhausted supply or satisfied demand. If both the supply is exhausted and the demand is satisfied, delete either the row or the column (this is the degenerate case). If a submatrix remains, then return to (i). Otherwise, assign all other $x_{ij}=0$, and stop.

Use this procedure to generate a BFS for the transportation that corresponds to the following tableau:

		Sinks			
		1	2	3	
	1	2	1	4	15
Sources	2	3	2	2	15
	3	4	1	5	15
		10	10	25	

Then solve the problem.

6.12 The following procedure can be used to obtain a BFS of the transshipment problem:

(i) Add an artificial node to the network.

(ii) Add an uncapacitated arc from each source node to this new artificial node. Set the corresponding cost equal to 0, and the corresponding flow equal to the supply at the source.

(iii) Add an uncapacitated arc from the artificial node to each sink node. Set the corresponding cost equal to M, where M is an unspecified large number. Set the flow equal to the demand at the source.

(iv) Apply the network simplex procedure to the resulting network, using the arcs from (ii), (iii), and the necessary arcs (with flow of zero) from the original network to determine a spanning tree of the modified network.

(v) Drive out the arcs from (iii).

Use this procedure to determine a BFS for the transshipment problem 8(a), where $b = [12 \ 18 \ -20 \ -10 \ 0]^T$, $c_{13} = c_{24} = 1$, $c_{41} = c_{43} = c_{52} = 2$, and $c_{45} = c_{51} = -1$. Then solve the problem.

6.13 Recall that the sufficiency conditions for an optimal solution of the minimal cost network flow problem (3) are

$$(i) \ z_{ij} - c_{ij} \leq 0, \text{ whenever } x_{ij} = 0$$

and

$$(ii) \ z_{ij} - c_{ij} \geq 0, \text{ whenever } x_{ij} = u_{ij}$$

(See Exercise 3.7). Use these conditions and the following procedure to solve the minimal cost network flow problems that follow the procedure. *Procedure:*

Step 1 Determine an initial BFS to the problem and identify the corresponding spanning tree.

Step 2 If conditions (i) and (ii) hold, then the problem is solved. Otherwise select a nonbasic variable, x_{ij}, that does not satisfy either (i) or (ii).

Step 3 Join the arc $a(i,j)$ to the tree that corresponds to the BFS. This forms a spanning network possessing a cycle containing $a(i,j)$. For example, see (21).

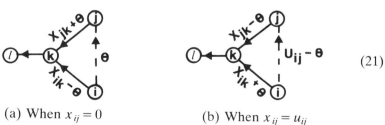

(a) When $x_{ij} = 0$ (b) When $x_{ij} = u_{ij}$

(21)

If $x_{ij} = 0$, then increase x_{ij} to $x_{ij} + \theta$, where $\theta \geq 0$ (see 21(a)). If $x_{ij} = u_{ij}$, then decrease x_{ij} to $x_{ij} - \theta$, where $\theta \geq 0$ (see 21(b)). In either case, increase θ until an arc in the cycle becomes saturated, i.e., the corresponding variable reaches its upper bound, or until the flow in one of the arcs of the cycle diminishes to 0. If a saturated or flowless arc corresponds to a basic variable, then we have a new BFS. Otherwise, the nonbasic variable, x_{ij}, takes on its other bound and the values of the basic variables change. In either case return to Step 2.

(a)

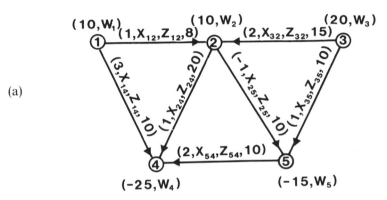

Here the labels on the arcs are of the form $(c_{ij}, x_{ij}, z_{ij}, u_{ij})$.

(b)

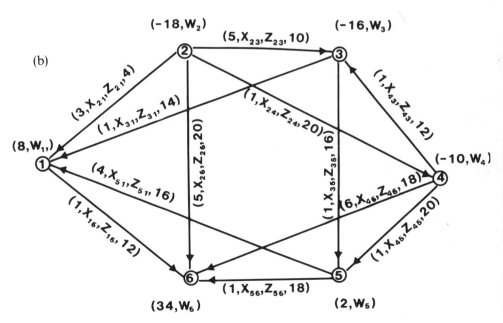

6.14   Solve the maximal flow problem for a network of five nodes and seven arcs having capacities of $u_{12} = 10$, $u_{13} = 4$, $u_{24} = u_{25} = u_{35} = 5$, and $u_{43} = u_{45} = 5$, where node 1 is the only source and node 5 is the only sink.

6.15   Show that the network flow problem

$$\text{Minimize } c^T x \text{ subject to } Ax = b \text{ and } 0 \leq \ell \leq x \leq u$$

can be converted into form (3) by the transformation $\bar{x} = x - \ell$.

6.16   Solve the assignment problem that corresponds to network (5) and the following values of the components of c: $c_{14} = 1$, $c_{15} = 4$, $c_{16} = 2$, $c_{24} = 3$, $c_{25} = 4$, $c_{26} = 1$, $c_{34} = 3$, $c_{35} = 2$, and $c_{36} = 1$.

4.   THE MAXIMAL FLOW PROBLEM

As indicated in the last sections, the maximal flow problem can be solved as a minimal cost network flow problem by using the network simplex procedure. However, the highly specialized nature of this important problem permits the development of more efficient procedures for resolving it. One such technique will be presented in this section.

   We shall assume that the problem has a unique source, node 1, and a unique sink, node m. The remaining nodes, 2, 3,..., $m-1$, are all intermediate nodes. The arcs are usually capacitated. Recall that a commodity enters the network at the source node, then flows through the network, and finally leaves at the sink. Multiple sources are handled by creating an artificial source, and artificial uncapacitated arcs which link the artificial source to the real sources. Multiple sinks are handled in a similar fashion. In order to simplify our notation, we shall assume that each arc $a(i,j)$ exists, and require that $u_{ij} = 0$ for those arcs which do not actually exist. The problem can then be expressed as:

$$\text{Maximize } f \text{ subject to}$$

$$\sum_{i=1}^{m} (x_{1i} - x_{i1}) = f$$

$$\sum_{i=1}^{m} (x_{ji} - x_{ij}) = 0, \text{ where } j \notin \{1, m\} \tag{22}$$

$$\sum_{i=1}^{m} (x_{mi} - x_{im}) = -f$$

$$\text{and each } 0 \leq x_{ij} \leq u_{ij}$$

where f is the amount of flow in the network. The constraints represent the conservation of flow at each node, i.e., any amount flowing into a node must flow out as well. The source is considered to have an unlimited supply of the commodity available. Notice that the trival flow where all $x_{ij} = 0$ is always a feasible solution of the problem.

Let S be a set of nodes from the network that contains the source, but not the sink. Let $\bar{S}$ be the set of nodes from the network that are not in S. Clearly, the sink is in $\bar{S}$. Let $(S,\bar{S}) = \{a(i,j): i \in S \text{ and } j \in \bar{S}\}$. Then $(S,\bar{S})$ is called a *cut* (or *cut-set*) that separates the source and the sink. The *capacity* of a cut $(S,\bar{S})$, denoted $\text{cap}(S,\bar{S})$, is defined to be the sum of the capacities of the arcs belonging to the cut. For example, in network (23) let $S = \{1,2,4\}$ and $\bar{S} = \{3,5\}$.

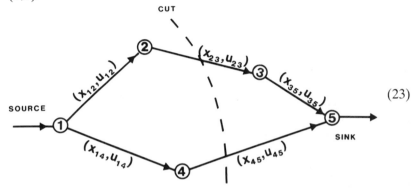

$$(23)$$

Then $(S,\bar{S}) = \{a(2,3), a(4,5)\}$ and $\text{cap}(S,\bar{S}) = u_{35} + u_{45}$. Notice that the arcs have labels (x_{ij}, u_{ij}).

We shall now present a sequence of fundamental results that relate the minimal cut capacity of a network to its maximal flow. Some of the proofs are left as exercises.

PROPOSITION 2   For any network, the value of any flow f does not exceed the capacity of any cut.

Proof: This follows from the following,

$$f = \sum_{i=1}^{m} (x_{1i} - x_{i1}) = \sum_{i=1}^{m} (x_{1i} - x_{i1}) + \sum_{\substack{j \in S \\ j \neq 1}} \sum_{i=1}^{m} (x_{ji} - x_{ij})$$

$$= \sum_{j \in S} \sum_{i \in S \cup \bar{S}} (x_{ji} - x_{ij}) = \sum_{j \in S} \sum_{i \in S} (x_{ji} - x_{ij}) + \sum_{j \in S} \sum_{i \in \bar{S}} (x_{ji} - x_{ij})$$

$$= \left(\sum_{j, i \in S} x_{ji} - \sum_{j, i \in S} x_{ij} \right) + \sum_{\substack{j \in S \\ i \in \bar{S}}} x_{ji} - \sum_{\substack{j \in S \\ i \in \bar{S}}} x_{ij}$$

$$= \sum_{\substack{j \in S \\ i \in \bar{S}}} x_{ji} - \sum_{\substack{j \in S \\ i \in \bar{S}}} x_{ij} \leq \sum_{\substack{j \in S \\ i \in \bar{S}}} x_{ji} \leq \sum_{\substack{j \in S \\ i \in \bar{S}}} u_{ji} = \text{cap}(S,\bar{S})$$

The proof of the following observation will be left as an exercise.

COROLLARY 4   For any network, the maximal flow does not exceed the minimal cut capacity. Moreover, if $f = \text{cap}(S,\bar{S})$, where f is the value of a flow and $(S,\bar{S})$ is a cut, then f is a maximal flow and $(S,\bar{S})$ is a minimal cut.

THEOREM 4 (Max-Flow Min-Cut)   For any network, the maximal flow is equal to the minimal cut capacity.

Proof: We shall construct a flow and a cut for which the value of the flow equals the capacity of the cut. The result will then follow from the second statement of Corollary 4. We start with a flow f that is maximal. The existence of such a flow follows from the first part of Corollary 4.

Define S recursively by the following statements:

(i) $1 \, \varepsilon \, S$,

(ii) If $i \, \varepsilon \, S$ and $x_{ij} < u_{ij}$, then place $j \in S$.

(iii) If $i \, \varepsilon \, S$ and $x_{ji} > 0$, then place $j \in S$

Suppose that the sink m is in S. Then there exists a chain that links 1 to m such that all the nodes in the chain belong to S. Moreover, the flow in each arc in the chain has the property that either $x_{ij} < u_{ij}$ or $x_{ji} > 0$. A typical situation is illustrated in (24) where $x_{12} < u_{12}$, $x_{42} > 0$, and $x_{45} < u_{45}$. There

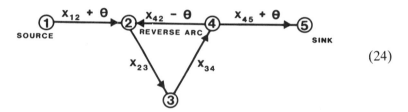

$$(24)$$

$\theta > 0$ such that $x_{12} + \theta \le u_{12}$, $x_{42} - \theta \ge 0$, and $x_{45} + \theta \le u_{45}$. The values x_{23} and x_{34} do not change. Notice that the flow to node 5 has been increased by θ units. As in Example (24), the flow from 1 through the chain in S to m can be increased by some positive amount. This contradicts the fact that f is a maximal flow. Thus, $m \notin S$. It follows that $(S,\bar{S})$ is a cut. By the way S, and $\bar{S}$, are defined it as follows that $x_{ij} = u_{ij}$ whenever $i \in S$ and $j \in \bar{S}$. Likewise, $x_{ji} = 0$ when $i \in S$ and $j \in \bar{S}$. Thus as in the proof of Proposition 2,

$$f = \sum_{\substack{i \in S \\ j \in \bar{S}}} x_{ij} - \sum_{\substack{i \in S \\ j \in \bar{S}}} x_{ji} = \sum_{\substack{i \in S \\ j \in \bar{S}}} u_{ij} = \text{cap}(S,\bar{S})$$

and the result follows.

A chain from 1 to m is a *flow augmenting chain* with respect to a flow f provided that $x_{ij} < u_{ij}$ over the forward arcs of the chain $x_{ji} > 0$ over the reverse arcs.

COROLLARY 5   A flow is maximal if and only if there does not exist a flow augmenting chain with respect to the flow.

The constructive proof of Theorem 4 leads to several algorithms for solving the maximal flow problem. One of the more common is the *labeling* procedure to be presented next. In this procedure we start with a flow f, and then the nodes beginning with the source are labeled in a fashion that is consistent with the generation of the cut in the proof of Theorem 4. The source, node 1, always receives the label $[-,\infty]$. If node i has been labeled and node j is linked to i by $a(i,j)$, where $x_{ij} < u_{ij}$, then node j receives a label $[i^{+}, \theta_{j}]$, where $\theta_{j} = \min\{\theta_{i}, u_{ij} - x_{ij}\}$. All of this means that node i has θ_{i} units of the commodity, and that node j can receive θ_{j} of these units via $a(i,j)$. If node i has been labeled and node j is linked to i by $a(j,i)$, where $x_{ji} > 0$, then node j receives a label $[i^{-}, \theta_{j}]$, where $\theta_{j} = \min\{x_{ji}, \theta_{i}\}$. Thus, the flow can be rerouted as in (24). When the nodes of a chain connecting 1 to m are labeled, a flow augmenting chain has been identified and the flow is not maximal. The flow is then increased as much as possible until one of the variables in the chain reaches its upper or lower bound. Then the procedure is repeated with the new flow and new labels, which start with $[-,\infty]$ at node 1. The procedure terminates when a cut is generated. The algorithm is summarized as follows

(i)   Start with each $x_{ij} = 0$.

(ii) Label node 1 with $[-,\infty]$.

(iii) If node i has been labeled and node j is linked to i by $A(i,j)$, where $x_{ij} < u_{ij}$, then label j with $[i^{+}, \theta_{j}]$, where $\theta_{j} = \min\{\theta_{i}, u_{ij} - x_{ij}\}$.

(iv) If node i has been labeled and node j is linked to i by $a(j,i)$, where $x_{ji} > 0$, then label j with $[i^{-}, \theta_{j}]$, where $\theta_{j} = \min\{\theta_{i}, x_{ji}\}$.

(v)   If the sink m has been labeled, then a chain links 1 to m and each node in the chain has a label. Let θ be the smallest θ_{i} in the labels. Then increase the flow in the forward arcs by θ and decrease the flow in the reverse arcs by θ. Discard the labels and return to (ii).

(vi) If the sink cannot be labeled, then a minimal cut $(S,\bar{S})$ has been generated, where S is the collection of labeled nodes. The maximum flow is the sum of the capacities of the arcs in the cut.

Clearly, when the u_{ij} are all positive integers, then there exists an optimal flow for which each x_{ij} is an integer. Rational arc capacities can be converted to integral capacities by enlarging all of the capacities equally with a suitable integral factor.

In the following example the labels are applied to the network in each iteration. The actual practice the labels can be generated without drawing the network. Notice that the set of labeled nodes in any particular iteration may contain more nodes than just the nodes of a chain from the source to the last labeled node. For example, in iteration (c) node 3 was labeled $[1^+, 4]$, next node 2 was labeled $[1^+, 2]$, and then nodes 4 and 5 were labeled. In any iteration the process of labeling a previously unlabeled node, which must be linked to some labeled node, continues as long as possible or until the sink is labeled.

EXAMPLE 4   Solve the following maximal flow problem.

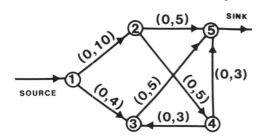

Solution:

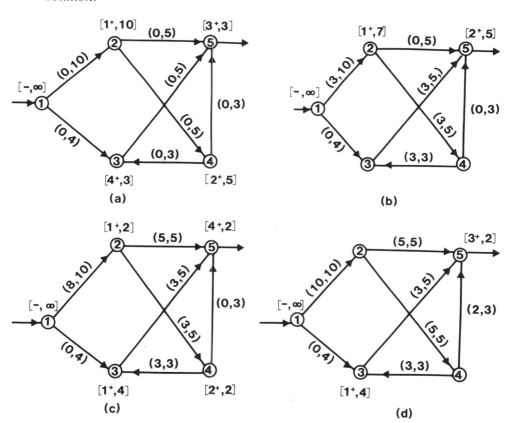

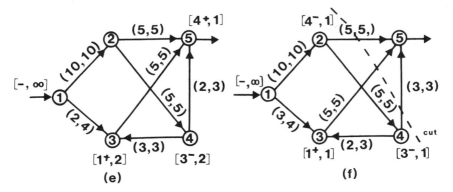

(e)   (f)

Thus, a minimal cut is $S = \{1,2,3,4\}$ and $\bar{S} = \{5\}$ and the capacity of this cut
is $5 + 5 + 3 = 13$.

Exercises

6.17   Solve the following maximal flow problems with the labeling
procedure.

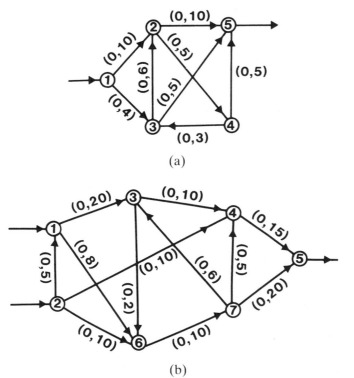

(a)

(b)

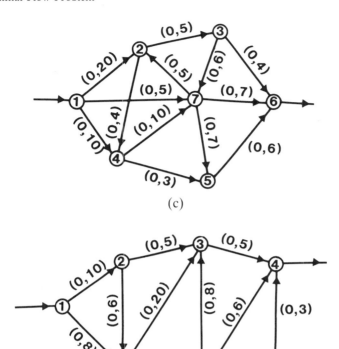

(c)

(d)

(e) Redo Exercise 6.14.

6.18 Prove Corollary 4.

6.19 Prove Corollary 5.

6.20 Prove that a cut $(S,\bar{S})$ is minimal if and only if every maximal flow saturates each arc of the cut $(S,\bar{S})$, and is flowless over each arc of $(\bar{S},S)$, where $(\bar{S},S) = \{a(i,j): i \varepsilon \bar{S} \text{ and } j \varepsilon S\}$.

6.21 (a) If $(S,\bar{S})$ and $(V,\bar{V})$ are minimal cuts, then prove that $(S \cup V, \overline{S \cup V})$ and $(S \cap V, \overline{S \cap V})$ are also minimal cuts.
(b) Let $(S,\bar{S})$ be the minimal cut that was generated in the proof of Theorem 4. Let $(V,\bar{V})$ be any other minimal cut. Prove that $S \subseteq V$.
(c) Prove that the set S of the minimal cut $(S,\bar{S})$ of Theorem 4 is the intersection of all the sets S_i for which $(S_i, \bar{S}_i)$ is a minimal cut.
(d) Prove that the cut $(S,\bar{S})$ generated in the proof of Theorem 4 does not depend on the selection of the maximal flow f.

6.22 (a) Show that the dual of the maximal flow problem (22) can be expressed in the form:

$$\text{Minimize} \sum_i \sum_j u_{ij} w_{ij} \text{ subject to}$$

$$-t_1 + t_m = 1$$

$$t_i - t_j + w_{ij} \geq 0, \text{ for all } i \text{ and } j$$

$$w_{ij} \geq 0, \text{ for all } i \text{ and } j$$

(Hint: Let the w_{ij} correspond to the constraints $x_{ij} \leq u_{ij}$, and the t_i correspond to the conservation of flow constraints.)

(b) Let $(S,\bar{S})$ be any cut and define $w_{ij} = 1$ whenever $(i,j) \in (S,\bar{S})$, and zero otherwise. Also define $t_i = 1$ when $i \in \bar{S}$, and zero otherwise. Show that this is a feasible solution to the dual problem. Finally, use these results to derive the max-flow min-cut theorem as a corollary to the fundamental duality theorem.

5. PRIMAL DUAL PROCEDURES FOR NETWORK FLOW PROBLEMS

As indicated in the last chapter, many of the techniques that are currently used to solve network flow problems are specialized applications of the primal-dual procedure. One common such specialization for the shortest path problem is next considered.

Let a network have m nodes and n arcs. For each arc $a(i,j)$ let the cost coefficient c_{ij} denote the length of the arc. For the present time assume that each $c_{ij} \geq 0$. Suppose that we wish to travel from node 1 to node m along a path of shortest length. The shortest path problem is to find such a path. Recall from Section 1 that the shortest path problem can be expressed as

$$\text{Minimize } c^T x \text{ subject to}$$

$$\sum_{j=1}^{m} (x_{1j} - x_{j1}) = 1$$

$$\sum_{j=1}^{m} (x_{ij} - x_{ji}) = 0, \text{ where } i \neq 1 \text{ and } i \neq m \qquad (25)$$

$$\sum_{j=1}^{m} (x_{mj} - x_{jm}) = -1$$

and each $0 \leq x_{ij} \leq \infty$.

The problem can also be viewed as an attempt to ship a unit of a commodity from source node 1 to sink node m at the smallest possible cost. In this case, the cost coefficient c_{ij} represents the cost of shipping one unit of the commodity along $a(i,j)$.

In either case, each component of an optimal solution x would be expected to be 0 or 1. When more than one optimal path exists, the components of x can differ from 0 or 1.

The dual of problem (25) is

$$\text{Maximize } w_1 - w_m \text{ subject to}$$
$$w_i - w_j \leq c_{ij}, \text{ for all } a(i,j) \text{ in the network} \tag{26}$$

Recall that the last constraint of Problem (25) is redundant and thus can be omitted. This would eliminate the variable w_m in (26). Hence, we shall assume that $w_m = 0$ in (26).

Also recall that the necessary and sufficient conditions for feasible solutions, x of (25) and w of (26), to be optimal are that each $x_{ij}(c_{ij} - w_i + w_j) = 0$ (see Exercise 4.25). Thus, when $x_{ij} > 0$, then $w_i - w_j = c_{ij}$. Also, when $w_i - w_j < c_{ij}$ then $a(i,j)$ is not in the optimal path.

Suppose now that w is a feasible solution of (26). Let $\Omega = \{(i,j): w_i - w_j = c_{ij}\}$. The restricted primal problem is then

$$\text{Minimize } \sum_{i=1}^{m} y_i \text{ subject to}$$

$$\sum_{j=1}^{m} (x_{1j} - x_{j1}) + y_1 = 1 \tag{27}$$

$$\sum_{j=1}^{m} (x_{ij} - x_{ji}) + y_i = 0, i = 2, \ldots, m-1$$

$x_{ij} \geq 0$ when $(i,j) \in \Omega$, $x_{ij} = 0$ when $(i,j) \notin \Omega$, and $y_i \geq 0$ for all i

The dual of this restricted problem is

$$\text{Maximize } r_1 \text{ subject to}$$
$$r_i - r_j \leq 0, \text{ for all } (i,j) \in \Omega \tag{28}$$
$$r_i \leq 1, \text{ for all } i$$

In the primal-dual algorithm (see Section 5 of Chapter 5), a solution, r, of (28) is found, and then w is replaced by $w + \theta r$, where

$$\theta = \min \{(c_{ij} - (w_i - w_j))/(r_i - r_j): r_i - r_j > 0\}$$

Next Ω is recomputed and the process is repeated. When $\sum_{i=1}^{m} y_i = 0$, the procedure terminates with an optimal solution of (25).

Solutions for (28), which enable the primal-dual procedure to be carried out efficiently, are easy to generate. For example, consider the following approach:

Step 1 Let $w = [0 \cdots 0]^T$. Notice that w is a feasible solution of (26) whenever $c \geq 0$.

Step 2 Let $\Omega = \{(i,j): w_i - w_j = c_{ij}\}$. Define $a(\Omega)$ to be the collection of $a(i,j)$ for which $(i,j) \in \Omega$.

Step 3 Define a solution r of (28) by

$$r_i = \begin{cases} 0, & \text{if a path of arcs from } a(\Omega) \text{ extends from } i \text{ to } m \\ 1, & \text{otherwise} \end{cases}$$

If a path of arcs belonging to $a(\Omega)$ extends from 1 to m, then go to Step 5.

Step 4 Replace w by $w + \theta r$, where

$$\theta = \min \{c_{ij} - (w_i - w_j): r_i - r_j > 0\}$$

and return to Step 2. Notice that when $r_i - r_j > 0$, then $r_i - r_j = 1$.

Step 5 Stop, the path extending from 1 to m is an optional solution of (25). To see this notice that $r_1 = 0$. It follows that the optimal value of the objective function of Problem (27) is 0. Thus, the arcs in the path from 1 to m constitute an optional path for problem (25).

Notice that $\{i : r_i = 0\} = \{i : \text{node } i \text{ is connected to node } m \text{ via the arcs}$ of $a(\Omega)\}$. In particular, when $c > 0$ and $w = [0 \ldots 0]^T$, then each $w_i - w_j < c_{ij}$, which implies that $a(\Omega) = \emptyset, \{i:r_i = 0\} = \{m\}$, and $r = [1 \ \ldots \ 1 \ 0]^T$. During each iteration at least one node enters the set $\{i : r_i = 0\}$ and at least one arc enters $a(\Omega)$, and they remain throughout the procedure. Thus, the process must produce a solution in no more than $m - 1$ steps. Notice also that x is not specifically computed.

EXAMPLE 5   Find the shortest path from node 1 to node 5 in the following network:

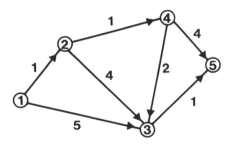

where the length of $a(1,3) = 5$ etc.

Solution: We shall attach a label (w_i, r_i) to each node i. Initially in (29), $(w_5, r_5) = (0,0)$ and $(w_i, r_i) = (0,1)$ for all $i = 1, \cdots, 4$. Then $\Omega = \emptyset$ and $a(\Omega) = \emptyset$.

Notice that $r_4-r_5 = 1$, $r_3-r_5 = 1$, and all other $r_i-r_j = 0$. Thus, $\theta = min \{4-(0-0), 1-(0-0)\} = 1$. Next replace w by $w+\theta r = [0\ 0\ 0\ 0\ 0]^T + [1\ 1\ 1\ 1\ 0]^T = [1\ 1\ 1\ 1\ 0]^T$. Notice that $w_3-w_5 = 1-0 = c_{35}$. In fact $\Omega = \{(3,5)\}$ and $a(\Omega) = \{a(3,5)\}$. Next compute a new $r = [1\ 1\ 0\ 1\ 0]^T$ using the rule in step 3. All of these steps can be carried out with or without the aid of drawing the network. However, we shall record all of the information in (30)

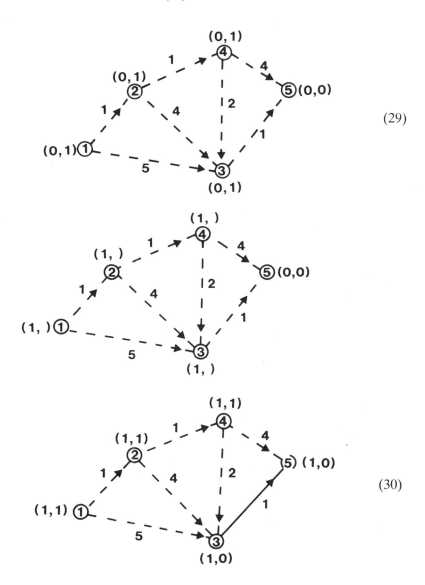

(29)

(30)

Next compute $\theta = \min \{5-(1-1), 4-(1-1), 2-(1-1)\} = 2$, and replace w by $w = [1\ 1\ 1\ 1\ 0]^T + 2[1\ 1\ 0\ 1\ 0]^T = [3\ 3\ 1\ 3\ 0]^T$. Then $\Omega = \{(4,3), (3,5)\}$ and $a(\Omega) = \{a(4,3), a(3,5)\}$. Hence, $r = [1\ 1\ 0\ 0\ 0]^T$. Again, all of this information is recorded in (31).

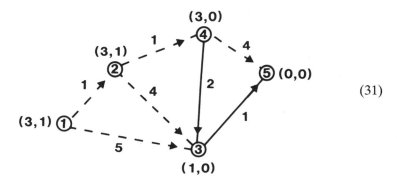

(31)

The final iterations are summarized in the following sequence of networks.

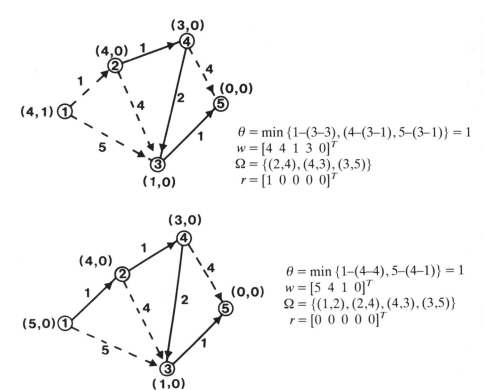

$\theta = \min \{1-(3-3), (4-(3-1), 5-(3-1)\} = 1$
$w = [4\ 4\ 1\ 3\ 0]^T$
$\Omega = \{(2,4), (4,3), (3,5)\}$
$r = [1\ 0\ 0\ 0\ 0]^T$

$\theta = \min \{1-(4-4), 5-(4-1)\} = 1$
$w = [5\ 4\ 1\ 0]^T$
$\Omega = \{(1,2), (2,4), (4,3), (3,5)\}$
$r = [0\ 0\ 0\ 0\ 0]^T$

Thus, a shortest path from 1 to 4 is $\{1, a(1,2), 2, a(2,4), 4, a(4,3), 3, a(3,5), 5\}$.

One of the more popular primal-dual procedures for solving the shortest path problem having all nonnegative costs is found in Exercise 6.24.

Situations do arise in which some of the c_{ij} are negative. If such a network contains a circuit where the sum of the cost coefficients is negative, then a path from source to sink of arbitrarily small length can be constructed by entering the negative circuit and traveling around the circuit a suitable number of times before exiting [see network (32)]. A primal-dual algorithm for a shortest path problem with some negative costs is found in Exercise 6.25.

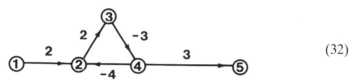

$$(32)$$

A generalization of the primal-dual procedure, known as the *out-of-kilter* algorithm, exists that will solve the more general minimal cost network flow problem. Unlike the network simple method, the out-of-kilter algorithm allows negative cost coefficients and positive lower bounds on the flow capacity through any arc. A detailed development of the algorithm can be found in several of the references at the end of this chapter.

Exercises

6.23 In each of the following networks find the shortest path from the first node to the last node.

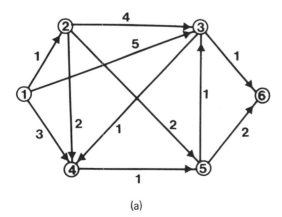

(a)

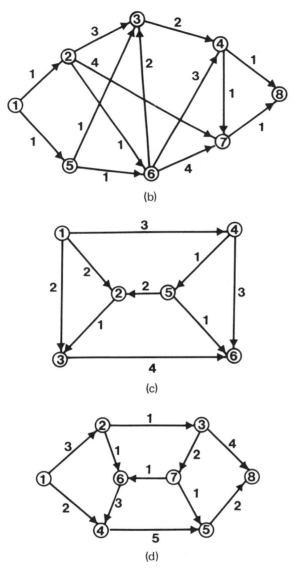

(b)

(c)

(d)

6.24 (Dijkstra's algorithm for the shortest path problem having nonnegative cost coefficients). Let m be the number of nodes in the network. Assume
that $c_{ij} \geq 0$ for each arc $a(i,j)$ in the network, and set $c_{ij} = \infty$ whenever $a(i,j)$
does not exists within the network. The following algorithm computes the
shortests paths from a fixed node, 1, to all of the other nodes in the network
by attaching a sequence of labels of the form $[p(i), \ell(i)]$ to each node i, where
$\ell(i)$ denotes the length of a path from 1 to j and $p(j)$ is the node immediately
preceding j in the path.

Step 1 Set $(p(1), \ell(1)) = (1,0)$ and $(p(i), \ell(i)) = (0,\infty)$ for all $i = 2, \ldots , m$. Also set $x = 1$ and $s = \{1\}$.

Step 2 For every $i \notin S$ such that $\ell(x) + c_{xi} < \ell(i)$, set $\ell(i) = \ell(x) + c_{xi}$ and $p(i) = x$.

Step 3 Select $j \notin S$ so that $\ell(j) = \min \{\ell(i) : i \notin S\}$. Set $S = S \cup \{j\}$ and $x = j$.

Step 4 If $S \neq \{1, \ldots , m\}$ return to Step 2.

Step 5 Stop. A shortest path from 1 to i has length $\ell(i)$, and the nodes in the path in reverse order are $i, p(i), p(p(i)), \ldots , 1$.

(a) Solve the shortest path problem in Example 5 using Dijkstra's algorithm.

(b) For each network in Exercise 6.23, use Dijkstra's algorithm to find the shortests paths from node 1 to the remaining nodes of the network.

(c) Verify that Dijkstra's algorithm is a primal-dual algorithm.

6.25 (A labeling algorithm for the shortest path problem possessing some negative cost coefficients). Again let m be the number of nodes in the network and let c be the sum of all the negative costs. The following algorithm produces a shortest path from node 1 to any other node provided that the network does not possess a circuit where the sum of the cost coefficients is negative. This algorithm attaches a sequence of labels of the form $[p(i), w_i]$ to each node, where $p(i)$ is the node immediately preceding node i in a given path and w_i is the i-th dual variable.

Step 1 Set $(p(1), w_1) = (1,0)$ and $(p(i), w_i) = (0,-\infty)$ when $i \neq 1$.

Step 2 If each $w_i - w_j \leq c_{ij}$ (or $w_i \leq w_j + c_{ij}$), then stop. The problem is solved.

Step 3 Select (i,j) such that $w_i - w_j > c_{ij}$. Set $p(j) = i$ and $w_j = w_i - c_{ij}$.

Step 4 If $w_j > -c$, then stop the problem possesses a circuit where the sum of the cost coefficients is negative.

Step 5 Return to Step 2.

When the algorithm does not terminate in Step 4, then the nodes of shortest path from 1 to any reachable node i listed in reverse order are $i, p(i), p(p(i)), \ldots ,1$.

Use this algorithm to solve the following shortest path problems:

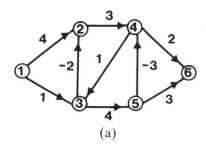

(a)

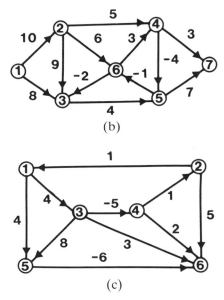

(b)

(c)

(d) Prove that for every $w_j > -\infty$, there exists a path from 1 to j for which w_j is the negative of the sum of the related cost coefficients.

(e) Prove that $w_j \le -c$ whenever the network does not possess a circuit for which the sum of the corresponding cost coefficients is negative.

6.26 The following algorithm is actually a primal-dual method specialized for the assignment problem. It is known as the *Hungarian method*. The method is applied to the coefficient matrix

$$c = \begin{bmatrix} c_{11} \ldots c_{1n} \\ \cdot \quad \quad \cdot \\ \cdot \quad \quad \cdot \\ \cdot \quad \quad \cdot \\ c_{n1} \ldots c_{nn} \end{bmatrix}$$

where c_{ij} is the value of the ith person in the jth job. Clearly, there are n jobs to be filled by n people.

   Step 1 In each row, subtract the smallest entry in the row from each entry in the row (including itself).

   Step 2 In each column subtract the smallest entry in the column from each entry in the column (including itself).

   Step 3 If the resulting matrix has n zeros (it may contain zeros other than these) such that each row and each column contains a unique one of these special zeros, then stop. An optimal assignment is obtained by assigning the ith individual to the jth job, whenever the enrty in row i and column j of the reduced matrix is one of these special zeros. Otherwise, continue to Step 4.

Step 4 Cover all zeros of the reduced matrix by drawing the minimal number of horizontal and vertical lines through the zeros of the reduced matrix. Find a smallest element that is not covered by one of these lines. Subtract this element from every element not covered by one of the lines. Add the element to every element that has been covered twice by the lines. Set C equal to the new matrix and return to Step 3.

(a) Solve the assignment problem where

$$c = \begin{bmatrix} 5 & 3 & 1 & 3 & 1 \\ 1 & 4 & 1 & 4 & 1 \\ 6 & 1 & 8 & 7 & 2 \\ 6 & 4 & 4 & 1 & 1 \\ 2 & 2 & 3 & 2 & 9 \end{bmatrix}$$

(b) Solve the assignment problem where

$$c = \begin{bmatrix} 1 & 0 & 1 & 3 & 1 \\ 0 & 4 & 1 & 4 & 6 \\ 1 & 1 & 2 & -1 & 2 \\ 2 & 0 & 1 & 3 & 0 \\ 3 & 6 & 0 & 4 & 0 \end{bmatrix}$$

(c) Verify that the Hungarian method is a primal-dual process.

REFERENCES

1. M. Bazaraa and J. J. Jarvis, *Linear Programming and Network Flows*, John Wiley and Sons, New York (1977).
2. R. Bronson, *Operations Research* (Schamn's Outline Series), McGraw-Hill Book Company, New York (1982).
3. V. Chvátal, *Linear Programming*, W. H. Freeman and Company, New York (1980).
4. L. R. Ford, Jr. and D. R. Fulkerson, *Flows in Networks*, Princeton University Press, Princeton, N.J. (1962).
5. R. V. Hartley, *Operations Research, A Managerial Approach*, Goodyear Publishing Company, Pacific Palisades, Ca. (1976).
6. J. L. Kennington and R. V. Helgason, *Algorithms for Network Programming*, John Wiley and Sons, New York (1980).
7. E. L. Lawler, *Combinatorial Optimization: Networks and Matroids*, Holt, Rinehart and Winston, New York (1976).

8. E. Minieka, *Optimization Algorithms for Networks and Graphs*, Marcel Dekker, Inc., New York (1978).

9. K. Murty, *Linear and Combinatorial Programming*, John Wiley and Sons, New York (1976).

10. C. H. Papadimitriou and K. Steiglitz, *Combinatorial Optimization, Algorithms and Complexity*, Prentice-Hall, Inc., Englewood Cliffs, N.J. (1982).

11. M. N. S. Swamy and K. Thulasiraman, *Graphs, Networks and Algorithms*, John Wiley and Sons, New York (1981).

7

Convex and Concave Functions

Chapter 2 contained an introduction to convex analysis and its applications to linear programming. This chapter is an extension of that introduction to include convex functions and their applications to mathematical programming. Section 1 begins with the basic definition and examples of a convex function. This is followed by a brief study of convex functions of a real variable. Then after a brief calculus review a similar study is made for functions of several variables. In Section 4 some basic results concerning the optimization of convex functions are presented. The last section deals with some of the popular generalizations of this topic and their applications.

1. INTRODUCTION

Let K be a convex subset of $\mathbf{R}_{n \times 1}$. A function of $f: K \to \mathbf{R}$ is a *convex* function if and only if $f(\alpha x + (1 - \alpha)y) \leq \alpha f(x) + (1 - \alpha)f(y)$ whenever $x, y \in K$ and $\alpha \in [0, 1]$. Figure 1(a) is an example of a convex function while Figure 1(b) is an example of a function that is not convex. Notice that in Figure 1(a), the point $(\alpha x_1 + (1 - \alpha)x_2, f(\alpha x_1 + (1 - \alpha)x_2))$, where $0 < \alpha < 1$, is not above the corresponding point $(\alpha x_1 + (1 - \alpha)x_2, \alpha f(x_1) + (1 - \alpha)f(x_2))$ on the chord that connects the points $(x_1, f(x_1))$ and $(x_2, f(x_2))$.

The following function are examples of convex functions.
1. Any linear function $f(x) = c^T x = \sum_{i=1}^{n} c_i x_i$ over $K = \mathbf{R}_{n \times 1}$. Here, $x = [x_1 \dots x_n]^T$ and $c = [c_1 \dots c_n]^T$ (see Chapter 2).
2. Any quadratic function $q(x) = ax^2 + bx + c$, where a, b and c are constants, $a > 0$ and $K = \mathbf{R}$.
3. $f(x) = |x|$, where $x \in K = \mathbf{R}$.
4. $f(x) = x^3$, where $x \in K = [0, \infty)$.
5. $f([x_1 x_2]^T) = 4x_1^2 + 25x_2^2$, where $[x_1 x_2]^T \in K = \mathbf{R}_{2 \times 1}$ (Notice that the graph of f is an elliptic paraboloid.)

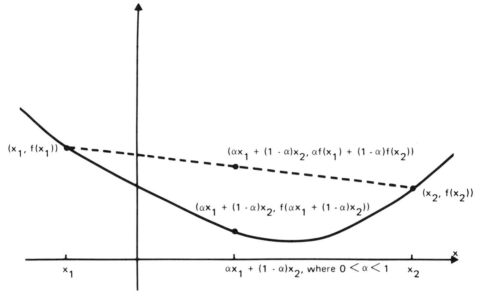

$(x_1, f(x_1))$

$(\alpha x_1 + (1 - \alpha)x_2, \alpha f(x_1) + (1 - \alpha)f(x_2))$

$(\alpha x_1 + (1 - \alpha)x_2, f(\alpha x_1 + (1 - \alpha)x_2))$

$(x_2, f(x_2))$

x_1

$\alpha x_1 + (1 - \alpha)x_2$, where $0 < \alpha < 1$ x_2

(a) f is a convex function over $K = \mathbf{R}$.

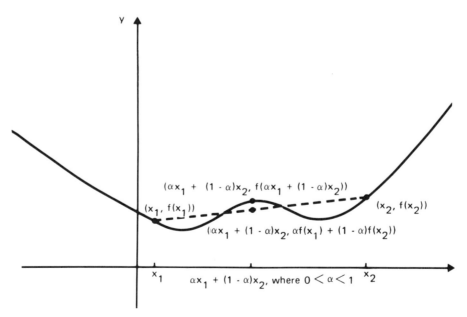

$(\alpha x_1 + (1 - \alpha)x_2, f(\alpha x_1 + (1 - \alpha)x_2))$

$(x_1, f(x_1))$

$(x_2, f(x_2))$

$(\alpha x_1 + (1 - \alpha)x_2, \alpha f(x_1) + (1 - \alpha)f(x_2))$

x_1

$\alpha x_1 + (1 - \alpha)x_2$, where $0 < \alpha < 1$ x_2

(b) f is not a convex function over $K = \mathbf{R}$.

Figure 1

Again consider a function $f: K \to \mathbf{R}$, where K is a convex subset of $\mathbf{R}_{n \times 1}$. The function f is concave if and only if $f(\alpha x_1 + (1 - \alpha)x_2) \geq \alpha f(x_1) + (1 - \alpha)f(x_2)$, whenever $x, y \in K$ and $\alpha \in K$ and $\alpha \in [0, 1]$. It is a simple exercise to show that f is a concave function if and only if $-f$ is a convex function. A function that is both convex and concave over K is said to be *affine*. Any linear function over K is both convex and concave and, hence, *affine*. An affine function is not necessarily linear.

Some examples of concave functions are
1. $f(x) = x^3$, where $x \in K = (-\infty, 0]$.
2.   Any quadratic function $f(x) = ax^2 + bx + c$, where a, b, and c are constants, $a < 0$ and $K = \mathbf{R}$.

The following proposition gives us an alternate way of defining a convex function.

PROPOSITION 1   Let $f: K \to \mathbf{R}$, where K is a convex subset of $\mathbf{R}_{n \times 1}$. Then f is a convex function if and only if

$$f\left(\sum_{i=1}^{n} \alpha_i x_i\right) \leq \sum_{i=1}^{n} \alpha_i f(x_i) \tag{1}$$

whenever each $x_i \in K$, each $\alpha_i \in [0, 1]$ and $\sum_{i=1}^{n} \alpha_i = 1$.
Proof: Clearly when f satisfies (1), then f is a convex function. Next suppose that f is a convex function. Then $f(a_1 x_1 + a_2 x_2) < a_1 f(x_1) + a_2 f(x_2)$ whenever $x_1, x_2 \in K$; $a_1, a_2 \in [0, 1]$ and $a_1 + a_2 = 1$. Suppose that the result holds when $n = k$. Let $\sum_{i=1}^{k+1} a_i x_i$ be a convex combination of vectors from K, i.e., each $x_i \in K$, each $\alpha_i \in [0, 1]$ and $\sum_{i=1}^{k+1} \alpha_i = 1$. Let $\alpha = \sum_{i=1}^{k} a_i$. Then $\alpha \neq 0$, $\alpha_{k+1} = 1 - \alpha$, and

$$\sum_{i=1}^{k+1} \alpha_i x_i = \sum_{i=1}^{k} \alpha_i x_i + \alpha_{k+1} x_{k+1} = \alpha \sum_{i=1}^{k} \frac{\alpha_i}{\alpha} x_i + (1 - \alpha)x_{k+1}$$

Since each $a_i/\alpha > 0$ and $\sum_{i=1}^{k} a_i/\alpha = 1$, then $\sum_{i=1}^{k} (\alpha_i/\alpha) x_i \in K$. Also, f is convex and we are assuming inductively that (1) holds when $n = k$. Thus, we have the following:

$$f\left(\sum_{i=1}^{k+1} \alpha_i x_i\right) = f\left(\alpha \sum_{i=1}^{k} \frac{\alpha_i}{\alpha} x_i + (1 - \alpha)x_{k+1}\right) \leq \alpha f\left(\sum_{i=1}^{k} \frac{\alpha_i}{\alpha} x_i\right) + (1 - \alpha)f(x_{k+1})$$

$$\leq \alpha\left(\sum_{i=1}^{k} \frac{\alpha_i}{\alpha} f(x_i)\right) + (1 - \alpha)f(x_{k+1})$$

$$= \sum_{i=1}^{k} \alpha_i f(x_i) + \alpha_{k+1} f(x_{k+1}) = \sum_{i=1}^{k+1} \alpha_i f(x_i)$$

It follows from the principle of mathematical induction that (1) holds for all n.

Convex function can often be combined in some fashion to determine other convex functions. The following propositions are examples.

PROPOSITION 2   Let K be a convex subset of $\mathbf{R}_{n \times 1}$. Then any conical combination of convex functions defined on K is again a convex function defined on K.

Proof: Let $f_1, \ldots, f_n$ be convex functions defined on K. Let $\alpha_1 \geq 0, \ldots, \alpha_n \geq 0$. If $x, y \in K$ and $\alpha \in [0,1]$, then

$$(\sum_{i=1}^{n} \alpha_i f_i)(\alpha x + (1 - \alpha)y) = \sum_{i=1}^{n} (\alpha_i f_i)(\alpha x + (1 - \alpha)y)$$

$$= \sum_{i=1}^{n} \alpha_i (f_i(\alpha x + (1 - \alpha)y)$$

$$\leq \sum_{i=1}^{n} \alpha_i (\alpha f_i(x) + (1 - \alpha)f_i(y))$$

$$= \alpha \sum_{i=1}^{n} (\alpha_i f_i)(x) + (1 - \alpha) \sum_{i=1}^{n} (\alpha_i f_i)(y)$$

$$= \alpha(\sum_{i=1}^{n} \alpha_i f_i)(x) + (1 - \alpha)(\sum_{i=1}^{n} \alpha_i f_i)(y)$$

and $\Sigma_{i=1}^{n} \alpha_i f_i$ is a convex function.

For example the function $f(x) = x^3 + 7x^2 - 2x + 3$ is a convex function over $K = [0, \infty)$ since both $g(x) = x^3$ and $h(x) = 7x^2 - 2x + 3$ are convex functions over $[0, \infty)$.

PROPOSITION 3   Let $f : K \to \mathbf{R}$ be a convex function defined on the convex set K. Let I be a convex set in $\mathbf{R}$ that contains $f(K) = \{f(x) : x \in K\}$. If $g : I \to \mathbf{R}$ is an increasing convex function, then $g \circ f : K \to \mathbf{R}$ is a convex function.

Proof: Let $x, y \in K$ and $\alpha \in [0, 1]$. Then

$$(g \circ f)(\alpha x + (1 - \alpha)y) = g(f(\alpha x + (1 - \alpha)y)) \leq g(\alpha f(x) + (1 - \alpha)f(y))$$

$$\leq \alpha g(f(x)) + (1 - \alpha)g(f(y))$$

$$= \alpha(g \circ f)(x) + (1 - \alpha)(g \circ f)y$$

For example, the function $h([x_1 x_2]^T) = (4x_1^2 + 25x_2^2)^3$ is a convex function on E_2 since both $f([x_1 x_2]^T) = 4x_1^2 + 25x_2^2$ and $g(x) = x^3$ are convex functions defined on the sets $K = \mathbf{R}_{2 \times 1}$ and $L = [0, \infty)$, respectively. Other examples of functional operations on a given collection of convex functions to obtain another convex function can be found in the exercises at the end of this section. We shall now direct our attention to a particular type of function $f : \mathbf{R}_{n \times 1} \to \mathbf{R}$.

Let $A = [a_{ij}]$ be a real $n \times n$ matrix. Also, let $x = [x_1 \ldots x_n]^T \in \mathbf{R}_{n \times 1}$. Then the function $f(x) = x^T A x = \sum_{i=1}^{n} \sum_{j=1}^{n} a_{ij} x_i x_j$ is called a *quadratic form* in n variables. Since $f(x)$ is a scalar, it follows that $x^T A x = f(x) = f(x)^T = (x^T A x)^T = x^T A^T x$. Hence,

$$f(x) = \frac{1}{2}(x^T A x + x^T A^T x) = \frac{1}{2} x^T (A + A^T) x$$

Notice that $A + A^T$ is a symmetric matrix, i.e., $(A + A^T)^T = A + A^T$. Thus, any quadratic form, f, can be expressed as $f(x) = x^T B x$, where B is a symmetric real matrix.

EXAMPLE 1 When $x = [x_1 \, x_2 \, x_3]^T$, the quadratic form $f(x) = x_1^2 + 2x_1 x_2 - x_1 x_3 + 4x_2 x_3 - x_2^2 + 7x_3^2$ can be expressed as

$$f(x) = [x_1 \, x_2 \, x_3] \begin{bmatrix} 1 & 2 & -1 \\ 0 & -1 & 4 \\ 0 & 0 & 7 \end{bmatrix} \begin{bmatrix} x_1 \\ x_2 \\ x_3 \end{bmatrix}$$

This, in turn, can be rewritten as

$$f(x) = [x_1 \, x_2 \, x_3] \frac{\begin{bmatrix} 1 & 2 & -1 \\ 0 & -1 & 4 \\ 0 & 0 & 7 \end{bmatrix} + \begin{bmatrix} 1 & 0 & 0 \\ 2 & -1 & 0 \\ -1 & 4 & 7 \end{bmatrix}}{2} \begin{bmatrix} x_1 \\ x_2 \\ x_3 \end{bmatrix}$$

$$= [x_1 \, x_2 \, x_3] \begin{bmatrix} 1 & 1 & -\frac{1}{2} \\ 1 & -1 & 2 \\ -\frac{1}{2} & 2 & 7 \end{bmatrix} \begin{bmatrix} x_1 \\ x_2 \\ x_3 \end{bmatrix} = x^T B x$$

where B is the symmetrix matrix

$$B = \begin{bmatrix} 1 & 1 & -\frac{1}{2} \\ 1 & -1 & 2 \\ -\frac{1}{2} & 2 & 7 \end{bmatrix}$$

The quadratic form $f(x) = x^T B x$, where B is a symmetric real square matrix, is said to be *positive definite* if and only if $f(x) = x^T B x > 0$ for all nonzero x in $\mathbf{R}_{n+1}$. Moreover, $f(x) = x^T B x$ is said to be *positive semidefinite* if and only if $f(x) = x^T B x \geq 0$ for all $x \in \mathbf{R}_{n \times 1}$. *Negative definite* and *negative semidefinite* are defined in a similar fashion by the reversal of the inequalities in the above definitions.

PROPOSITION 4   Any positive semidefinite quadratic form $f(x) = x^T B x$, where B is a square symmetric real matrix, is a convex function.

Proof: Let $x, y \in \mathbf{R}_{n+1}$ and $\alpha \in [0,1]$. Since f is positive semi-definite it follows that

$$0 \le f(x-y) = (x-y)^T B(x-y) = (x^T - y^T) B(x-y) = (x^T B - y^T B)(x-y)$$
$$= x^T B x - x^T B y - y^T B x + y^T B y$$

Hence,

$$x^T B y + y^T B x \le x^T B x + y^T B y$$

Now consider $f(\alpha x + (1-\alpha)y)$,

$$f(\alpha x + (1-\alpha)y) = (\alpha x + (1-\alpha)y)^T B(\alpha x + (1-\alpha)y)$$
$$= (\alpha x^T + (1-\alpha)y^T) B(\alpha x + (1-\alpha)y)$$
$$= (\alpha x^T B + (1-\alpha)y^T B)(\alpha x + (1-\alpha)y)$$
$$= \alpha^2 x^T B x + \alpha(1-\alpha)(x^T B y + y^T B x) + (1-\alpha)^2 y^T B y$$
$$\le \alpha^2 x^T B x + \alpha(1-\alpha)(x^T B x + y^T B y) + (1-\alpha)^2 y^T B y$$
$$= \alpha^2 x^T B x + \alpha x^T B x - \alpha^2 x^T B x + (1-\alpha)(\alpha + (1-\alpha))y^T B y$$
$$= \alpha x^T B x + (1-\alpha)y^T B y = \alpha f(x) + (1-\alpha)f(y)$$

A similar result holds for concave quadratic forms, namely $f(x) = x^T B x$ is concave whenever B is negative semidefinite.

We shall close this section with two results relating a convex function $f : K \to \mathbf{R}$, where K is a convex subset of $\mathbf{R}_{n \times 1}$, with some convex subsets of $\mathbf{R}_{n \times 1}$ and $\mathbf{R}_{(n+1) \times 1}$, respectively. First let $\lambda \in \mathbf{R}$ and $Lev_\lambda f = \{x : f(x) \le \lambda\}$. $Lev_\lambda f$ is called a *level* set.

PROPOSITION 5   If $f : K \to \mathbf{R}$, where K is a convex subset of $\mathbf{R}_{n+1}$, is a convex function, then $Lev_\lambda f$ is a convex set for every $\lambda \in \mathbf{R}$.

Proof: Let $x, y \in Lev_\lambda f$ and $\alpha \in [0,1]$. Then

$$f(\alpha x = (1-\alpha)y) \le \alpha f(x) + (1-\alpha)f(y) \le \alpha\lambda + (1-\alpha)\lambda = \lambda$$

which implies that $\alpha x + (1-\alpha)y \in Lev_\lambda f$. Hence, $Lev_\lambda f$ is a convex set.

The converse of this proposition is not true; i.e., examples of functions

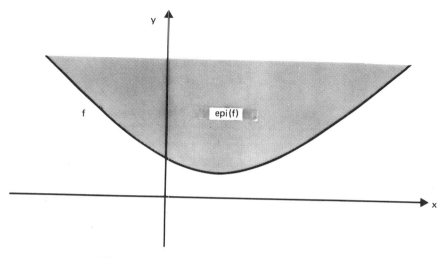

Figure 2 f is a convex function defined on **R**.

that are not convex but having all convex level sets can be constructed. In the last section of this chapter we shall introduce a class of functions for which the proposition and its converse both hold.

Our final set is related to the graph of f in $\mathbf{R}_{(n+1)\times 1}$, i.e., the set of points $\{(x,f(x)) : x \in K\}$, where $[x^T f(x)]^T$ is denoted by $(x,f(x))$ in order to simplify notation. The *epigraph* of f is denoted by epi(f) and defined by

$$\text{epi}(f) = \{(x,y) : x \in K, y \in \mathbf{R}, f(x) \leq y\}$$

Clearly, epi(f) is the collection of points in $\mathbf{R}_{(n+1)\times 1}$ which lies on or above the graph, $\{(x,f(x)) : x \in K\}$, of f. Figure 2 illustrates this concept. Notice in Figure 2 that epi(f) is a convex set.

PROPOSITION 6   Let $f : K \to \mathbf{R}$, where K is a convex subset of $\mathbf{R}_{n\times 1}$. Then f is a convex function if and only if epi(f) is a convex set.
Proof: First suppose that f is a convex function. Let (x,y) and (u,v) both belong to epi(f) and let $\alpha \in [0,1]$. Then

$$f(\alpha x + (1-\alpha)u) \leq \alpha f(x) + (1-\alpha)f(u) \leq \alpha y + (1-\alpha)v$$

Since K is a convex, then $\alpha x + (1-\alpha)u \in K$. This implies that $(\alpha x + (1-\alpha)u, \alpha y + (1-\alpha)v) \in \text{epi}(f)$. But

$$(\alpha x + (1-\alpha)u, \ \alpha y + (1-\alpha)v) = \alpha(x,y) + (1-\alpha)(u,v)$$

Hence, $\alpha(x,y) + (1-\alpha)(u,v) \in \text{epi}(f)$ and epi(f) is a convex set.

Next suppose that epi(f) is a convex set.   Let $x, u \in K$ and $\alpha \in [0, 1]$. Clearly, $(x, f(x))$ and $(u, f(u))$ both belong to epi(f). Since epi(f) is convex, then

$$(\alpha x + (1 - \alpha)u, \alpha f(x) + (1 - \alpha)f(u)) = \alpha(x, f(x)) + (1 - \alpha)(u, f(u)) \in \text{epi}(f)$$

Hence, $f(\alpha x + (1 - \alpha)u) \le \alpha f(x) + (1 - \alpha)f(u)$ and f is a convex function.

In more advanced work Proposition 6 is often used as the definition of a convex function. In particular, this is often done when the function f is allowed to have infinite values. In this text the functions are all finite valued.

Exercises

7.1   Show that the linear function $f(x) = c^T x$, where $c = [c_1 \ldots c_n]^T$ and $x = [x_1 \ldots x_n]^T$ is both a convex and a concave function.

7.2   Show that f is a concave function if and only if $-f$ is a convex function.

7.3   (a) Show that f is a concave function over a convex set K if and only if $f(\sum_{i=1}^n \alpha_i x_i) \ge \sum_{i=1}^n \alpha_i f(x_i)$ whenever each $x_i \in K$, each $\alpha_i \in [0, 1]$, $n \in N$ (the set of natural numbers), and $\sum_{i=1}^n \alpha_i = 1$.
(b) Show that f is an affine function over K if and only if $f(\sum_{i=1}^n \alpha_i x_i) = \sum_{i=1}^n \alpha_i f(x_i)$ whenever each $x_i \in K, n \in N$ and $\sum_{i=1}^n \alpha_i = 1$.

7.4   Show that f is a concave function over the convex set K if and only if the set hyp(f) $= \{(x, y) : x \in K, \ y \in \mathbf{R}, f(x) \ge y\}$ is a convex set. (This set is called the *hypograph* of f.)

7.5   A function $f : K \to \mathbf{R}$, where K is a convex subset of $\mathbf{R}_{n \times 1}$ is said to be *strictly convex* if and only if $f(\alpha x + (1 - \alpha)y) < \alpha f(x) + (1 - \alpha)f(y)$, for every distinct pair of points $x, y \in K$ and $\alpha \in (0, 1)$. *Strictly concave* functions are defined in a similar manner. Show that any positive definite quadratic form is a strictly convex function.

7.6   Prove that any negative semidefinite quadratic form is a concave function. Also, prove that any negative definite quadratic form is a strictly concave function.

7.7   Show that the set $\{x : f(x) < \lambda\}$ is convex whenever f is a convex function. Also show that both $\{x : f(x) \ge \lambda\}$ and $\{x : f(x) > \lambda\}$ are convex whenever f is a concave function.

7.8   Let $f_\alpha : K_\alpha \to \mathbf{R}$, where K_α is a convex subset of $\mathbf{R}_{n \times 1}$, be a convex function for every $\alpha \in \Omega$. Let $K = \bigcap_{\alpha \in \Omega} K_\alpha$. If $K \ne \emptyset$ define $f : K \to \mathbf{R}$ by $f(x) = \sup\{f_\alpha(x) : \alpha \in \Omega\}$. Let $S = \{x \in K : f(x) < \infty\}$. Then prove that S is a convex set and f is a convex function on S.

7.9   Determine whether the following functions are convex or not:

(a) $f(x,y) = \dfrac{x^2}{a^2} - \dfrac{y^2}{b^2}$ over $K = \mathbf{R}_{2 \times 1}$

(b) $f(x) = x^3 - 3x^2 + 4$ over $K = [1, \infty)$

(c) $f(x) = x^2 + \dfrac{1}{x}$ over $K = (0, \infty)$

(d) $f(x) = x^2 + \dfrac{1}{x}$ over $K = (-\infty, \infty)$

(e) $f(x) = 3x^5 - 10x^3$ over $K = (0, \infty)$

7.10   Let f_1 and f_2 be convex functions on $K = \mathbf{R}_{n \times 1}$. Define $f_1 \square f_2(x) = \inf \{f_1(x_1) + f_2(x_2): x_1, x_2 \in \mathbf{R}_{n \times 1}$ and $x = x_1 + x_2\}$. Show that $f_1 \square f_2$ is a convex function over $\mathbf{R}_{n \times 1}$. $f_1 \square f_2$ is called the *infimal convolution* of f_1 and f_2.

7.11   Let C be a nonempty convex set in $\mathbf{R}_{n \times 1}$. Show that each of the following functions is convex over $\mathbf{R}_{n \times 1}$.

(a) $\delta(x) = \sup \{y^T x: y \in C\}$ (δ is called the *support* function of C).
(b) $\gamma(x) = \inf \{\lambda \geq 0: x \in \lambda C\}$ (γ is called the *gauge* function of C).
(c) $d(x) = \inf \{|x - y|: y \in C\}$ (d is called the *distance* function of C).

7.12   Give an example of an affine function that is not a linear function.

7.13   Let $f(x) = c^T x + b$ where $x = [x_1 \ldots x_n]^T$, $c = [c_1 \ldots c_n]^T$ and $b \in \mathbf{R}$. Show that f is an affine function. Determine when f is linear.

2.   CONVEX FUNCTIONS OF ONE REAL VARIABLE

This section contains a brief study of functions $f: K \to \mathbf{R}$, where K is a convex subset of $\mathbf{R}$, that are convex (or concave). Some of the basic properties developed in this section will be extended in the next section to convex functions of several variables.

   When working with convex functions of one real variable the concept of a one-sided derivative is useful. Recall that for any function $f: \mathbf{R} \to \mathbf{R}$ the *derivative from the right* is defined at $x = x_0$ to be the following limit when it exists

$$f_R'(x_0) = \lim_{h \to 0^+} \frac{f(x_0 + h) - f(x_0)}{h}$$

Likewise, the derivative from the left at $x = x_0$ is

$$f_L'(x_0) = \lim_{h \to 0^-} \frac{f(x_0 + h) - f(x_0)}{h}$$

Also, the derivative $f'(x_0)$ exists if and only if $f'_R(x_0) = f'_L(x_0)$ and when this happens $f'(x_0)$ is equal to the one-sided derivatives.

PROPOSITION 7 Let I be an interval and $f:I \to \mathbf{R}$ be a convex function. If x_0 is in the interior of I, then
(a) $f'_R(x_0)$ and $f'_L(x_0)$ both exist.
(b) f is continuous at x_0.
Proof: (a) First let $a,b,c \in I$ such that $a < b < c$. We shall first establish the following inequality (see Figure 3).

$$\frac{f(c) - f(b)}{c - b} \geq \frac{f(c) - f(a)}{c - a} \geq \frac{f(b) - f(a)}{b - a} \tag{2}$$

Since $a < b < c$, there exist $\alpha \in (0, 1)$ such that $b = \alpha a + (1 - \alpha)c$. Solving for α and $1 - \alpha$ gives

$$\alpha = \frac{b - c}{a - c} \quad \text{and} \quad 1 - \alpha = \frac{a - b}{a - c}$$

Hence,

$$f(b) = f(\alpha a + (1 - \alpha)c) \leq \alpha f(a) + (1 - \alpha)f(c) = \frac{b - c}{a - c}f(a) + \frac{a - b}{a - c}f(c)$$

which implies that

$$f(b) - f(a) \leq \left(\frac{b - c}{a - c} - 1 \right) f(a) + \frac{a - b}{a - c}f(c) = \frac{b - a}{a - c}f(a) + \frac{a - b}{a - c}f(c)$$

$$f(b) - f(a) \leq \frac{b - a}{a - c}(f(a) - f(c))$$

$$\frac{f(b) - f(a)}{b - a} \leq \frac{f(a) - f(c)}{a - c} = \frac{f(c) - f(a)}{c - a}$$

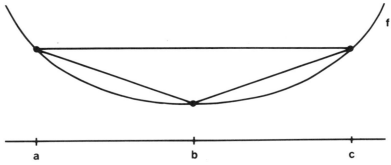

Figure 3   One-sided derivatives.

The other inequality is established in a similar fashion and is left as an exercise.

Now suppose that x_0 is in the interior of I. This means that there exist points $a,c \in I$ such that $x_0 \in (a,c) \subseteq I$. Consider the interval (x_0,c). Let $x,z \in (x_0,c)$ such that $x_0 < x < z < c$. Then applying inequality (2) to $x_0 < x < z$ and $a < x_0 < x$ gives

$$\frac{f(z)-f(x)}{z-x} \geq \frac{f(z)-f(x_0)}{z-x_0} \geq \frac{f(x)-f(x_0)}{x-x_0}$$

and

$$\frac{f(x)-f(x_0)}{x-x_0} \geq \frac{f(x)-f(a)}{x-a} \geq \frac{f(x_0)-f(a)}{x_0-a}$$

so,

$$\frac{f(z)-f(x_0)}{z-x_0} \geq \frac{f(x)-f(x_0)}{x-x_0} \geq \frac{f(x_0)-f(a)}{x_0-a}$$

As $x \to x_0^+$. the expression $(f(x)-f(x_0))/(x-x_0)$ is decreasing and bounded from below. It follows that $f_R'(x_0) = \lim_{x \to x_0} (f(x)-f(x_0))/(x-x_0)$ exists. Likewise in a similar fashion $f_L'(x_0)$ exists. This completes the proof of (a).
(b) Now,

$$\lim_{x \to x_0^+} f(x) = \lim_{x \to x_0^+} \left[\frac{f(x)-f(x_0)}{x-x_0}(x-x_0)+f(x_0) \right] = f_R'(x_0)\cdot 0 + f(x_0) = f(x_0)$$

Likewise, $\lim_{x \to x_0^-} f(x) = f(x_0)$. Hence, $\lim_{x \to x_0} f(x) = f(x_0)$ and f is continuous at $x = x_0$.

Our next proposition establishes a convenient test for the convexity of a twice continuously differentiable function.

PROPOSITION 8   Let $f:I \to \mathbf{R}$ be a twice continuously differentiable function over the interval I. Then f is convex if and only if $f''(x) \geq 0$ for every $x \in I$.
Proof: First suppose that $f''(x) \geq 0$ for every $x \varepsilon I$. Then f' is nondecreasing over I. Let $z = \lambda x + (1-\lambda)y$, where $\lambda \in (0,1)$, $x < y$ and $x,y \in I$. Now

$$f(z)-f(x) = \int_x^z f'(t)\, dt \leq f'(z)(z-x)$$

and

$$f(y)-f(z) = \int_z^y f'(t)\, dt \geq f'(z)(y-z)$$

Also

$$z-x = \lambda x + (1-\lambda)y - x = (1-\lambda)(y-x)$$

and

$$y-z = y - \lambda x - (1-\lambda)y = \lambda(y-x)$$

Therefore,

$$f(z) \leq f(x) + f'(z)(z - x) = f(x) + (1 - \lambda)f'(z)(y - x)$$

and

$$f(z) \leq f(y) - f'(z)(y - z) = f(y) - \lambda f'(z)(y - x)$$

Hence,

$$f(z) = \lambda f(z) + (1 - \lambda)f(z) \leq \lambda[f(x) + (1 - \lambda)f'(z)(y - x)]$$
$$+ (1 - \lambda)[f(y) - \lambda f'(z)(y - x)]$$
$$= \lambda f(x) + (1 - \lambda)f(y)$$

and f is a convex function over I.

Next suppose that f is convex over I. If there exists $x_o \in (a,b)$ such that $f''(x_o) < 0$, then since f'' is continuous over I there must exists $\varepsilon > 0$ such that $f''(x) < 0$ whenever $x \in (x_o - \varepsilon, x_o + \varepsilon)$. But then using an argument identical to that in the first part of the proof, it would then follow that f is strictly concave over $(x_o - \varepsilon, x_o + \varepsilon)$. This is a contradiction and hence $f''(x) \geq 0$ for all $x \in I$.

For example the function $f(x) = -\ln x$ is convex over $(0,\infty)$ since $f''(x) = x^{-2} \geq 0$ over $(0, \infty)$. The function $g(x) = x^3$ is convex over $[0, \infty)$ since $g''(x) = 6x \geq 0$ over $[0, \infty)$. The function g is not convex over $(-\infty, \infty)$.

Exercises

7.14   Let $f: I \to \mathbf{R}$ be a twice continuously differentiable function on the interval I. Prove the following:
(a) f is strictly convex function over I if $f''(x) > 0$ over I.
(b) f is concave over I if and only if $f''(x) \leq 0$ for every $x \in I$.
(c) f is strictly concave over I if $f''(x) < 0$ for every $x \in I$.

7.15   Use Proposition 8 and Problem 7.14 to determine where the following functions are convex, strictly convex, concave and strictly concave.
(a) $f(x) = e^{ax}$, where a is a real constant
(b) $f(x) = x^p$, where $1 \leq p < \infty$
(c) $f(x) = (a^2 - x^2)^{-1/2}$
(d) $f(x) = x \ln x$
(e) $f(x) = x^3 - 3x$
(f) $f(x) = 3 - \dfrac{4}{x} + \dfrac{4}{x^2}$
(g) $f(x) = x^2 + \dfrac{1}{x}$

7.16   Complete the verification of inequality (2) in the proof of Proposition 7.

7.17 If f is a convex function on I, then prove that both f'_R and f'_L have at most a countable number of discontinuities in I.

7.18 If f is a convex function on I, then prove that $f'_L(x) \leq f'_R(x)$ for all x in the interior of I.

7.19 Prove that a function $f: (a,b) \to \mathbf{R}$ is convex (strictly convex) if and only if there exists an increasing (strictly increasing) function $g: (a,b) \to \mathbf{R}$ such that $f(x) = f(c) + \int_c^x g(t)dt$, where $c \in (a,b)$.

3. SOME TOPICS FROM CALCULUS

In this section we shall briefly review some concepts from the differential calculus of several variables. Only the basic definitions and results will be stated. More detailed discussions can be found in any of the calculus books listed in the references found at the end of this chapter.

A function $f: S \to \mathbf{R}$, where $S \subseteq \mathbf{R}_{n \times 1}$ is *differentiable* at any interior point $\bar{x} \in S$ if and only there exist functions $\varepsilon_i = \varepsilon_i([\Delta x_1 \dots \Delta x_n]^T)$ so that

$$\Delta f = \frac{\partial f(\bar{x})}{\partial x_1} \Delta x_1 + \cdots + \frac{\partial f(\bar{x})}{\partial x_n} \Delta x_n + \varepsilon_1 \Delta x_1 + \cdots + \varepsilon_n \Delta x_n$$

where $\varepsilon_1 \to 0, \cdots, \varepsilon_n \to 0$ as $\Delta x_1 \to 0, \cdots, \Delta x_n \to 0$. (Here $\Delta f = f(\bar{x} + \Delta x) - f(\bar{x})$, where $\bar{x} = [\bar{x}_1 \dots \bar{x}_n]^T$ and $\Delta x = [\Delta x_1 \dots \Delta x_n]^T$). If f is differentiable at $\bar{x}$, then f is continuous at $\bar{x}$ and all the first order partials $\partial f(\bar{x})/\partial x_i$ exist. If all the first order partials exist and are continuous over some neighborhood of $\bar{x}$, then f is differentiable at $\bar{x}$. The function f is *continuously differentiable* at $\bar{x}$ whenever all of the first order partials exist and are continuous throughout a neighborhood of $\bar{x}$. A function having continuous second partial derivatives throughout a neighborhood of a point is said to be *twice continuously differentiable* at the point.

Let $v = [v_1 \cdots v_n]^T$ be a nonzero vector in $\mathbf{R}_{n \times 1}$. Let $L = \{\bar{x} + tv: t \in \mathbf{R}\}$. Then L is the line in $\mathbf{R}_{n \times 1}$ that passes through $\bar{x}$ in the direction v. If

$$\lim_{t \to 0} \frac{f(\bar{x} + tv) - f(\bar{x})}{t}$$

exists, then the limit is called the *derivative of f at $\bar{x}$ in the direction v* (*directional derivative*). In this case we write

$$D_v f(\bar{x}) = \lim_{t \to 0} \frac{f(\bar{x} + tv) - f(\bar{x})}{t}$$

Unless otherwise stated in this section we shall assume that f is continuous and has continuous partial derivatives throughout some neighborhood of x. In this case

$$D_V f(x) = \lim_{t \to 0} \frac{f(\bar{x} + tv) - f(\bar{x})}{t} = \sum_{i=1}^{n} \frac{\partial f(\bar{x})}{\partial x_i} v_i$$

Now suppose that v is a unit vector, i.e., $\|v\| = 1$. Then $D_v f(\bar{x})$ is the slope of the tangent line at $(\bar{x}, f(\bar{x}))$ in the cross section of the graph of f with the vertical plane passing through L (see Figure 4). Also, $D_V f(\bar{x})$ gives the instantaneous rate of change of f at $\bar{x}$ in the direction of v. If $v = e_i = [0 \cdots 0 \ 1 \ 0 \cdots 0]^T$, where the only nonzero entry of e_i is in the i-th position, then $D_V f(x) = \partial f(\bar{x})/\partial x_i$.

The *gradient* of f at $\bar{x}$ is denoted by $\Delta f(\bar{x})$ (or $(\partial f/\partial x)(\bar{x})$ and defined by

$$\nabla f(\bar{x}) = \left[\frac{\partial f(\bar{x})}{\partial x_1} \cdots \frac{\partial f(\bar{x})}{\partial x_n} \right]^T$$

Whenever f is differentiable at $\bar{x}$, then the directional derivative can be computed by $D_V f(\bar{x}) = f(\bar{x})^T v$. Also,

$$\max_{\|v\|=1} D_V f(\bar{x}) = \|\nabla f(\bar{x})\|$$

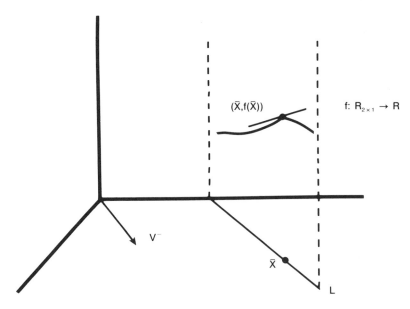

Figure 4   Directional derivative

and if $\nabla f(\bar{x}) \neq 0$, then $\|\nabla f(\bar{x})\|$ occurs whenever

$$v = \frac{\nabla f(\bar{x})}{\|\nabla f(\bar{x})\|}$$

Likewise,

$$\min_{\|v\|=1} D_v f(\bar{x}) = -\|\nabla f(\bar{x})\|$$

which occurs when

$$v = \frac{-\nabla f(\bar{x})}{\|\nabla f(\bar{x})\|}$$

provided $\nabla f(\bar{x}) \neq 0$. Thus, the gradient vector $\nabla f(\bar{x})$ points in the direction of steepest ascent and the direction of steepest descent is the opposite direction.

If a differentiable function f has either a relative maximum or a relative minimum value at any interior point, $\bar{x}$, of its domain then it necessarily follows that $\nabla f(\bar{x}) = 0$. Such points are called *critical* points. all boundary points must be investigated as well.

When $f: S \rightarrow \mathbf{R}$, where S is a closed and bounded subset of $\mathbf{R}_{n \times 1}$, is continuous, then there exist points x^* and y^* in S such that $f(x^*) = \min\{f(x): x \in S\}$ and $f(y^*) = \max\{f(x): x \in S\}$. This result is known as the *Weierstrass theorem*.

When $f(\bar{x}) = c$, each $\partial f(\bar{x})/\partial x_i$ is continuous at $\bar{x}$, and $\nabla f(\bar{x})$ is not the zero vector, then $\nabla f(\bar{x})$ is perpendicular to the set $\{x: f(x) = c\}$, where c is a constant. Also, an equation of the tangent plane to the surface $f(x) = c$ at $\bar{x}$ is

$$0 = \nabla f(\bar{x})^T (x - \bar{x}) = \sum_{i=1}^{n} \frac{\partial f(\bar{x})}{\partial x_i} (x_i - \bar{x}_i)$$

Now, identify (x,y) with the column matrix $[x^T y]^T = [x_1 \ldots x_n y]^T$. If $y = f(x)$, then let $F(x,y) = f(x) - y$. Since $\nabla F(x,y) = [\nabla f(x)^T \ -1]^T$, then an equation of the tangent plane to the surface $\{(x,y): y = f(x)\} = \{(x,y): F(x,y) = 0\}$ at the point $(\bar{x}, \bar{y})$, where $\bar{y} = f(\bar{x})$, is

$$0 = \sum_{i=1}^{n} \frac{\partial f(\bar{x})}{\partial x_i} (x - \bar{x}_i) - (y - \bar{y})$$

If we denote the matrix product $z^T x$ by $<z|x>$, then the above equation can be expressed in the form $0 = <(\nabla f(\bar{x}),1) | (x,y) - (\bar{x},\bar{y})>$. Thus, the tangent plane to the surface at $(\bar{x},\bar{y})$ is the hyperplane

$$\{(x,y): <(\nabla f(\bar{x}),-1)|(x,y) - (\bar{x},\bar{y})> = 0\} = \{(x,y): y = \bar{y} + \nabla f(\bar{x})^T (x - \bar{x})\}$$

When f has second partial derivatives throughout a neighborhood of $\bar{x}$, then the *Hessian* matrix of f at $\bar{x}$ is defined by

$$H_f(\bar{x}) = \left[\frac{\partial^2 f(\bar{x})}{\partial x_i \partial x_j}\right] = \begin{bmatrix} \dfrac{\partial^2 f(\bar{x})}{\partial^2 x_1} & \cdots & \dfrac{\partial^2 f(x)}{\partial x_1 \partial x_n} \\ & \vdots & \\ \dfrac{\partial^2 f(\bar{x})}{\partial x_n \partial x_1} & \cdots & \dfrac{\partial^2 f(\bar{x})}{\partial^2 x_n} \end{bmatrix}$$

Notice that $H_f(\bar{x})$ is a symmetric matrix, i.e., $H_f(\bar{x})^T = H_f(\bar{x})$, whenever the second partial derivatives are continuous throughout the neighborhood of $\bar{x}$. When f is understood, the notation $H_f(x^*)$ will be simplified to read $H(x^*)$.

Using both the gradient and the Hessian, *Taylor's theorem* can be extended to real valued functions of several real variables. Formally, let $f: S \to \mathbf{R}$ be twice continuously differentiable over an open convex subset S of $\mathbf{R}_{n \times 1}$. Then for every $x \in \mathbf{R}_{n \times 1}$ and $\bar{x} \in S$ it follows that

$$f(x) = f(\bar{x}) + \nabla f(\bar{x})^T(x - \bar{x}) + \frac{1}{2}(x - \bar{x})^T H(\lambda x + (1 - \lambda)\bar{x})(x - \bar{x})$$

for some $\lambda \in (0, 1)$.

Moreover, there exists a function $\delta(x,\bar{x})$ so that $\delta(x,\bar{x}) \to 0$ as $x \to \bar{x}$ and

$$f(x) = f(\bar{x}) + \nabla f(\bar{x})^T(x - \bar{x}) + \frac{1}{2}(x - \bar{x})^T H(\bar{x})(x - \bar{x}) + \delta(x,\bar{x})\|x - \bar{x}\|^2$$

Finally, if $u = f(x_1, \ldots x_n)$ is a differentiable function and each x_j is a function of the independent variables $r_1, \ldots, r_m$ such that each $\partial x_j/\partial r_k$ exists, then

$$\frac{\partial u}{\partial r_k} = \sum_{j=1}^{n} \frac{\partial f}{\partial x_j} \frac{\partial x_j}{\partial r_k}$$

This is known as the generalized chain rule.

4.   CONVEX FUNCTIONS OF SEVERAL VARIABLES

Our first result gives necessary and sufficient conditions for a differentiable function to be convex. The second result generalizes Proposition 8 for functions of several variables.

PROPOSITION 9   Let $f:K \to \mathbf{R}$, where K is an open convex subset of $\mathbf{R}_{n \times 1}$, be differentiable. Then f is convex if and only if $f(x) - f(y) \geq \nabla f(y)^T(x - y)$, for all $x,y \in K$.

Proof: First suppose that f is convex over K. Then for all $x, y \in K$ and $\lambda \varepsilon (0,1)$ it follows that $\lambda f x) + (1 - \lambda)f(y) \geq f(\lambda x + (1 - \lambda)y)$. Thus,

$$\lambda[f(x) - f(y)] \geq f(\lambda x + (1 - \lambda)y) - f(y)$$

$$f(x) - f(y) \geq \frac{f(\lambda x + (1 - \lambda)y) - f(y)}{\lambda}$$

and

$$f(x) - f(y) \geq \lim_{\lambda \to 0^+} \frac{f(\lambda x + (1 - \lambda)y) - f(y)}{\lambda} = D_{x-y}f(y) = \nabla f(y)^T(x - y)$$

Next let us suppose that $f(x) - f(y) \geq \nabla f(y)^T(x - y)$ for all $x, y \in K$. Let $z = \lambda x + (1 - \lambda)y$ where $\lambda \in (0, 1)$. Then $f(x) - f(z) \geq \nabla f(z)^T(x - z)$ and $f(y) - f(z)) \geq \nabla f(z)^T(y - z)$. Therefore,

$$\lambda(f(x) - f(z)) + (1 - \lambda)(f(y) - f(z)) \geq \lambda \nabla f(z)^T(x - z) + (1 - \lambda)\nabla f(z)^T(y - z),$$

$$\lambda f(x) + (1 - \lambda)f(y) - \lambda f(z) - f(z) + \lambda f(z) \geq$$
$$\nabla f(z)^T(\lambda x + (1 - \lambda)y) - (\lambda + (1 - \lambda))\nabla f(z)^T z$$

and

$$\lambda f(x) + (1 - \lambda)f(y) - f(z) \geq \nabla f(z)^T z - \nabla f(z)^T z = 0$$

Thus, $f(\lambda x + (1 - \lambda)y) = f(z) \leq \lambda f(x) + (1 - \lambda)f(y)$ which implies that f is convex.

PROPOSITION 10 Let $f : K \to \mathbf{R}$ be a twice continuously differentiable function, where K is an open convex subset of $\mathbf{R}_{n \times 1}$. Then f is convex over K if and only if $z^T H_f(y + \lambda z)z \geq 0$ for every point $y + \lambda z \in K$, where $y \in K, z \in \mathbf{R}_{n \times 1}$ and $\lambda \in \mathbf{R}$.

Proof: Let $y \in K$, $z \in \mathbf{R}_{n \times 1}$ and $\lambda \in \mathbf{R}$. Then the line passing through y in the direction z is the set of points $\ell = \{y + \lambda z : \lambda \in \mathbf{R}\}$. Since K is an open convex set, then $K \cap \ell$ is an open line segment contained in K. Now f is convex over K if and only if f is convex over all such line segments. Suppose that both $y + \lambda_1 z$ and $y + \lambda_2 z$ are points on $K \cap \ell$ and $\lambda_1 < \lambda_2$. Consider the interval $I = (\lambda_1, \lambda_2) = \{\lambda : \lambda_1 < \lambda < \lambda_2\}$. Define $g : I \to \mathbf{R}$ by $g(\lambda) = f(y + \lambda z)$. It is a simple problem to show that g is a convex function over I if and only if f is a convex function over $K \cap \ell$. But g is convex over I if and only if $g''(\lambda) \geq 0$, for every $\lambda \in I$ (see Exercise 7.29). Moreover,

$$g'(\lambda) = \lim_{h \to 0} \frac{g(\lambda + h) - g(\lambda)}{h} = \lim_{h \to 0} \frac{f(y + \lambda z + hz) - f(y + \lambda z)}{h} = D_z f(y + \lambda z)$$

$$= \nabla f(y + \lambda z)^T z = \sum_{i=1}^{n} \frac{\partial f(y + \lambda z)}{\partial x_i} z_i$$

Using the chain rule it follows that

$$g''(\lambda) = \sum_{j=1}^{n} \sum_{i=1}^{n} \frac{\partial^2 f(y + \lambda z)}{\partial x_j \partial x_i} z_i z_j \tag{2}$$

But a simple calculation shows that the right hand side of equation (2) is $z^T H_f(y + \lambda z)z$. Thus, f is convex over $K \cap \ell$ if and only if $z^T H_f(y + \lambda z)z \geq 0$ for all $z \in \mathbf{R}_{n \times 1}$.

We should mention that since K is an open convex set then every point in K can be expressed in the form $y + \lambda z$ where $y \in K, z \in \mathbf{R}_{n \times 1}$ and $\lambda \in \mathbf{R}$. Thus, f is convex over K if and only if the Hessian matrix is positive semidefinite at each point in K.

EXAMPLE 2   Examine the convexity of $f(x) = x_1^3 + x_2^2 + x_3$, where $x = [x_1 \ x_2 \ x_3]^T$. Since

$$H_f(x) = \begin{vmatrix} 6x_1 & 0 & 0 \\ 0 & 2 & 0 \\ 0 & 0 & 0 \end{vmatrix}$$

then $z^T H_f(x)z = 6x_1 z_1^2 + 2z_2^2$. Clearly, $z^T H_f(x)z \geq 0$, for all $z = [z_1 \ z_2 \ z_3]^T \in \mathbf{R}_{3 \times 1}$, only when $x_1 \geq 0$. Thus, f is convex over $K = \{x = [x_1 \ x_2 \ x_3]^T : x_1 \geq 0\}$.

It can be shown by a proof similar to that of Proposition 10, that a twice continuously differentiable function defined on a convex subset K of $\mathbf{R}_{n \times 1}$ is strictly convex whenever its Hessian matrix is positive definite over K, i.e., $z^T H_f(x)z > 0$ for all $x \in K$ and $z \neq 0$. The converse statement is not true, i.e., it is possible to construct examples of strictly convex functions that do not have a positive definite Hessian matrix over K. However, any strictly convex function is a convex function and hence has a positive semidefinite Hessian matrix at each point of its domain K.

The following result is often used to establish that a matrix is positive definite. The proof can be found in [9] of the references at the end of the chapter.

PROPOSITION 11   Each of the following conditions is a necessary and sufficient condition for a square real symmetric matrix $A = [a_{ij}] \in \mathbf{R}_{n \times n}$ to be positive definite:
(a)   All the eigenvalues of A are positive
(b) $a_{11} > 0$ and each determinant (also called principal minor)

$$\begin{vmatrix} a_{11} & a_{12} \\ a_{21} & a_{22} \end{vmatrix} > 0, \begin{vmatrix} a_{11} & a_{12} & a_{13} \\ a_{21} & a_{22} & a_{23} \\ a_{31} & a_{32} & a_{33} \end{vmatrix} > 0, \dots, |A| > 0$$

EXAMPLE 3   The function $f(x) = x_1^2 + 2x_2^2 + x_3^2 + 3x_1 - x_2 + 6x_3 + x_2x_3 - 9$, where $x = [x_1 \ x_2 \ x_3]^T$ is strictly convex over $\mathbf{R}_{3\times1}$ since

$$H_f(x) = \begin{vmatrix} 2 & 0 & 0 \\ 0 & 4 & 1 \\ 0 & 1 & 2 \end{vmatrix}$$

$$2 > 0, \quad \begin{vmatrix} 2 & 0 \\ 0 & 4 \end{vmatrix} > 0 \text{ and } \begin{vmatrix} 2 & 0 & 0 \\ 0 & 4 & 1 \\ 0 & 1 & 2 \end{vmatrix} = 14 > 0$$

for all $x \in \mathbf{R}_{3\times1}$.

Analogous results also hold true for concave functions.
In particular, a twice continuously differentiable function over a convex set K is concave if and only if the Hessian matrix $H_f(x)$ is negative semidefinite for each $x \in K$. More will be said about other such results in the following exercises.

Exercises

7.20   Let $f:K \to \mathbf{R}$ where K is on open convex subset of $\mathbf{R}_{n\times1}$, be twice continuously differentiable.
(a)   Prove that f is strictly convex over K whenever the Hessian matrix, $H_f(x)$, is positive definite at each $x \in K$.
(b)   Give an example of a strictly convex function defined over a convex set K that does not have a positive definite Hessian matrix, $H_f(x)$, at each $x \in K$.
(c)   Prove that f is concave of K if and only if the Hessian matrix, $H_f(x)$, is negative semidefinite for each $x \in K$.
(d)   Prove that f is strictly concave over K whenever the Hessian matrix, $H_f(x)$, is negative definite for each $x \in K$.
(e)   Give an example of a strictly concave function over a convex set K that does not have a negative definite Hessian matrix, $H_f(x)$, at each $x \in K$.

7.21   Prove that a matrix M is negative definite if and only if the principal minors in Proposition 11 of odd order are negative and those of even order are positive.

7.22   Determine if the following functions are convex, strictly convex, concave or strictly concave over the indicated sets.
(a) $f(x) = e^{x_1^2 + x_2^2}$, where $x = [x_1 \ x_2]^T \in K = \mathbf{R}_{2\times1}$.
(b) $f(x) = 2(x_1 + 2)^2 + x_1x_2 + e^{x_1x_2 + x_3}$, where $x = [x_1 \ x_2 \ x_3] \in K = \mathbf{R}_{3\times1}$.
(c) $f(x) = 4(x_1x_2)^2 - x_1x_2$, where $x = [x_1 \ x_2]^T \in K = \mathbf{R}_{2\times1}$.
(d) $f(x) = -(x_1x_2x_3)^{1/3}$, where $x = [x_1 \ x_2 \ x_3]^T \in K = \mathbf{R}_{3\times1}$.
(e) $f(x) = 8x_1x_2 + 4x_1 + 12x_2 - 16x_1^2$, where $x = [x_1 \ x_2]^T \in K = \mathbf{R}_{2\times1}$.
(f) $f(x) = x_1 + x_2 + 3x_3 - \sqrt{x_1^2 + x_2^2 + x_3^2}$, where $x = [x_1 \ x_2 \ x_3]^T \in K = \mathbf{R}_{3\times1}$.

(g) $f(x) = -10x_1 - 20x_2 - x_1x_2 - 2x_1^2 - 2x_2^2$, where $x = [x_1 \; x_2]^T \in K = \mathbf{R}_{2 \times 1}$.
(h) $f(x) = 6x_1 + 2x_1x_2 - 2x_1^2 - 2x_2^2$, where $x = [x_1 \; x_2]^T \in K = \mathbf{R}_{2 \times 1}$.
(j) $f(x) = \ln(x_1^2 + x_2^2)$, where $x = [x_1 \; x_2]^T \in K = \mathbf{R}_{2 \times 1}$.

(k) $f(x) = \sqrt{x_1^2 + x_2^2 + x_3^2}$, where $x = [x_1 \; x_2 \; x_3]^T \in K = \mathbf{R}_{3 \times 1}$.

7.23   (a) If $f(x) = (1/2)x^T A x$, then prove that $\nabla f(x) = (1/2)(A + A^T)x$. When A is symmetric, show that $\nabla f(x) = Ax$.
(b)   Let $f(x) = (1/2)(x^T A x) + b^T x + c$, where A is a $n \times n$ symmetric matrix, b is a $n \times 1$ matrix and c is a constant. Prove that f is a convex function if and only if A is positive semidefinite by showing that $H_f(x) = A$.

7.24   Let $f : K \to \mathbf{R}$ be a convex function, where K is convex subset of $\mathbf{R}_{n \times 1}$. Then the vector $\lambda = [\lambda_1 \ldots \lambda_n]^T$ is called a *subgradient* of f at $\bar{x} \in K$ if and only if $f(x) - f(\bar{x}) \geq \lambda^T(x - \bar{x})$, for every $x \in K$. (Compare with Proposition 9.)
(a)   Prove that the collection of all subgradients of f at $\bar{x}$ is a convex set.
(b)   If $\bar{x} \in \text{int } K$, show that there exists a subgradient λ of f at $\bar{x}$ (Hint: Show that epi(f) is contained in a closed half space of a hyperplane $H(c, \beta)$, where $c = [\lambda^T \; -1]^T$ and $\beta = \lambda^T \bar{x} - f(\bar{x})$.)
(c)   Find an example of a function that has a subgradient at each point in the interior of its domain but is not a convex function.
(d)   If $g : K \to \mathbf{R}$ and g has a subgradient λ at each point $x \in \text{int } K$, then prove that g is a convex function over $\text{int } K$.
(e)   If f is differentiable at $\bar{x} \in \text{int } K$ and λ is any subgradient of f at $\bar{x}$, then $\lambda = \nabla f(\bar{x})$.

7.25   Let K be a convex set in $\mathbf{R}_{n \times 1}$. Define the *relative interior* of K to be $ri(k) = \{x \in K : \text{there exists } \varepsilon > 0 \text{ such that } N_\varepsilon(x) \cap \text{aff}(K) \subseteq K\}$. Then prove the following propositions:
(a) $ri(K)$ is a convex set. (Hint: Let $x, y \in ri(K)$. Since $y \in ri(K)$, there exists $\varepsilon > 0$ such that $N_\varepsilon(y) \cap \text{aff}(K) \subseteq K$. Let $\lambda \in (0,1)$ and denote $z = \lambda x + (1 - \lambda)y$. Show that $N_{(1-\lambda)\varepsilon}(z) \cap \text{aff}(K) \subseteq K$.)
(b)   A convex function $f : K \to E_1$ is continuous over $ri(K)$.

7.26   If $x = [x_1 \ldots x_n]^T \geq 0$, $\alpha = [\alpha_1 \ldots \alpha_n]^T > 0$ and $\sum_{i=1}^n \alpha_i = 1$, then prove that

$$x_1^{\alpha_1} x_2^{\alpha_2} \cdots x_n^{\alpha_n} \leq \alpha_1 x_1 + \cdots + \alpha_n x_n$$

This is sometimes called the *geometric mean-arithmetic mean inequality*. (Hint: The result is trivial when any $x_i = 0$. If each $x_i > 0$ rewrite

$$x_i^{\alpha_i} = e^{\alpha_i \ln x_i}$$

and use the fact that $f(t) = e^t$ is a convex function.)

7.27 If $x = [x_1 \ldots x_n]^T \geq 0$, $y = [y_1 \ldots y_n]^T \geq 0$, $p > 1$ and $1/p + 1/q = 1$, then

$$\sum_{i=1}^{n} x_i y_i \leq \left(\sum_{i=1}^{n} x_i^p \right)^{1/p} \left(\sum_{i=1}^{n} y_i^q \right)^{1/q}$$

This is called *Hölder's inequality*. (Hint: Apply Problem 7.26 on the terms x_i/u and y_i/v, where $u = (\Sigma x_i^p)^{1/p}$, $v = (\Sigma y_i^q)^{1/q}$)

7.28 If $x = [x_1 \ldots x_n]^T \geq 0$, $y = [y_1 \ldots y_n]^T \geq 0$ and $p \geq 1$, then

$$\left(\sum_{i=1}^{n} (x_i + y_i)^p \right)^{1/p} \leq \left(\sum_{i=1}^{n} x_i^p \right)^{1/p} + \left(\sum_{i=1}^{n} y_i^p \right)^{1/p}$$

This is called *Minkowski's inequality*. (Hint: Apply Problem 7.27 where $q = p/(p-1)$ when $p \neq 1$.)

7.29 Complete the proof of Proposition 10 by showing that
(a) $g(\lambda) = f(y + \lambda z)$ is convex over I if and only if f is convex over $K \cap \ell$.
(b) $g''(\lambda) = z^T H_f(y + \lambda z) z$.

7.30 Let $f: K \to R$ where K is an open convex subset of $R_{n \times 1}$, be differentiable. Then prove that f is concave over K if and only if $f(x) - f(y) \leq \nabla f(y)^T(x - y)$ for all $x, y \in K$.

7.31 Let $f: K \to R$, where K is a convex subset of $R_{n \times 1}$, be differentiable over int K. If f is convex, $y \in$ int K and $x \in K$, then prove that $\nabla f(y)^T(x - y) \leq f(x) - f(y)$.

5. OPTIMIZATION OF CONVEX FUNCTIONS

In this section we shall derive some important results concerning the minimization and maximization of a convex function. Our first result deals with the problem of minimizing a convex function of several variables over a convex set. Quite often when minimizing a function of several variables, we are only able to determine the relative minima. Our first result will show that when the function is convex any relative minimum is a global minimum.

THEOREM 1   Let $f: K \to R$ be a convex function, where K is a convex subset of $R_{n \times 1}$. If there exists $\bar{x} \in K$ such that f has a relative minimum at $\bar{x}$, then f has a global minimum at $\bar{x}$.
Proof: Since $\bar{x}$ gives a relative minimum value of f then there exists $\varepsilon > 0$ such that $f(\bar{x}) \leq f(z)$ for every $z \in K \cap N_\varepsilon(\bar{x})$. Let $x \in K \backslash N_\varepsilon(\bar{x})$. Since $x \notin N_\varepsilon(\bar{x})$ then $\|x - \bar{x}\| > \varepsilon$. Pick $\alpha \in (0, 1]$ such that $0 < \alpha < \varepsilon/(\|x - \bar{x}\|) \leq 1$ and let $z = \alpha x + (1 - \alpha)\bar{x}$ (see Figure 5). Since K is convex, $z \in K$. Also,

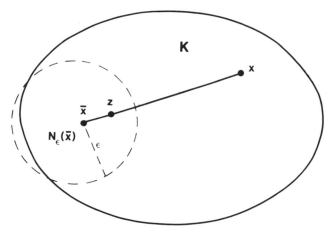

Figure 5   A global minimum

$$\|z - \bar{x}\| = \|\alpha x + (1 - \alpha)\bar{x} - \bar{x}\| = \|\alpha(x - \bar{x})\|$$

$$= |\alpha| \, \|x - \bar{x}\| < \frac{\varepsilon}{\|x - \bar{x}\|} \|x - \bar{x}\| = \varepsilon$$

which implies that $z \in N_\varepsilon(\bar{x})$. Therefore,

$$f(\bar{x}) \leq f(z) = f(\alpha x + (1 - \alpha)\bar{x}) \leq \alpha f(x) + (1 - \alpha)f(\bar{x})$$

and

$$0 \leq \alpha f(x) + (1 - \alpha)f(\bar{x}) - f(\bar{x}) = \alpha(f(x) - f(\bar{x}))$$

But $\alpha > 0$ and hence $0 \leq f(x) - f(\bar{x})$. Thus, $f(\bar{x}) \leq f(x)$ for all $x \in K$ and f has a global minimum of $f(\bar{x})$ at $\bar{x}$.

EXAMPLE 4   Minimize $f(x) = e^{x^2}$ over $x \in K = \mathbf{R}$.
Solution: Since $f'(x) = 2xe^{x^2}$, then $f''(x) = 2e^{x^2} + 4x^2 e^{x^2} \geq 0$ for all $x \in \mathbf{R}$. Thus it follows from Proposition 8 that f is convex over $\mathbf{R}$. It follows from elementary calculus that any relative minimum must occur at a value of x for which $f'(x) = 0$ (the function is differentiable over $\mathbf{R}$). Since $0 = f'(x) = 2xe^{x^2}$ implies that $x = 0$, it follows that a possible relative minimum could exist at $x = 0$. Again from elementary calculus it follows that a relative minimum does exist at $x = 0$ since $f''(0) = 2$. It follows from Theorem 1 that f has a global minimum of $f(0) = 1$ when $x = 0$.

Our next result gives necessary and sufficient conditions for a convex continuously differentiable function to have a global minimum at a point of its domain. In the next section we shall derive some necessary and sufficient conditions for a function of several variables to have a relative extrema at a point of its domain.

PROPOSITION 12   Let $f:K\to R$ be a convex continuously differentiable function defined on a convex subset K of $\mathbf{R}_{n\times 1}$. Then f has a global minimum at $\bar{x}\in K$ if and only if $\nabla f(\bar{x})^T(x-\bar{x})\geq 0$ for every $x\in K$.

Proof: First suppose that $\nabla f(\bar{x})^T(x-\bar{x})\geq 0$, for every $x\in K$. Since f is a convex function over K, Proposition 9 implies that $f(x)-f(\bar{x})\geq \nabla f(\bar{x})^T(x-\bar{x})$, for each $x\in K$. Hence, $f(x)-f(\bar{x})\geq \nabla f(\bar{x})^T(x-\bar{x})\geq 0$ and $f(x)\geq f(\bar{x})$, for each $x\in K$. It follows that f has a global minimum at $\bar{x}$.

Next let us suppose that the function f has a global minimum at $\bar{x}$. Let $x\in K$. Since f has a global minimum at $\bar{x}$ and since K is a convex set it follows that $f(\bar{x})\leq f(\lambda x+(1-\lambda)\bar{x})$, for each $\lambda\in(0,1)$. Thus

$$\frac{f(\lambda(x-\bar{x})+\bar{x})-f(\bar{x})}{\lambda}=\frac{f(\lambda x+(1-\lambda)\bar{x})-f(\bar{x})}{\lambda}\geq 0$$

for each $\lambda\ \varepsilon\ (0,1)$. Since f is continuously differentiable, then

$$\nabla f(\bar{x})^T(x-\bar{x})=D_{x-\bar{x}}f(\bar{x})=\lim_{\lambda\to 0+}\frac{f(\lambda(x-\bar{x})+\bar{x})-f(\bar{x})}{\lambda}\geq 0$$

Geometrically, Proposition 12 says that a continuously differentiable convex function f has a global minimum at a boundary point $\bar{x}$ of K where $\nabla f(\bar{x})\neq 0$ if and only if the set K is contained in the upper half space $H^+(\nabla f(\bar{x}),\nabla f(\bar{x})^T\bar{x})$ where

$$H^+(\nabla f(\bar{x}),\nabla f(\bar{x})^T\bar{x})=\{x:\nabla f(\bar{x})^Tx\geq \nabla f(\bar{x})^T\bar{x}\}$$

$$=\{x:\nabla f(\bar{x})^T(x-\bar{x})\geq 0\}$$

(see Figure 6). Notice that the hyperplane $H(\nabla f(\bar{x}),\nabla f(\bar{x})^T\bar{x})$ is a subset of $\mathbf{R}_{n\times 1}$. When $\bar{x}\in int(k)$, a global minimum would only be possible whenever $\nabla f(\bar{x})=0$. In fact, when $\nabla f(\bar{x})=0$, then f has a global minimum at $\bar{x}$.

Our next theorem deals with the problem of finding a global maximum of a convex function over a convex polytope.

THEOREM 2   Let $f:K\to R$ be a convex function, where K is a convex polytope in $\mathbf{R}_{n\times 1}$. Then f has a global maximum which is achieved at one or more of the extreme points of K.

Proof: Since K is a convex polytope, the finite basis theorem (Theorem 6 of Chapter 2) implies that there exists a finite number, say m, of extreme points of K whose convex hull is all of K. Let these extreme points be denoted by $x_1,\ldots,x_i,\ldots,x_m$, where each $x_i=[x_{i_1}\ldots x_{i_n}]^T$. Then $K=conv\{x_1,\ldots,x_m\}$. Let $\bar{x}$ be any element in $\{x_1,\ldots,x_m\}$ for which

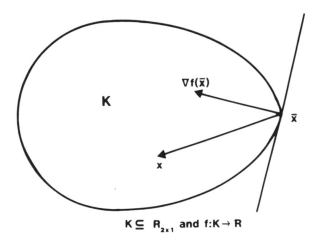

$$K \subseteq R_{2 \times 1} \text{ and } f:K \to R$$

Figure 6   In this figure $K \subseteq \mathbf{R}_{2\times 1}$ and $f:K\to\mathbf{R}$.

$f(\bar{x}) = \max\{f(x_i):i = 1,\dots,m\}$.   Let $x \varepsilon K$.   Then $x = \Sigma_{i=1}^{m} \alpha_i x_i$,   where $\Sigma_{i=1}^{m} \alpha_i = 1$ and each $0 \leq \alpha_i \leq 1$. Since

$$f(x) = f(\sum_{i=1}^{m} \alpha_i x_i) \leq \sum_{i=1}^{m} \alpha_i f(x_i) \leq \sum_{i=1}^{m} \alpha_i f(\bar{x}) = f(\bar{x}) \sum_{i=1}^{m} \alpha_i = f(\bar{x})$$

it follows that f has a global maximum of $f(\bar{x})$ at $\bar{x}$.

It should be noted that the maximum could occur at several, not just one, of the extreme points of K. Also, Theorem 2 has several generalizations, one of which is to the case where K is a convex polyhedron, which will be developed in the exercises at the end of this chapter. For example, the Weierstrass theorem from calculus says that a continuous function over a compact (i.e., closed and bounded) set K attains a global maximum at some point in K. (The function also attains a global minimum at some point in K.) Recall that a convex function is continuous over $ri(K)$ [Problem 7.25(b) of the last exercise set]. Suppose in addition that the function is continuous over all of K. It can be shown that any set K in $\mathbf{R}_{n\times 1}$ that is both convex and compact (i.e., closed and bounded) is the convex hull of its extreme points. (This is a special case of a more general result known as the Krein-Millman theorem, see [6].) Thus, using an argument similar to that in the proof of Theorem 2 it follows that any function that is continuous and convex over a compact convex subset K of $\mathbf{R}_{n\times 1}$ attains a global maximum at some extreme point of K.

EXAMPLE 5   Maximize $2x_1^2 + x_2^4$ subject to the constraints

$$- x_1 + 2x_2 \leq 10$$
$$5x_1 + 3x_2 \leq 41$$
$$2x_1 - 3x_2 \leq 8$$
$$x_1 \geq 0 \text{ and } x_2 \geq 0$$

Solution: Let $f(x) = 2x_1^2 + x_2^4$, where $x = [x_1 \ x_2]^T$. Since

$\dfrac{\partial f}{\partial x_1} = 4x_1$ and $\dfrac{\partial f}{\partial x_2} = 4x_2^3$ it follows that

$$H_f(x) = \begin{vmatrix} 4 & 0 \\ 0 & 12x_2^2 \end{vmatrix}$$

Since $z^T H_f(x)z = 4z_1^2 + 12x_2^2 z_2^2 \geq 0$ for every $z = [z_1 \ z_2]^T \in \mathbf{R}_{2\times 1}$ it follows that f is a convex function over $\mathbf{R}_{2\times 1}$. Let K be the set of points $x = [x_1 \ x_2]^T$ that satisfies the above constraints. Then K is a convex polytope with extreme points $[7 \ 2]^T$, $[4 \ 7]^T$, $[4 \ 0]^T$, $[0 \ 5]^T$ and $[0 \ 0]^T$ (see Example 18, Chapter 2). Theorem 2 implies that f has a global maximum that must occur at one of these extreme points. Since $f([7 \ \ 2]^T) = 114$, $f([4 \ \ 7]^T) = 2433$, $f([4 \ \ 0]^T) = 32$, $f([0 \ \ 5]^T) = 625$ and $f([0 \ \ 0]^T) = 0$, it follows that f has a global maximum value of 2433 which occurs at $[4 \ \ 7]^T$.

Exercises

7.32   Let $f: K \to \mathbf{R}$ be a concave function, where K is a convex subset of $\mathbf{R}$. Prove the following:
(a) If f has a relative maximum at $\bar{x} \in K$, then f has a global maximum at $\bar{x}$.
(b) If f is also continuously differentiable, then f has a global maximum at $\bar{x} \varepsilon K$ if and only if $\nabla f(x)^T(x - \bar{x}) \leq 0$ for every $x \in K$.
(c) If K is a convex polytope, then f has a global minimum which is achieved at one or more of the extreme points of K.
(d) The subset S of K that contains the points at which f attains a relative maximum is either empty or convex.
(e) If f attains a global minimum at point $\bar{x} \in ri(K)$, then f is constant over K.

7.33   If $f: K \to \mathbf{R}$ is a convex function, where K is a convex subset of $\mathbf{R}_{n\times 1}$ then prove the following:
(a) The subset S of K that contains the points at which f attains a relative minimum is either empty or convex.
(b) If f attains a global maximum at a point $\bar{x} \in ri(K)$, then f is constant over K.

7.34   (a) Prove that any relative minimum of a strictly convex function, defined over a convex set K, is the unique global minimum over K.
(b) Prove that if a convex function f defined over a convex set K has a relative minimum at $\bar{x} \in K$ and if f is strictly convex over a neighborhood of $\bar{x}$, then $\bar{x}$ is the unique point in K where f has a global minimum.

7.35   Let $f: K \rightarrow \mathbf{R}$ be a convex function, where K is a convex subset of $\mathbf{R}_{n \times 1}$.
(a) If $\nabla f(\bar{x}) = 0$, where $\bar{x} \in K$, then prove that $f(\bar{x})$ is a global minimum for f over K.
(b) Additionally, if $\bar{x} \in \text{int } K$, f is continuously differentiable throughout a neighborhood of $\bar{x}$, $H_f(x)$ exists and is positive definite throughout the neighborhood, then prove that $\bar{x}$ is the unique minimizing point for f in K.
(c) Also, if K is open and f is differentiable, then prove that f has a global minimum at $\bar{x} \in K$ if and only if $\nabla f(\bar{x}) = 0$.

7.36   Maximize $f(x) = x_1^2 + 3x_1x_2 + 5x_2^2$, where $x = [x_1 \ x_2]^T$ and

$$2x_1 - x_2 + 2 \geq 0$$
$$2x_1 - 4x_2 \leq 6$$
$$x_1 + 2x_2 - 11 \leq 0$$
$$x_1 \geq 0, x_2 \geq 0$$

7.37   Minimize $f(x) = -10x_1 - 20x_2 - x_1x_2 - 2x_1^2 - 2x_2^2$, where $x = [x_1 \ x_2 \ x_3 \ x_4]^T$ and

$$x_2 + x_3 \qquad = 8$$
$$x_1 + x_2 \qquad + x_4 = 10$$
$$x_1 \geq 0, x_2 \geq 0, x_3 \geq 0, x_4 \geq 0$$

7.38   Maximize $f(x) = \sqrt{x_1^2 + x_2^2 + x_3^2 + x_4^2 + x_5^2}$ where $x = [x_1 \ x_2 \ x_3 \ x_4 \ x_5]^T$ and

$$-2x_1 + x_2 + x_3 \qquad = 2$$
$$x_1 - 2x_2 \qquad + x_4 \qquad = 3$$
$$x_1 + 2x_2 \qquad + x_5 = 11$$
$$\text{all } x_i \geq 0$$

7.39   Minimize $f(x) = -(4x_1^2 + 25x_2^2)$ where $x = [x_1 \ x_2]^T$ and

$$-x_1 + 2x_2 \leq 10$$
$$5x_1 + 3x_2 \leq 41$$
$$2x_1 - 3x_2 \leq 8$$
$$x_1 \geq 0, x_2 \geq 0$$

7.40   Let $f:K \rightarrow \mathbf{R}$ be a convex function, where K is a convex subset of $\mathbf{R}_{n \times 1}$.
(a) Suppose that K does not contain any lines. Then prove that if f has a global maximum at $\bar{x} \in K$, then f also attains its maximum at some extreme point of K.
(b) If K is a convex polyhedron and f is bounded from above over K, then f does have a global maximum which occurs at an extreme point of K.

7.41   Let $f:K \rightarrow \mathbf{R}$ be a convex function, where K is a convex subset of $\mathbf{R}_{n \times 1}$. Prove that f has a global minimum at $\bar{x} \in K$ if and only if f has a subgradient $\lambda = [\lambda_1 \dots \lambda_n]^T$ at $\bar{x}$ such that $\lambda^T(x - \bar{x}) \geq 0$, for all $x \in K$ (Compare this result with Proposition 12.)

7.42   Let $f:K \rightarrow \mathbf{R}$ be a convex function, where K is a convex subset of $\mathbf{R}_{n \times 1}$.
(a) If f has a relative maximum at $\bar{x} \in K$, then $\lambda^T(x - \bar{x}) \leq 0$ for each $x \in K$ and each subgradient $\lambda = [\lambda_1 \dots \lambda_n]^T$ of $\bar{x}$.
(b) If f is differentiable and f has a relative maximum at $\bar{x} \in K$, then $\nabla f(\bar{x})^T(x - \bar{x}) \leq 0$, for every $x \in K$.
(c) Give an example showing that the converse of (b) is not true.

7.43   Let $f:\mathbf{R}_{n \times 1} \rightarrow \mathbf{R}$ be a twice continuously differentiable function. The *linear approximation of f at $\bar{x}$* is defined to be $\ell(x) = f(\bar{x}) + \nabla f(\bar{x})^T(x - \bar{x})$. The *quadratic approximation of f at $\bar{x}$* is defined to be $q(x) = f(\bar{x}) + \nabla f(\bar{x})^T(x - \bar{x}) + (1/2)(x - \bar{x})^T H_f(\bar{x})(x - \bar{x})$. Show that ℓ is an affine function over $\mathbf{R}_{n \times 1}$. Determine when q is convex, concave or affine.

7.44   Maximize $h(x) = e^{x_1^4 + x_2^2 + x_3^4}$, where $x = [x_1 \; x_2 \; x_3]^T$ belongs to the convex polytope having $[0 \;\; 0 \;\; 0]^T, [0 \;\; 4 \;\; 2]^T, [5 \;\; 6 \;\; 1]^T, [9 \;\; 0 \;\; 3]^T$ and $[7 \;\; 8 \;\; 1]^T$ as its extreme points.

6.   QUASICONVEX FUNCTIONS AND OTHER GENERALIZATIONS

In general the problem of determining the minima or maxima of a function of several variables is difficult. Often we must be contented to determine only the relative minima or relative maxima. However, if the function is convex (or concave), then Theorem 1 says that any relative minimum (maximum) is also a global minimum (maximum). Moreover, Theorem 2 tells us that a convex function defined on a convex polytope has a global maximum at some extreme point of the polytope. In this section we shall study some less restrictive types of functions for which similar results hold.

Let $f:K \rightarrow \mathbf{R}$, where K is a convex subset of $\mathbf{R}_{n \times 1}$. The function f is *quasiconvex* if and only if $f(\alpha x + (1 - \alpha)y) \leq \max\{f(x), f(y)\}$, whenever $x, y \in K$ and $\alpha \varepsilon (0,1)$. The function f is *quasiconcave* if and only if $-f$ is quasiconvex. since $\max\{-f(x), -f(y)\} = -\min\{f(x), f(y)\}$, it follows that f is

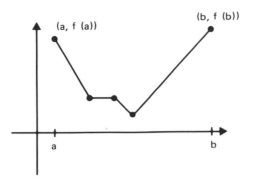

(a)   A quasiconvex function that is not strongly quasiconvex.

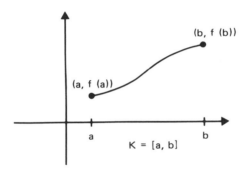

(b)   A strongly quasiconvex function.

Figure 7

quasiconcave if and only if $f(\alpha x + (1 - \alpha)y) \geq \min\{f(x), f(y)\}$. When the inequality is strict, i.e., $f(\alpha x + (1 - \alpha)y) < \max\{f(x), f(y)\}$, for $x \neq y$ and $\alpha \in (0,1)$, the function f is said to be *strongly quasiconvex*. Again, f is *strongly quasiconcave* if and only if $-f$ is strongly quasiconvex. Every strong quasiconvex function is a quasiconvex function. Moreover, any strictly convex function is a strongly quasiconvex function. Also, any convex function is also quasiconvex. Figure 7 illustrates some of the concepts.

Recall that the level sets, Lev$_\lambda f$, of a convex function f are always convex (Proposition 5). The converse is not true. For example in Figure 7(a) the level sets of f are all convex intervals, but f is certainly not a convex function. Our next proposition shows that convexity of the level sets of a function is a necessary and sufficient condition for the function to be quasiconvex.

PROPOSITION 13   Let $f:K \to \mathbf{R}$, where K is a convex subset of $\mathbf{R}_{n \times 1}$. Then f is quasiconvex if and only if Lev$_\lambda f$ is convex for each $\lambda \in \mathbf{R}$.

Proof: First suppose that f is quasiconvex and $\lambda \in \mathbf{R}$. Let $x,y \in \mathrm{Lev}_\lambda f$. Then $f(x) \leq \lambda$ and $f(y) \leq \lambda$. Let $\alpha \in (0,1)$. Since $f(\alpha x + (1 - \alpha)y) \leq \max\{f(x),f(y)\} \leq \lambda$, then $\alpha x + (1 - \alpha)y \in \mathrm{Lev}_\lambda f$ and hence $\mathrm{Lev}_\lambda f$ is convex.

Next suppose that each $\mathrm{Lev}_\lambda f$ is convex. Let $x,y \in K$ and $\alpha \in (0,1)$. Let $\lambda = \max\{f(x),f(y)\}$ and consider $\mathrm{Lev}_\lambda f$. Since $x,y \in \mathrm{Lev}_\lambda f$ and $\mathrm{Lev}_\lambda f$ is convex it follows that $\alpha x + (1 - \alpha)y \in \mathrm{Lev}_\lambda f$. Thus, $f(\alpha x + (1 - \alpha)y) \leq \lambda = \max\{f(x),f(y)\}$ and f is quasiconvex.

COROLLARY 1   Let $f:K \rightarrow \mathbf{R}$, where K is a convex subset of $\mathbf{R}_{n \times 1}$. Then f is quasiconvex if and only if $f(\Sigma^n_{i=1} \alpha_i x_i) \leq \max\{f(x_i): i = 1, \dots, n\}$ where each $x_i \varepsilon K$, each $\alpha_i \geq 0$, $\Sigma^n_{i=1} \alpha_i = 1$ and n is a natural number.
Proof: Suppose that f is quasiconvex and consider $\Sigma^n_{i=1} \alpha_i x_i$, where each α_i, x_i and n satisfy the above conditions. Let $\lambda = \max\{f(x_i): i = 1, \dots, n\}$. Since each $x_i \in \mathrm{Lev}_\lambda f$, a convex set, then $\Sigma^n_{i=1} \alpha_i x_i \in \mathrm{Lev}_\lambda f$. Hence, $f(\Sigma^n_{i=1} \alpha_i x_i) \leq \lambda = \max\{f(x_i): i = 1, \dots, n\}$. The proof of the reverse implication is trivial.

The next two propositions generalize the results of Theorem 2 and Theorem 1 to quasiconvex and strongly quasiconvex functions.

PROPOSITION 14   Let $f:K \rightarrow \mathbf{R}$, where K is a convex polytope in $\mathbf{R}_{n \times 1}$, be a quasiconvex function. Then f has a global maximum at one of the extreme points of K.
Proof: Let $x_1, \dots, x_n$ denote the extreme points of the convex polytope K. Let $x \in K$. Then $x = \Sigma^n_{i=1} \alpha_i x_i$, where each $\alpha_i \geq 0$ and $\Sigma^n_{i=1} \alpha_i = 1$. By Corollary 1, $f(x) \leq \max f(x_i)$. Thus, f clearly has a global maximum at any extreme point x_k, where $f(x_k) = \max f(x_i)$.

PROPOSITION 15   Let $f:K \rightarrow \mathbf{R}$, where K is a convex subset of $\mathbf{R}_{n \times 1}$, be a strongly quasiconvex function. If f has a relative minimum at $\bar{x} \in K$, then f has a global minimum at $\bar{x}$.
Proof: Since f has a relative minimum at $\bar{x}$, these exists $\varepsilon > 0$ such that $f(\bar{x}) \leq f(x)$ for every $x \in N_\varepsilon(\bar{x}) \cap K$. Suppose that there is some point $z \in K$ for which $z \neq \bar{x}$ and $f(z) \leq f(\bar{x})$. Consider $f(\alpha z + (1 - \alpha)\bar{x})$, where $\alpha \in (0,1)$ (see Figure 8). Since f is strongly quasiconvex, $f(\alpha z + (1 - \alpha)\bar{x}) < f(\bar{x})$. But for some α, $x = \alpha z + (1 - \alpha)\bar{x} \in N_\varepsilon(x) \cap K$ and hence $f(\bar{x}) \leq f(x)$. This is a contradiction and thus $f(\bar{x}) < f(z)$ for each $z \in K$, and f has a global minimum at $\bar{x}$.

Another simple restriction can be placed on f to insure that any relative minimum is also a global minimum. First we need the following definition. Let K be a subset of $\mathbf{R}_{n \times 1}$. The set K is said to be *star-shaped* at $\bar{x} \in K$

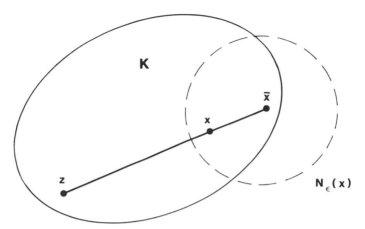

Figure 8   A global minimum

provided that $\alpha x + (1-\alpha)\bar{x} \; \varepsilon \; K$, whenever $x \in K$ and $\alpha \in (0,1)$. Clearly, any convex set is star-shaped at each of its points. Figure 9 gives several examples of star-shaped sets that are not convex.

   Suppose that K is star-shaped at $\bar{x}$ and $f:K \rightarrow \mathbf{R}$. Then f is *convex at* $\bar{x}$ if and only if $f(\alpha x + (1-\alpha)\bar{x}) \le \alpha f(x) + (1-\alpha)f(\bar{x})$ whenever $x \in K$ and $\alpha \in (0,1)$. Any convex function is convex at each point of its domain.

PROPOSITION 16   Let $f:K \rightarrow \mathbf{R}$, where K is a subset of $\mathbf{R}_{n \times 1}$, have a relative minimum at $\bar{x} \in K$. If K is star-shaped at $\bar{x}$ and f is convex at $\bar{x}$, then f has a global minimum at $\bar{x}$.
Proof: Since f has a relative minimum at $\bar{x}$, there exists $\varepsilon > 0$ such that $f(\bar{x}) \le f(x)$, whenever $x \in N_\varepsilon(\bar{x}) \cap K$. Suppose that $z \in K$ and $f(z) < f(\bar{x})$.

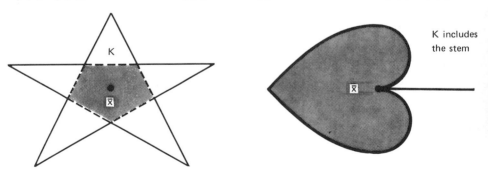

(a) The set K is star-shaped to   (b) The set K is star-shaped to x̄
any x̄ in the shaded area

Figure 9   Examples of star-shaped sets.

Consider $\alpha z + (1-\alpha)\bar{x}$ where $\alpha \in (0,1)$. Since K is star-shaped at $\bar{x}$ and f is convex at $\bar{x}$, it follows that $f(\alpha z + (1-\alpha)\bar{x}) \le \alpha f(z) + (1-\alpha)f(\bar{x}) < f(\bar{x})$. But $x = \alpha z + (1-\alpha)\bar{x} \in N_\varepsilon(\bar{x}) \cap K$ for some $\alpha \in (0,1)$ and hence $f(x) \ge f(\bar{x})$. This is a contradiction and hence $f(z) \ge f(\bar{x})$ for every $z \in K$.

These are only a few of the generalizations of convex functions which can be found in the literature. Additional generalizations are mentioned in the following exercises. Detailed discussions can be found in references [1], [6], and [9] at the end of this chapter.

Exercises

7.45 Let $f: K \to \mathbf{R}$ be a quasiconvex function over a convex subset K of $\mathbf{R}_{n \times 1}$. Prove the following results.

(a) Every relative minimum is a global minimum or f is constant in a neighborhood of the relative minimum.

(b) The subset of K where f has a global minimum is convex.

7.46 Let K be an open convex subset of $\mathbf{R}_{n \times 1}$ and $f: K \to \mathbf{R}$. Then prove that f is quasiconvex over K if and only if $f(x) \le f(w)$ always implies that $\nabla f(w)^T(x-w) \le 0$.

7.47 Let K be a convex subset of $\mathbf{R}_{n \times 1}$ and $f: K \to \mathbf{R}$. The function f is said to be *strictly quasiconvex* if and only if $f(\alpha x + (1-\alpha)z) < \max\{f(x),f(z)\}$ whenever $x,z \in S$, $\alpha \in (0,1)$ and $f(x) \ne f(z)$. Prove the following results.

(a) Every strongly quasiconvex function is strictly quasiconvex.

(b) Find an example of a strictly quasiconvex function that is not quasiconvex.

(c) Any relative minimum of a strictly quasiconvex function is a global minimum (not necessarily unique).

7.48 Let K be an open convex subset of $\mathbf{R}_{n \times 1}$ and $f: K \to \mathbf{R}$ be differentiable over K. Then f is *pseudoconvex* over K if and only if $x,z \in K$ and $\nabla f(z)^T(x-z) \ge 0$ implies that $f(x) \ge f(z)$. Prove the following results.

(a) Any differentiable convex function is a pseudoconvex function.

(b) Any pseudoconvex function is a strictly quasiconvex function.

(c) Any pseudoconvex function is a quasiconvex function.

(d) If f is a pseudoconvex function and $\nabla f(\bar{x}) = 0$, then f has a global minimum at $\bar{x}$.

7.49 Let $f: K \to \mathbf{R}$, where K is a convex subset of $\mathbf{R}_{n \times 1}$, be convex at $\bar{x} \in K$. If f is differentiable at $\bar{x}$, then $f(x) - f(\bar{x}) \ge \nabla f(\bar{x})^T(x-\bar{x})$.

REFERENCES

1. M. S. Bazaraa and C. M. Shetty, *Nonlinear Programming*, John Wiley and Sons, New York (1969).
2. P. S. Clark, Jr., *Calculus and Analytic Geometry*, D. C. Heath and Company, Lexington, Mass. (1974).
3. M. J. Forray, *Calculus and Analytic Geometry*, Macmillan, New York (1978).
4. S. I. Gass, *Linear Programming, Methods and Applications*, McGraw Hill, New York (1969).
5. W. H. Marlow, *Mathematics For Operations Research*, John Wiley and Sons, New York (1978).
6. A. W. Roberts and D. E. Varberg, *Convex Functions*, Academic Press, New York (1975).
7. R. T. Rockafellar, *Convex Analysis*, Princeton University Press, Princeton, N.J. (1970).
8. J. Stoer and C. Witzgall, *Convexity and Optimization in Finite Dimensions I*, Springer-Verlag, Berlin (1970).
9. G. Strang, *Linear Algebra and Its Applications*, Academic Press, New York (1976).

8
Optimality Conditions

The first five sections of this chapter contain conditions that are necessary and conditions that are sometimes sufficient for a relative extreme value of a real valued function of several real variables which satisfy various types of constraints. In particular, the Kuhn-Tucker necessary conditions for a relative maximum of a function of several nonnegative real variables that are subject to equality constraints are found in Section 4. Section 6 contains the Kuhn-Tucker theorem which establishes the relationship that exists between solutions of the type of problem just mentioned and the solutions of the corresponding saddle-point problem of the related Lagrangran function. Section 6 also contains necessary and sufficient conditions for a maximum value of a differentiable concave function of several real variables subject to inequality constraints. Finally, the Kuhn-Tucker conditions are applied to the quadratic programming problem resulting in a restatement of the problem as a linear complementarity problem.

1. UNCONSTRAINED PROBLEMS

We begin this section with a study of the following problem:

$$\text{Maximize } f(x), \text{ where } f: \mathbf{R}_{n \times 1} \to \mathbf{R} \tag{1}$$

As stated earlier in Chapter 1, we will probably have to be content with finding those points x^* where f has a relative maximum. Let x^* be such a point and assume that f is twice continuously differentiable throughout a neighborhood $N_\varepsilon(x^*)$ of x^*, where $x \in N_\varepsilon(x^*)$ implies that $f(x) \le f(x^*)$. Then for every direction vector v, where $\|v\| = 1$, it follows that

$$0 \ge D_v f(x^*) = \lim_{t \to 0} \frac{f(x^* + tv) - f(x^*)}{t} = \nabla f(x^*)^T v$$

Thus, it follows by considering the cases where $v = e_i$ and $v = -e_i$ that $\partial x/\partial x_i(x^*) = 0$. Doing this for each i, we have our *first necessary condition*,

$$\nabla f(x^*) = 0 \tag{2}$$

for a relative maximum at x^*.

Notice that every point x in $N_\varepsilon(x^*)$ can be expressed in the form $x = x^* + tv$, where $t > 0$ (in fact $0 < t < \varepsilon$, since $t = \|tv\| = \|x - x^*\| < \varepsilon$). Thus, applying Taylor's theorem we obtain

$$f(x^*) \geq f(x) = f(x^* + tv) = f(x^*) + t\nabla f(x^*)^T v + \frac{t^2}{2} v^T H(x^* + \lambda tv)v$$

where $\lambda \in (0,1)$. But $\nabla f(x^*) = 0$ and $\dfrac{t^2}{2} > 0$. Hence, a *second necessary condition* for f to have a relative maximum at x^* is

$$0 \geq v^T H(x^* + \lambda tv)v, \text{ where } \lambda \in (0,1) \tag{3}$$

It follows by continuity that (3) holds when $\lambda = 0$. In summary we have the following result.

THEOREM 1   Let $f: \mathbf{R}_{n \times 1} \to \mathbf{R}$ be twice continuously differentiable throughout a neighborhood of x^*. If f has a relative maximum at x^*, then it necessarily follows that
    (i) $\nabla f(x^*) = 0$.
    (ii) $H(x^*)$ is negative semidefinite.

Sufficient conditions for a relative maximum are given in the next theorem. The proof is left as an exercise [Exercise 8.1 (c)].

THEOREM 2   Let $f: \mathbf{R}_{n \times 1} \to \mathbf{R}$ be twice continuously differentiable throughout a neighborhood of x^*. Then a sufficient condition for $f(x)$ to have a strict relative maximum at x^*, where (2) holds, is that $H(x^*)$ be negative definite.

Comparable results hold for relative minimums (see Exercise 8.1).

EXAMPLE 1   Let $f: \mathbf{R}_2 \to \mathbf{R}$ be defined by $f(x) = 3x_1^3 + x_2^2 - 9x_1 + 4x_2$, where $x = [x_1 \quad x_2]^T$. Find the relative extreme values of the function.
Solution: Notice that

$$\nabla f(x) = [9x_1^2 - 9 \quad 2x_2 + 4]^T = [0 \quad 0]^T$$

implies that $x_1 = \pm 1$ and $x_2 = -2$. Checking $x = [1 \quad -2]^T$, we have that

$$H([1 \quad -2]^T) = \begin{bmatrix} 18 & 0 \\ 0 & 2 \end{bmatrix}$$

which is positive definite since $v^T H([1 \quad -2]^T)v = 18v_1^2 + 2v_2^2 > 0$ when $v \neq 0$. Thus, $[1 \quad -2]^T$ gives a strict relative minimum (see Exercise 9.1). Next consider $[-1 \quad -2]^T$. Since

$$H([-1 \quad -2]^T) = \begin{bmatrix} -18 & 0 \\ 0 & 2 \end{bmatrix}$$

then $v^T H([-1 \quad -2]^T)v = -18v_1^2 + 2v_2^2$, which may equal to zero when $v \neq 0$. Thus, the sufficient condition for either a relative maximum or minimum is not satisfied. Actually, the second necessary condition for either relative maximum or minimum is not satisfied. Therefore, $[1 \quad -2]^T$ gives the only relative extreme value of f. Notice that the first order necessary condition for a relative minimum of f at x^* is (2) [Exercise 8.1 (a)].

2. NONNEGATIVE VARIABLES

Again let $f: \mathbf{R}_{n \times 1} \to \mathbf{R}$ and consider the problem

$$\text{Maximize } f(x) \text{ subject to } x \geq 0 \qquad (4)$$

Suppose that f has a relative maximum at x^*, where $x^* \geq 0$. Then there exists a neighborhood $N_\varepsilon(x^*)$ of x^* so that whenever $x \in N_\varepsilon(x^*)$ and $x \geq 0$, then $f(x) \leq f(x^*)$. Represent x in the form $x = x^* + tv$, where v is a direction vector and $t > 0$. Assuming that f is twice continuously differentiable throughout $N_\varepsilon(x^*)$, Taylor's theorem implies that

$$f(x^*) \geq f(x) = f(x^*) + t\nabla f(x^*)^T v + \frac{t^2}{2} v^T H(x^* + \lambda tv)v$$

where $\lambda \in (0, 1)$. Therefore, $0 \geq f(x^*)^T v + t/2\, v^T H(x^* + \lambda tv)v$. Taking the limit as $t \to 0$, we have $0 \geq \nabla f(x^*)^T v$. If $x^* > 0$, we know that $\nabla f(x^*) = 0$. (Why?) Suppose that some $x_i^* = 0$. Let $v = e_i$. Then $0 \geq \nabla f(x^*)^T e_i = \partial f/\partial x_i(x^*)$. Finally, if $x^* \not> 0$ but $x_i^* > 0$, then consider the restriction of f to $T = \{[x_1^* \ldots x_{i-1}^* z_i x_{i+1}^* \ldots x_n^*]^T : z_i \geq 0\}$. Then f restricted to T has a relative maximum at x^*. It follows that $\partial f/\partial x_i = 0$. Thus, necessary conditions for a relative maximum of f at x^* include

$$\frac{\partial f}{\partial x_i}(x^*) = 0, \quad \text{if } x_i^* > 0$$

$$\frac{\partial f}{\partial x_i}(x^*) \leq 0, \quad \text{if } x_i^* = 0$$

These results are summarized in the following proposition.

PROPOSITION 1   Necessary conditions for a relative maximum of f in problem (4) to occur at x^* include

$$\nabla f(x^*) \leq 0, \nabla f(x^*)^T x^* = 0, x^* \geq 0 \qquad (5)$$

where f is twice continuously differentiable throughout a neighborhood of x^*.

EXAMPLE 2   Maximize $f(x) = -3x_1^2 - x_2^2 - x_3^2 + 2x_1 x_2 + 2x_1 x_3 + 2x_1 + 7$ subject to $x \geq 0$.
Solution: From (5) we have the following necessary conditions for a relative maximum,

(1) $0 \geq \dfrac{\partial f}{\partial x_1} = -6x_1 + 2x_2 + 2x_3 + 2$

(2) $0 = x_1 \dfrac{\partial f}{\partial x_1} = x_1(-6x_1 + 2x_2 + 2x_3 + 2)$

(3) $0 \geq \dfrac{\partial f}{\partial x_2} = -2x_2 + 2x_1$

(4) $0 = x_2 \dfrac{\partial f}{\partial x_2} = x_2(-2x_2 + 2x_1)$

(5) $0 \geq \dfrac{\partial f}{\partial x_3} = -2x_3 + 2x_1$

(6) $0 = x_3 \dfrac{\partial f}{\partial x_3} = x_3(-2x_3 + 2x_1)$

(7) $x_1 \geq 0, x_2 \geq 0$ and $x_3 \geq 0$

   Now (4) implies that $x_2 = 0$ or $x_1 = x_2$. When $x_2 = 0$, then (3) and (7) imply that $x_1 = 0$. So, (6) implies that $x_3 = 0$. But this contradicts (1). Thus, $x_2 \neq 0$. So, $x_1 = x_2$.
   Condition (6) implies that $x_3 = 0$ or $x_1 = x_2 = x_3$. If $x_3 = 0$, then (5), (4) and (7) imply that $x_1 = x_2 = x_3 = 0$. But this possibility has been ruled out. Thus, $x_1 = x_2 = x_3$. Hence, (2) implies that $x_1 = 0$ or $x_1 = 1$. Since $x_1 \neq 0$, then the only possible relative maximum of f occurs when $x_1 = x_2 = x_3 = 1$. The Hession at $x^* = [1 \quad 1 \quad 1]^T$ is

$$\begin{bmatrix} -6 & 2 & 2 \\ 2 & -2 & 0 \\ 2 & 0 & -2 \end{bmatrix}$$

H can be shown to be negative definite (see Exercise 7.21). Thus, f is strictly concave [Exercise 7.20 (d)] and f has a relative maximum at x^*. (Why?) It follows that $f(x^*) = 8$ is a global maximum [Exercise 7.32(a)].

Analogous conditions for a relative minimum can be found in Exercise 8.3(c).

Exercises

8.1   Let $f: \mathbf{R}_{n\times1} \to \mathbf{R}$ be twice continuously differentiable throughout a neighborhood of x^*.
(a) If f has a relative minimum at x^*, prove that it necessarily follows that
(i) $\nabla f(x^*)^T = 0$ (first necessary condition).
(ii) $H(x^*)$ is positive semidefinite (second necessary condition).
(b) A sufficient condition for f to have a strict relative minimum at x^* is that $H(x^*)$ be positive definite.
(c) Prove Theorem 2.

8.2   Find the relative extreme values of the functions below.
(a) $f(x) = x_1^2 + 6x_2^2 + 2x_3^2 + 18x_1 + x_2 - x_3 + 50$
(b) $f(x) = x_1^3 + x_2^3 - 6x_1x_2$
(c) $f(x) = (1/4)x_1^4 - (2/3)x_1^3 + (1/3)x_2^3 + x_2^2 + (1/2)x_3^2 - 3x_2 - 5x_3$
(d) $f(x) = -x_1^2 + x_2^2 + x_3^2 + 2x_1x_2 - x_2x_3 - 2x_1$

8.3   Let $f: \mathbf{R}_{n\times1} \to \mathbf{R}$ be twice continuously differentiable and let $S \subseteq \mathbf{R}_{n\times1}$.
(a) Let f have a relative maximum at $x^* \in S$, i.e., there exists an $\varepsilon > 0$ so that $f(x) \le f(x^*)$ whenever $x \in S \cap N_\varepsilon(x^*)$. Let $v \ne 0$ be any direction vector such that $x^* + tv \in S$ whenever t is a sufficiently small negative scalar. Prove that it necessarily follows that $\nabla f(x^*)^T v \le 0$. Moreover, $v^T H(x^*) v \le 0$ whenever equality holds.
(b) Prove that when f has a relative minimum at $x^* \in S$, then it necessarily follows that $\nabla f(x^*)^T v \ge 0$ and $v^T H(x^*) v \ge 0$, whenever $\nabla f(x^*)^T v = 0$.
(c) Let f have a relative minimum at $x^* \in S$, where

$$S = \{x: x \ge 0\} \quad \text{Prove that } \nabla f(x^*) \ge 0, \nabla f(x^*)^T x^* = 0, \text{ and } x^* \ge 0$$

8.4   Find the relative extreme values of the following problems:
(a) Maximize $f(x) = -x_1^2 - x_2^2 - x_3^2 + 2x_1 + 6x_2 - 8$, all $x_i \ge 0$.
(b) Maximize $f(x) = -x_1^2 - 2x_2^2 + 4$, where each $x_i \ge 0$.
(c) Minimize $f(x) = 2x_1^2 + x_2^2 - 8x_1 + 8x_2 + 24$, all $x_i \ge 0$.
(d) Minimize $f(x) = x_1^4 + x_2^2 + 2x_3^2 - 2x_2 - 16x_3 + 33$, all $x_i \ge 0$.

3.   EQUALITY CONSTRAINTS

Now let $f: \mathbf{R}_{n\times1} \to \mathbf{R}$ and consider the problem

$$\text{Maximize } f(x) \text{ subject to } g(x) = b \tag{6}$$

where $g(x) = [g_1(x) \dots g_m(x)]^T$ and each $g_i: \mathbf{R}_{n\times1} \to \mathbf{R}$. Here, $b \in \mathbf{R}_{m\times1}$ and

$m < n$. Let problem (6) have a relative maximum at x^*. Then there exists an $\varepsilon > 0$ such that $f(x^*) \geq f(x)$ whenever $x \in N_\varepsilon(x^*)$ and $g(x) = b$. Suppose that f is twice continuously differentiable throughout $N_\varepsilon(x^*)$, and assume that each $\partial g_i/\partial x_j$ is twice continuously differentiable throughout this neighborhood of x^*. Next, let the rank of the following matrix, known as the *Jacobian matrix*,

$$\frac{\partial g}{\partial x}(x^*) = \begin{bmatrix} \dfrac{\partial g_1}{\partial x_1}(x^*) & \cdots & \dfrac{\partial g_1}{\partial x_n}(x^*) \\[2ex] \vdots & & \vdots \\[2ex] \dfrac{\partial g_m}{\partial x_1}(x^*) & \cdots & \dfrac{\partial g_m}{\partial x_n}(x^*) \end{bmatrix} \tag{7}$$

be m. To simplify the following argument we shall assume that the last m columns of (7) are linearly independent. According to the implicit function theorem from calculus, given m continuously differentiable functions of n variables, where $m < n$, for which the Jacobian matrix has rank m, it is possible to solve for m of the variables in a neighborhood of x^* in terms of the remaining $n - m$ variables, i.e., $x_{n-m+i} = h_i(x_1, \ldots, x_{n-m})$ for $i = 1, \ldots, m$. Moreover, the functions h_i are unique and continuously differentiable throughout the neighborhood of x^*. Now let $x(1) = [x_1 \ldots x_{n-m}]^T$ and $x(2) = [x_{n-m+1} \ldots x_n]^T$. Then $x(2) = [h_1(x(1)) \ldots h_m(x(1))]^T$. Denote this last matrix by $h(x(1))$. Then Problem (6) can be restated as

$$\text{Maximize} f\left(\begin{bmatrix} x(1) \\ x(2) \end{bmatrix}\right) = \text{maximize} f\left(\begin{bmatrix} x(1) \\ h(x(1)) \end{bmatrix}\right) = \text{maximize } H(x(1)) \tag{8}$$

where $H(x(1)) = f([x(1)^T \ h(x(1))^T]^T)$. Since Problem (8) must have a relative maximum at $x^*(1) = [x_1^* \ldots x_{n-m}^*]^T$, then it follows from Theorem 1 of this chapter that $\nabla H(x^*(1)) = 0$. But

$$\frac{\partial H}{\partial x_i} = \sum_{j=1}^{n-m} \frac{\partial f}{\partial x_j} \frac{\partial x_j}{\partial x_i} + \sum_{j=n-m+1}^{n} \frac{\partial f}{\partial x_j} \frac{\partial h_j}{\partial x_i} = \frac{\partial f}{\partial x_i} + \begin{bmatrix} \dfrac{\partial f}{\partial x_{n-m+1}} & \cdots & \dfrac{\partial f}{\partial x_n} \end{bmatrix} \begin{bmatrix} \dfrac{\partial h_1}{\partial x_i} \\[1ex] \vdots \\[1ex] \dfrac{\partial h_m}{\partial x_i} \end{bmatrix}$$

Thus, letting
$$\nabla f(x(1))^T = [\partial f/\partial x_1 \ldots \partial f/\partial x_{n-m}], \quad \nabla f(x(2))^T = [\partial f/\partial x_{n-m+1} \ldots \partial f/\partial x_n]$$

and $\partial h / \partial x(1)$ be an $m \times (n-m)$ matrix composed of columns of the form $[\partial h_1 / \partial x_i \dots \partial h_m / \partial x_i]^T$, it follows that

$$0 = \nabla H(x(1))^T = \nabla f(x(1))^T + \nabla f(x(2))^T \frac{\partial h}{\partial x(1)} \tag{9}$$

must hold at $x^*(1)$. Also,

$$b = g(x) = g\left(\begin{bmatrix} x(1) \\ h(x(1)) \end{bmatrix}\right)$$

Hence, a similar calculation shows that (see Exercise 8.5)

$$0 = \frac{\partial g}{\partial x(1)} + \frac{\partial g}{\partial x(2)} \frac{\partial h}{\partial x(1)} \tag{10}$$

where $\partial g / \partial x(1)$ and $\partial g / \partial x(2)$ are the submatrices of the first $n-m$ and the last m columns of (7) respectively, must hold at $x^*(1)$. Since at $z^*(1)$ we have that

$$\frac{\partial h}{\partial x(1)} = -\left(\frac{\partial g}{\partial x(2)}\right)^{-1} \frac{\partial g}{\partial x(1)}$$

then it follows from equation (9) that

$$\nabla f(x(1))^T - \lambda^T \frac{\partial g}{\partial x(1)} = 0 \tag{11}$$

where

$$\lambda^T = \nabla f(x(2))^T \left(\frac{\partial g}{\partial x(2)}\right)^{-1}$$

at $x^*(1)$. Clearly,

$$\nabla f(x(2))^T - \lambda^T \frac{\partial g}{\partial x(2)} = 0 \tag{12}$$

Combining equations (11) and (12), and recalling the initial constraints of Problem (6), we have the following the first order necessary conditions,

$$\nabla f(x^*)^T - \lambda^T \frac{\partial g}{\partial x}(x^*) = 0 \quad \text{and} \quad b - g(x^*) = 0$$

for a relative maximum to occur at x^*. Letting $L(x,\lambda) = f(x) + \lambda^T(b - g(x))$, these necessary conditions can be stated in the following mannner.

THEOREM 3   In Problem (6), let f have a relative maximum at x^*. If f and each component g_i of g are twice continuously differentiable throughout a neighborhood of x^*, and if the Jacobian matrix (17) has full row rank of m, then there exists a λ^* so that

$$\frac{\partial L}{\partial x}(x^*, \lambda^*)^T = \nabla f(x^*)^T - \lambda^{*T}\frac{\partial g}{\partial x}(x^*) = 0$$

$$\frac{\partial L}{\partial \lambda}(x^*, \lambda^*) = b - g(x^*) = 0$$
(13)

The function L is known as the *Lagrangian* function and the components λ_i of λ are called *Lagrange multipliers*. Each Lagrange multiplier satisfies $\lambda_i = \partial f/\partial b_i (x^*)$ (see Exercise 8.11) and, hence, measures the sensitivity of the optimal value of $f(x)$ to changes in the constants b_i.

Second order necessary and second order sufficiency conditions for a relative maximum are also developed in the exercises. It should be noted that the first order necessary conditions for a relative minimum of $f(x)$ subject to $g(x) = b$ are that there exists a λ^* so that

$$\nabla f(x^*)^T + \lambda^{*T}\frac{\partial g}{\partial x}(x^*) = 0 \text{ and } b - g(x^*) = 0$$
(14)

Notice that the necessary conditions (14) can be obtained from $\partial L/\partial x = 0$ and $\partial L/\partial \lambda = 0$ by using either $L(x, \lambda) = -f(x) + \lambda^T(b - g(x))$ or $L(x, \lambda) = f(x) - \lambda^T(b - g(x))$.

EXAMPLE 2 Find the relative extrema of the function $f(x) = 2x_1x_2$ subject to $x_1^2 + x_2^2 = 1$.
Solution: Let $L(x_1, x_2, \lambda) = 2x_1x_2 + \lambda(1 - x_1^2 - x_2^2)$. Then conditions (12) become

$$0 = \frac{\partial L}{\partial x_1} = 2x_2 - 2x_1\lambda$$

$$0 = \frac{\partial L}{\partial x_2} = 2x_1 - 2x_2\lambda$$

$$0 = \frac{\partial L}{\partial \lambda} = 1 - x_1^2 - x_2^2$$

Thus, $x_2 = x_1\lambda$ and $x_1 = x_2\lambda$. Hence, $x_2 = x_1\lambda = x_2\lambda^2$. It follows that $x_2 = 0$ or $\lambda = \pm 1$. If $x_2 = 0$, then $x_1 = x_2\lambda = 0$. But this contradicts $x_1^2 + x_2^2 = 1$. Thus, $x_2 \neq 0$ and either $\lambda = 1$ or $\lambda = -1$. When $\lambda = 1$, then $x_1 = x_2$. Thus, $2x_1^2 = x_1^2 + x_2^2 = 1$ and $x_1 = \pm\sqrt{2}/2$. Likewise when $\lambda = -1$, then $x_2 = -x_1$ and $x_1 = \pm\sqrt{2}/2$. Thus, we have the following four possible solutions: $\{[\sqrt{2}/2 \ \sqrt{2}/2 \ 1]^T, [-\sqrt{2}/2 \ -\sqrt{2}/2 \ 1]^T, [\sqrt{2}/2 \ -\sqrt{2}/2 \ -1]^T, [-\sqrt{2}/2 \ \sqrt{2}/2 \ -1]^T\}$ for $[x_1 \ x_2 \ \lambda]^T$.

So far we have only worked with first order necessary conditions for a relative extrema. Thus, we only know that if optimal solutions exist, then they must occur at some of these points. Since the set $K = \{x: x_1^2 + x_2^2 = 1\}$ is closed and bounded and since $f(x) = 2x_1x_2$ is continuous, then the Weierstrass theorem implies that f has both a maximum and a minimum over k. It follows easily that the maximum of f over k is 1 and occurs at both $x_1 = x_2 = \sqrt{2}/2$ and $x_1 = x_2 = -\sqrt{2}/2$. With a similar argument using the necessary conditions in (14) for a relative minimum, it can be shown that the minimum value of f is –1 and this occurs at both $[\sqrt{2}/2 \ -\sqrt{2}/2]^T$ and $[-\sqrt{2}/2 \ \ \sqrt{2}/2]^T$.

4. NONNEGATIVE VARIABLES AND EQUALITY CONSTRAINTS

Next consider the problem

$$\text{Maximize } f(x) \text{ subject to } g(x) = b \text{ and } x \geq 0 \qquad (15)$$

where f and g satisfy the conditions of Section 3. Let f have a relative maximum at x^*. If $x^* > 0$, then it easily follows from Section 3 that there exists a λ^* so that $(\partial L/\partial x)(x^*,\lambda^*) = 0$ and $(\partial L/\partial \lambda)(x^*,\lambda^*) = 0$. Next suppose that $x^* \not> 0$. Then without loss of generality suppose that $x_i{}^* > 0$ for $i = 1, \ldots, k$ and $x_i^* = 0$ for $i = k + 1, \ldots, n$. Let $\bar{b}_i = 0$ for $i = k + 1, \ldots, n$ and consider the problem

$$\text{Maximize } f(x) \text{ subject to } g(x) = b, x_i = \bar{b}_i \text{ for } i = k + 1, \ldots, n,$$
$$\text{and } x_1 > 0, \ldots, x_k > 0 \qquad (16)$$

Clearly, this problem has a relative maximum at x^*. The Lagrangian function for this last problem is $\bar{L}(x,\lambda,\beta) = f(x) + \lambda^T(b - g(x)) + \sum_{i=k+1}^{n} \beta_i(\bar{b}_i - x_i)$. Since $x_1 > 0, \ldots, x_k > 0$, then it again follows [Exercise 8.6 (a)] that there exist $\beta_{k+1}^*, \ldots, \beta_n^*$ and λ^* so that

$$0 = \frac{\partial \bar{L}}{\partial \beta_{i+1}} = \cdots = \frac{\partial \bar{L}}{\partial \beta_n}, \quad 0 = \frac{\partial \bar{L}}{\partial \lambda}(x^*,\lambda^*), \quad \text{and} \quad 0 = \frac{\partial \bar{L}}{\partial x}(x^*, \lambda^*)$$

This last equation means that

$$0 = \frac{\partial f}{\partial x_i}(x^*) - \lambda^{*T} \frac{\partial g}{\partial x_i}(x^*), \text{ for } i = 1, \ldots, k$$

and $\qquad 0 = \frac{\partial f}{\partial x_i}(x^*) - \lambda^{*T} \frac{\partial g}{\partial x_i}(x^*) - \beta_i^* \text{ for } i = k+1, \ldots, n \qquad (17)$

Recall (also see Exercise 8.11) that β_i^* measures the sensitivity of optimal values f to changes in the parameter $\bar{b}_i$. Let each $\bar{b}_i$ decrease to a negative number near 0. Since this enlarges the set of feasible solutions, it follows that the optimal values of f either increase or remain constant, i.e., as $\bar{b}_i$ decreases, the optimal values of f are nondecreasing. It follows that $\beta_i^* \leq 0$.

Hence, it follows from (17) that

$$\frac{\partial f}{\partial x_i}(x^*) - \lambda^{*T}\frac{\partial g}{\partial x_i}(x^*) \leq 0$$

whenever $x_i^* = 0$.

Summarizing the results of this section in terms of the Lagrangian $L(x,\lambda) = f(x) + \lambda^T(b - g(x))$ of Problem (15) we have the following theorem.

THEOREM 4   If Problem (15) has a relative maximum at x^*, then there exists a λ^* so that

$$0 \geq \frac{\partial L}{\partial x}(x^*,\lambda^*)^T = \nabla f(x^*)^T - \lambda^{*T}\frac{\partial g}{\partial x}(x^*)$$

$$0 = \frac{\partial L}{\partial x}(x^*,\lambda^*)^T x^* = (\nabla f(x^*)^T - \lambda^{*T}\frac{\partial g}{\partial x}(x^*))x^* \qquad (18)$$

$$0 = \frac{\partial L}{\partial \lambda}(x^*,\lambda^*) = b - g(x^*) \text{ and } x^* \geq 0$$

(Note: f, each g_i, and the Jacobian $\partial g/\partial x$ satisfy the conditions of Theorem 2.)

Conditions (18) are known as the *Kuhn-Tucker* necessary conditions for Problem (15). Notice that the expression $0 = (\partial L/\partial x)(x^*,\lambda^*)^T x^* = \Sigma_{i=1}^n x_i^*(\partial L/\partial x_i)(x^*,\lambda^*)$ can be replaced by $x_i^*(\partial L/\partial x_i)(x^*,\lambda^*) = 0$ for $i = 1, ..., n$, since $x^* \geq 0$ and $(\partial L/\partial x)(x^*,\lambda^*) \leq 0$.

5.   NONNEGATIVE VARIABLES AND INEQUALITY CONSTRAINTS

Now consider the problem

$$\text{Maximize } f(x) \text{ subject to } g(x) \leq b \text{ and } x \geq 0 \qquad (19)$$

where again f and g satisfy the conditions of Section 3. Let $s = b - g(x)$. Notice that $s \geq 0$. Then Problem (19) can be related in the form of (15) as follows:

$$\text{Maximize } f(x) \text{ subject to } g(x) + s = b, x \geq 0 \text{ and } s \geq 0 \qquad (20)$$

If (19) has a relative maximum at x^*, then (20) has a relative maximum at (x^*,s^*) where $s^* = b - g(x^*)$. The Lagrangian function for (20) is $L(x,s,\lambda) = f(x) + \lambda^T(b - g(x) - s)$. Hence, it follows from the Kuhn-Tucker conditions (18) of Section 4 that there exists a λ^* so that

$$0 \geq \frac{\partial L}{\partial x}(x^*,s^*,\lambda^*)^T = \nabla f(x^*)^T - \lambda^{*T}\frac{\partial g}{\partial x}(x^*)$$

$$0 = \frac{\partial L}{\partial x}(x^*, s^*, \lambda^*)^T x^* = (\nabla f(x^*)^T - \frac{\partial g}{\partial x}(x^*))x^*$$

$$0 \geq \frac{\partial L}{\partial s}(x^*, s^*, \lambda^*)^T = -\lambda^{*T}$$

$$0 = \frac{\partial L}{\partial s}(x^*, s^*, \lambda^*)^T s^* = -\lambda^{*T} s^* \tag{21}$$

$$0 = \frac{\partial L}{\partial \lambda}(x^*, s^*, \lambda^*) = b - g(x^*) - s^*$$

$$x^* \geq 0 \text{ and } s^* \geq 0$$

Eliminating s^*, these Kuhn-Tucker necessary conditions can be restated in terms of the Lagrangian function $L(x,\lambda) = f(x) + \lambda^T(b - g(x))$ of problem (19).

THEOREM 5   If Problem (19) has a relative maximum at x^*, where f and each g_i are twice continuously differentiable throughout a neighborhood of x^*, and if $(\partial g/\partial x)(x^*)$ has full row rank m, then there exists a λ^* so that

$$\frac{\partial L}{\partial x}(x^*, \lambda^*)^T = \nabla f(x^*)^T - \lambda^{*T} \frac{\partial g}{\partial x}(x^*) \leq 0$$

$$\left(\frac{\partial L}{\partial x}(x^*, \lambda^*)\right)^T x^* = (\nabla f(x^*)^T - \lambda^{*T} \frac{\partial g}{\partial x}(x^*))x^* = 0$$

$$\frac{\partial L}{\partial \lambda}(x^*, \lambda^*) = b - g(x^*) \geq 0 \tag{22}$$

$$\left(\frac{\partial L}{\partial \lambda}(x^*, \lambda^*)\right)^T \lambda^* = (b - g(x^*))^T \lambda^* = 0$$

$$x^* \geq 0 \text{ and } \lambda^* \geq 0$$

Unfortunately, it is usually not practical to solve (22) for x^*. In the sections that follow it will be shown that under special circumstances the Kuhn-Tucker conditions are sufficient for a relative maximum at x^*.

EXAMPLE 3   Maximize $f(x) = -x_1^2 - 3x_2^2 + 4x_1 + 6x_2$ subject to $x_1 + 2x_2 \leq 4$, $x_1 \geq 0$ and $x_2 \geq 0$.
Solution: Let $L(x_1, x_2, \lambda) = -x_1^2 - 3x_2^2 + 4x_1 + 6x_2 + \lambda(4 - x_1 - 2x_2)$. Then the Kuhn-Tucker necessary conditions for a relative maximum are

$$0 \geq \frac{\partial L}{\partial x_1} = -2x_1 + 4 - \lambda, \quad 0 = x_1 \frac{\partial L}{\partial x_1} = x_1(-2x_1 + 4 - \lambda)$$

$$0 \geq \frac{\partial L}{\partial x_2} = -6x_2 + 6 - 2\lambda, \quad 0 = x_2 \frac{\partial L}{\partial x_2} = x_2(-6x_2 + 6 - 2\lambda)$$

$$0 \le \frac{\partial L}{\partial \lambda} = 4 - x_1 - 2x_2, \ 0 = \lambda \frac{\partial L}{\partial \lambda} = \lambda(4 - x_1 - 2x_2)$$

$$x_1 \ge 0, x_2 \ge 0, \text{ and } \lambda \ge 0$$

From $0 = x_1(-2x_1 + 4 - \lambda)$, it follows that $x_1 = 0$ or $-2x_1 + 4 - \lambda = 0$. If $x_1 = 0$, then $0 = \lambda(4 - x_1 - 2x_2)$ implies that $\lambda = 0$ or $x_2 = 2$. If $\lambda = 0$, then $0 \ge -2x_1 + 4 - \lambda = 4$ which is impossible. Thus, $x_2 = 2$, whenever $x_1 = 0$. But then $0 = x_2(-6x_2 + 6 - 2\lambda)$ implies that $\lambda = -3$. Since this is impossible it follows that $x_1 \ne 0$. So, $-2x_1 + 4 - \lambda = 0$.

With a similar argument it follows from $0 = x_2(-6x_2 + 6 - 2\lambda)$ that $-3x_2 + 3 - \lambda = 0$. Moreover, it follows that $0 = \lambda(4 - x_1 - 2x_2)$ implies that $4 - x_1 - 2x_2 = 0$ or that $\lambda = 0$, $x_1 = 2$ and $x_2 = 1$.

Checking, $x_1 = 2$, $x_2 = 1$, and $\lambda = 0$ satisfy the Kuhn-Tucker conditions and, hence, is a possible solution. Solving the equations $-2x_1 + 4 - \lambda = 0$, $-3x_2 + 3 - \lambda = 0$, and $4 - x_1 - 2x_2 = 0$ yields this same solution.

Finally the Weierstrass theorem implies that f has a maximum value of 7 when $x = [2 \quad 1]^T$.

Exercises

8.5 Verify equation (10).

8.6 (a) Verify in Problem (16) that there exists a λ^* so that $0 = (\partial \bar{L}/\partial \lambda)(x^*, \lambda^*)$ and $0 = (\partial \bar{L}/\partial x)(x^*, \lambda^*)$.
(b) In Problem (19), any constraint $g_i(x) \le b_i$ is said to be *active at x* whenever $g_i(x) = b_i$. Moreover, when $g_i(x) < b_i$, i.e., $s_i > 0$, then the constraint is said to be *inactive* at x. Suppose that the i-th constraint of (19) is inactive at x^*. Prove that $\lambda_i^* = 0$.
(c) Let Problem (19) have a relative maximum at x^*. Let I be the collection of indices that correspond to the active constraints of (19) relative to x^*. Determine whether Theorem 5 remains true whenever the assumptions concerning g are reduced to those g_i for which $i \in I$, i.e., g_i is twice continuously differentiable near x^* when $i \in I$ and the Jacobian matrix of the active constraints has full row rank.

8.7 Show that conditions (14) are necessary for the problem

$$\text{Minimize } f(x) \text{ subject to } g(x) = b$$

to have a relative minimum at x^*.

8.8 Minimize $2x_1^2 + 8x_2^2$ subject to $3x_1 + 6x_2 = 10$.

8.9 Minimize $(x_1 + 2)^2 + 2(x_2 - 1)^2 + x_3^2$ subject to $x_1^2 + 4x_1x_2 + x_2^2 = 1$ and $x_2 + x_3 = 6$.

8.10   Determine the relative extrema of the function $f(x) = x_1^2 + x_2^2 + x_1x_2$ subject to $x_1^2 + x_2^2 = 4$.

8.11   Verify that $\lambda^*_i = (\partial f/\partial b_i)(x^*)$, where (6) has a relative maximum at x^*.   (Hint: Use   the   chain   rule   to   obtain $(\partial f/\partial b_i)(x^*) =$ $\lambda^*_i + (\ f(x^*)^T - \lambda^T(\partial g/\partial x)(x^*))[(\partial x_1/\partial b_i)(x^*) \ldots (\partial x_n/\partial b_i)(x^*)]^T.)$

8.12   (Second order necessary conditions). Let
$$H_L(x^*,\lambda^*) = [(\partial^2 L/\partial x_i \partial x_j)(x^*,\partial \lambda^*)]$$
denote the Hessian matrix of L, where Problem (6) has relative maximum at x^*. Show that $z^T H_L(x^*,\lambda^*)z < 0$ whenever z satisfies $((\partial g/\partial x)(x^*))z = 0$.

8.13   As in Problem 8.12 let $H_L(x^*,\lambda^*)$ denote the Hessian matrix of L.
(a) (Sufficiency conditions). Show that Problem (6) has a relative maximum at x^* whenever $((\partial g/\partial x)(x^*))z = 0$ implies that $z^T H_L(x^*,\lambda^*)z < 0$.
(b) (Sufficiency conditions). Show that Problem (6) has a relative maximum at x^* whenever the eigenvalues of the matrix

$$\begin{bmatrix} H_L(x^*,\lambda^*) & \dfrac{\partial g}{\partial x}(x^*) \\ \hline \dfrac{\partial g}{\partial x}(x^*) & 0 \end{bmatrix}$$

are all negative.

8.14   Prove that the Kuhn-Tucker necessary conditions for the problem

$$\text{Minimize } f(x) \text{ subject to } g(x) \le b \text{ and } x \ge 0$$

are

$$\nabla f(x^*)^T + \lambda^{*T}\frac{\partial g}{\partial x}(x^*) \ge 0, \ (\nabla f(x^*)^T + \lambda^{*T}\frac{\partial g}{\partial x}(x^*))x^* = 0, x^* \ge 0$$

$$b - g(x^*) \ge 0, (b - g(x^*))^T \lambda^* = 0, \text{ and } \lambda^* \ge 0$$

(Hint: Maximize $-f(x)$ subject to $g(x) \le b$ and $x \ge 0$. In this case the Lagrangian is $L(x,\lambda) = -f(x) + \lambda^T(b - g(x))$. Notice that the necessary conditions in this exercise could also be obtained from $0 \le \partial L/\partial x$, $0 = (\partial L/\partial x)^T x$, $x > 0$, $\partial L/\partial \lambda > 0$, $(\partial L/\partial \lambda)^T \lambda = 0$ and $\lambda > 0$ where the Lagrangian L is defined by $L(x,\lambda) = f(x) + \lambda^T(b - g(x))$. Also notice that the sufficiency conditions for a relative minimum at x^* of $f(x)$ subject to the constraints $g(x) = b$ are that there exists a λ^* such that

$$\nabla f(x^*)^T + \lambda^{*T}\frac{\partial g}{\partial x}(x^*) = 0, \ g(x^*) = b$$

and $z^T H_L(x^*,z^*)z > 0$ whenever $((\partial g/\partial x)(x^*))z = 0$.

8.15   Determine the Kuhn-Tucker necessary conditions for a relative maximum at x^* for the problem

$$\text{Maximize } f(x) \text{ subject to } g(x) \leq b, h(x) = 0 \text{ and } x \geq 0$$

where $h(x) = [h_1(x) \ldots h_k(x)^T]$ and each $h_i: \mathbf{R}_{n \times 1} \rightarrow \mathbf{R}$.

8.16   Determine the Kuhn-Tucker necessary conditions for the following problems:
(a) Maximize $x_1 x_2 x_3$ subject to $5x_1 + x_2 \leq 4$, $3x_2 - x_3 \leq 5$ and each $x_i \geq 0$.
(b) Minimize $x_1^2 + x_1 x_2$ subject to $x_1 + 6x_2 \leq 3$, $x_1 \sin x_2 = 1$, $x_1 \geq 0$ and $x_2 \geq 0$.
(c) Maximize $x_1 x_2 x_3$ subject to $x_1^2 + x_2^2 \leq 9$, $4x_1 - x_3 \geq 6$, and each $x_i \geq 0$.
(d) Minimize $c^T x$ subject to $Ax \leq b$ and $x \geq 0$.
(e) Maximize $(\frac{1}{2})x^T Bx + c^T x$ subject to $Ax \leq b$ and $x \geq 0$, where, B is symmetric.
(f) Maximize $f(x)$ subject to $g(x) \leq b$.

6.   THE KUHN-TUCKER THEOREM

Consider now the problem

$$\text{Maximize } f(x) \text{ subject to } g(x) \leq b \text{ and } x \geq 0 \qquad (23)$$

Problem (23) differs from Problem (19) in that no special assumption concerning the functions f or g are made in (23). As in (19), define the Lagrangian function as $L(x,\lambda) = f(x) + \lambda^T(b - g(x))$.
    Also consider the problem: find nonnegative vectors x^* and λ^* such that

$$L(x, \lambda^*) \leq L(x^*, \lambda^*) \leq L(x^*, \lambda) \qquad (24)$$

for every $x \geq 0$ and $\lambda \geq 0$. Such a pair (x^*, λ^*) will be called a *saddle point* of L. Notice that

$$L(x^*, \lambda^*) = \max\{L(x,\lambda^*): x \geq 0\} = \min\{L(x^*,\lambda): \lambda \geq 0\} \qquad (25)$$

Observe that no explicit assumption was made in either (24) or (25) about x or x^* satisfying the constraint $g(x) \leq b$. Also, notice that λ is nonnegative.
    In proving the following theorem we shall make use of a well known fact from convex analysis that two nonempty disjoint convex sets, S_1 and S_2, in $\mathbf{R}_{n \times 1}$ can be separated by a hyperplane, i.e., there exists a hyperplane $H(c,\beta)$, where $c \neq 0$, such that $S_1 \subseteq H^+(c,\beta)$ and $S_2 \subseteq H^-(c,\beta)$. A proof can be found in Valentine [11]. Notice that $c^T x \geq \beta \geq c^T y$ whenever $x \in S_1$ and $y \in S_2$.

THEOREM 6 (KUHN-TUCKER)   If (x^*, λ^*) solves Problem (24), then x^* solves problem (23). If in addition f is a concave function, each g_i is a convex function, there exists an $\bar{x} \geq 0$ such that $g(\bar{x}) < b$, and if x^* is a solution to Problem (23), then there exists a $\lambda^* \geq 0$ so that (x^*, λ^*) is a saddle point of Problem (24).

Proof: First suppose that (x^*, λ^*) solves problem (24). Then from the second inequality in (24) it follows that

$$f(x^*) + \lambda^{*T}(b - g(x^*)) \leq f(x^*) + \lambda^T(b - g(x^*))$$
$$(\lambda - \lambda^*)^T(b - g(x^*)) \geq 0 \qquad (26)$$

where $\lambda \geq 0$. Let $\lambda = [\lambda_1^* \ldots \lambda_{i-1}^* \; \lambda_i^* + 1 \; \lambda_{i+1}^* \ldots \lambda_n^*]^T$. Then $\lambda - \lambda^* = e_i$. Hence, $b_i - g_i(x^*) \geq 0$. Doing this for each i, we have that $b - g(x^*) \geq 0$ and x^* satisfies the constraints of (23). Also, if $\lambda = 0$ in (26), then $-\lambda^{*T}(b - g(x^*)) \geq 0$. Thus $\lambda^{*T}(b - g(x^*)) \leq 0$. But $\lambda^* \geq 0$ and $b - g(x^*) \geq 0$. Hence, $\lambda^{*T}(b - g(x^*)) = 0$. Thus, the first inequality in (24) implies that

$$f(x) + \lambda^{*T}(b - g(x)) \leq f(x^*) + \lambda^{*T}(b - g(x^*)) = f(x^*)$$

Suppose that x satisifies $g(x) \leq b$ and $x \geq 0$. Then $\lambda^{*T}(b - g(x)) \geq 0$. Thus, $f(x) \leq f(x^*)$ and x^* solves (23).

Now suppose that F is concave, each g_i is convex, there exists an $\bar{x} \geq 0$ so that $g(\bar{x}) < b$, and x^* solves (23). Define two sets in $\mathbf{R}_{m+1}$ by

$$A = \{[a_0 \; a^T]^T : [a_0 \; a^T]^T \leq [f(x) \; (b - g(x))^T]^T, \text{ for some } x \geq 0\}$$

and

$$B = \{[b_0 \; b^T]^T : [b_0 \; b^T]^T > [f(x^*) \; 0^T]^T\},$$

where a_0 and b_0 are scalars and $a, b \in \mathbf{R}_{m \times 1}$. It is a simple exercise to show that A and B are disjoint convex sets (see Exercise 8.17). It follows from the remarks preceding the theorem that A and B can be separated by a hyperplane. Thus, there exists a $\bar{c} = [c_0 \; c^T]^T \neq 0$, where c_0 is a scalar and $c \in \mathbf{R}_{m \times 1}$, such that

$$[c_0 \; c^T]\begin{bmatrix} a_0 \\ a \end{bmatrix} \leq [c_0 \; c^T]\begin{bmatrix} b_0 \\ b \end{bmatrix} \qquad (27)$$

whenever $[a_0 \; a^T]^T \in A$ and $[b_0 \; b^T]^T \in B$. Rewriting (27), we have

$$c_0 a_0 + \sum_{i=1}^{m} c_i a_i \leq c_0 b_0 + \sum_{i=1}^{m} c_i b_i \qquad (28)$$

In (28) hold each a_i and all b_i except b_j fixed. Then

$$c_0 a_0 + \sum_{i=1}^{m} c_i a_i - \sum_{\substack{i=0 \\ i \neq j}}^{m} c_i b_i \leq c_j b_j \qquad (29)$$

holds for every b_j as $b_j \to \infty$. It follows from (29) that $c_j b_j$ is bounded from below as $b_j \to \infty$. Hence, $c_j \geq 0$. Repeating the argument for $j = 0, 1, \ldots, m$, we have that $\bar{c} \geq 0$. Since the inner products in (27) are continuous and since $[f(x^*)\ 0^T]^T$ is in the boundary of B, it follows that

$$c_0 f(x) + c^T(b - g(x)) = [c_0 \quad c^T]\begin{bmatrix} f(x) \\ b - g(x) \end{bmatrix}$$

$$\leq [c_0 \quad c^T]\begin{bmatrix} f(x^*) \\ 0 \end{bmatrix} = c_0 f(x^*)$$

(30)

whenever $x \geq 0$. Suppose that $c_0 = 0$. Then $c^T(b - g(x)) \leq 0$ for every $x \geq 0$. But recall that $g(\bar{x}) < b$ and $\bar{x} \geq 0$. Since $0 \neq c \geq 0$, then $c^T(b - g(\bar{x})) > 0$, which is a contradiction. Thus, $c_0 > 0$. Hence, dividing both sides of (30) by c_0 gives

$$f(x) + (1/c_0)c^T(b - g(x)) \leq f(x^*)$$

(31)

whenever $x \geq 0$. Let $\lambda^* = (1/c_0)c$. When $x = x^*$, it follows from (31) that $\lambda^{*T}(b - g(x^*)) \leq 0$. But $\lambda^* \geq 0$ and $g(x^*) \leq b$. Thus, $\lambda^{*T}(b - g(x^*)) \geq 0$. Hence, $\lambda^{*T}(b - g(x^*)) = 0$. It follows using (31) that $L(x, \lambda^*) \leq L(x^*, \lambda^*)$, whenever $x \geq 0$. Also, $L(x^*, \lambda^*) = f(x^*) \leq f(x^*) + \lambda^T(b - g(x^*)) = L(x^*, \lambda)$, whenever $\lambda \geq 0$. Hence, (x^*, λ^*) is a saddle point of L (Uzawa [10]).

The condition that there exists an $\bar{x} \geq 0$ such that $g(\bar{x}) < b$ was used to ensure the existence of a $\lambda^* \geq 0$ so that (x^*, λ^*) was a saddle point of Problem (24). Such a condition is known as a *constraint qualification*. In Section 3, the constraint qualification was that the Jacobian matrix (7) have full row rank.

Adding some of the conditions of problem (19), we can establish that the Kuhn-Tucker conditions are both necessary and sufficient for a solution of problem (23). In particular, consider the problem

$$\text{Maximize } f(x) \text{ subject to } g(x) \leq b \text{ and } x \geq 0$$

(32)

where f is a twice continuously differentiable concave function, each component g_i of g is a twice continuously differentiable convex function, and there exists an $\bar{x} \geq 0$ such that $g(\bar{x}) < b$.

THEOREM 7   Necessary and sufficient conditions for x^* to solve problem (32) (i.e., $(x^* \ \lambda^*)$ to be a saddle point of (24)) are that

$$\frac{\partial L}{\partial x}(x^*, \lambda^*) \leq 0, \left(\frac{\partial L}{\partial x}(x^*, \lambda^*)\right)^T x^* = 0, \ x^* \geq 0$$

$$\frac{\partial L}{\partial \lambda}(x^*, \lambda^*) \geq 0, \left(\frac{\partial L}{\partial \lambda}(x^*, \lambda^*)\right)^T \lambda^* = 0, \ \lambda^* \geq 0$$

(33)

Proof: It is left as a simple exercise (Exercise 8.18) for the reader to show that conditions (33) hold, whenever (x^*, λ^*) is a saddle point of (24).

Next suppose that x^* and λ^* satisfy conditions (33). From Exercise 8.19, $L(x, \lambda^*)$ is a concave function in x. Thus,

$$L(x, \lambda^*) - L(x^*, \lambda^*) \le \frac{\partial L}{\partial x}(x^*, \lambda^*)^T(x - x^*)$$

(see Exercise 7.30). Therefore,

$$L(x^*, \lambda^*) \ge L(x, \lambda^*) + \frac{\partial L}{\partial x}(x^*, \lambda^*)^T x^* - \frac{\partial L}{\partial x}(x^*, \lambda^*)^T x \ge L(x, \lambda^*)$$

Since $((\partial L/\partial x)(x^*,\lambda^*))^T x^* = 0$, $x \ge 0$ and $(\partial L/\partial x)(x^*,\lambda^*) \le 0$. Moreover since L is linear in λ, then it also follows that

$$L(x^*, \lambda^*) - L(x^*, \lambda) = \frac{\partial L}{\partial \lambda}(x^*, \lambda^*)^T(\lambda^* - \lambda)$$

or

$$L(x^*, \lambda^*) = L(x^*, \lambda) + \frac{\partial L}{\partial x}(x^*, \lambda^*)^T \lambda^* - \frac{\partial L}{\partial \lambda}(x^*, \lambda^*)^T \lambda \le L(x^*, \lambda)$$

(see Exercise 7.30 and Proposition 9 of Chapter 7), since

$$\frac{\partial L}{\partial \lambda}(x^*, \lambda^*)^T \lambda^* = 0, \lambda \ge 0 \text{ and } \frac{\partial L}{\partial \lambda}(x^*, \lambda^*) \ge 0$$

Hence, (x^*, λ^*) is a saddle point of (24) and x^* solves (32).

Using the Kuhn-Tucker conditions, we shall now reformulate a quadratic programming problem (as was promised in Chapter 4) as a linear complementarity problem. As seen in the following result, it follows from Theorem 6 that in many important situations it suffices to solve a linear complementarity problem in order to solve a quadratic programming problem.

THEOREM 8   If a relative maximum of the quadratic programming problem

$$\text{Maximize } c^T x + \frac{1}{2}x^T B x \text{ subject to } Ax \le b \text{ and } x \ge 0 \qquad (34)$$

where B is a symmetric matrix, occurs at x^*, then it necessarily follows that there exist $\lambda^* \ge 0, u \ge 0$ and $v \ge 0$ that satisfy

$$\begin{bmatrix} -c \\ b \end{bmatrix} = \begin{bmatrix} u \\ v \end{bmatrix} - \begin{bmatrix} -B & A^T \\ -A & 0 \end{bmatrix} \begin{bmatrix} x^* \\ \lambda^* \end{bmatrix} \text{ and } u^T x^* = v^T \lambda^* = 0 \qquad (35)$$

Moreover, when B is negative semidefinite and x^*, λ^*, u and v are nonnegative vectors that satisfy conditions (35), then x^* solves Problem (34).

Proof: Let $L(x,\lambda) = c^T x + (\frac{1}{2})x^T Bx + \lambda^T(b - Ax)$. Then the Kuhn-Tucker conditions for a relative maximum at x^* can easily be shown to be (Exercise 8.20)

$$0 \geq L_x(x^*, \lambda^*) = c + Bx^* - A^T\lambda^*, 0 \leq L_\lambda(x^*, \lambda^*) = b - Ax^*$$
$$0 = L_x(x^*, \lambda^*)^T x^*, 0 = L_\lambda(x^*, \lambda^*)^T\lambda \tag{36}$$
$$x^* \geq 0 \text{ and } \lambda^* \geq 0$$

Using slack variables, the first two conditions can be expressed as

$$0 = c + Bx^* - A^T\lambda^* + u \text{ and } 0 = b - Ax^* - v \tag{37}$$

where $u \geq 0$ and $v \geq 0$. The next two Kuhn-Tucker conditions imply that $u^T x^* = 0 = v^T\lambda^*$. But conditions (37) can be represented as the matrix equation in (35).

When B is negative semidefinite, then the objective function is a concave function. Since the constraints are linear it follows from Theorem 7 that any solution x^*, λ^*, u and v of the linear complementarity problem (35) determines a solution x^* of Problem (34).

The reader should recall Lemke's algorithm for solving a linear complementarity problem. An example of the algorithm can be found in Section 6 of Chapter 4.

Exercises

8.17   Prove that sets A and B in Theorem 3 are disjoint convex sets.

8.18   Show that conditions (33) hold whenever (x^*, λ^*) is a saddle point of (24).

8.19   Prove that $L(x, \lambda)$ in Theorem 7 is a concave function in x. Also, prove that $L(x, \lambda)$ is linear in λ.

8.20   Verify that the six conditions in (36) are the Kuhn-Tucker condition for problem (34).

8.21   Represent the Kuhn-Tucker conditions for the following problems as linear complementarity problems.
(a) Minimize $c^T x$ subject to $Ax = b$ and $x \geq 0$
(b) Maximize $c^T x + \frac{1}{2}x^T Bx$ subject to $Ax = b$ and $x \geq 0$
(c) Minimize $c^T x + \frac{1}{2}x^T Bx$ subject to $Ax = b$ and $x \geq 0$
(d) Minimize $c^T x + \frac{1}{2}x^T Bx$ subject to $Ax \geq b$ and $x \geq 0$

where B is a symmetric matrix in (b), (c), and (d).

8.22   State and prove an analogous result to Theorem 8 for the problem in Exercise 8.21 (d).

8.23   Solve the quadratic programming problem (34) for the following cases:

$$\text{(a)} \quad B = \begin{bmatrix} -1 & 1 \\ 1 & -1 \end{bmatrix}, A = \begin{bmatrix} 1 & 4 \\ 3 & 2 \end{bmatrix}, b = \begin{bmatrix} 7 \\ 8 \end{bmatrix}$$

and $c = [-1 \quad 3]^T$

$$\text{(b)} \quad B = \begin{bmatrix} 2 & -1 \\ -1 & 1 \end{bmatrix}, A = \begin{bmatrix} 3 & 5 \\ 1 & 6 \end{bmatrix}, b = \begin{bmatrix} 12 \\ 9 \end{bmatrix}$$

and $c = [2 \quad 2]^T$

$$\text{(c)} \quad B = \begin{bmatrix} -1 & 1 & 0 \\ 1 & -3 & 1 \\ 0 & 1 & -1 \end{bmatrix}, A = \begin{bmatrix} 1 & -1 & 2 \\ 0 & 5 & 4 \\ 3 & -2 & 0 \end{bmatrix}, b = \begin{bmatrix} 1 \\ 2 \\ 1 \end{bmatrix}$$

and $c = [1 \quad 0 \quad 2]^T$

$$\text{(d)} \quad B = \begin{bmatrix} 2 & -1 & 0 \\ -1 & 2 & 1 \\ 0 & 1 & 2 \end{bmatrix}, A = \begin{bmatrix} 1 & -1 & 2 \\ 0 & 5 & 4 \\ 3 & -2 & 0 \end{bmatrix}, b = \begin{bmatrix} 1 \\ 2 \\ 1 \end{bmatrix}$$

and $c = [1 \quad 0 \quad 2]^T$.

8.24   Find the point $x = [x_1 \ x_2]^T$ satisfying $x_1 + x_2 \le 10, x_1 - 2x_2 \le 4, x_1 \ge 0$, and $x_2 \ge 0$ that is nearest $[6 \quad 8]^T$.

8.25   First consider the problem

$$\text{Minimize } f(x) \text{ subject to } g(x) \le b \text{ and } x \ge 0$$

Define $L(x, \lambda) = f(x) + \lambda^T(b - g(x))$. Then consider a second problem of finding $x^* \ge 0$ and $\lambda^* \ge 0$ so that

$$L(x^*, \lambda) \le L(x^*, \lambda^*) \le L(x, \lambda^*)$$

for every $x \ge 0$ and $\lambda \ge 0$. Prove that when (x^*, λ^*) solves the second problem, then x^* solves the first problem. If in addition $f(x)$ is a convex function, each component $g_i(x)$ of $g(x)$ is a convex function, there exists an $\bar{x} \ge 0$ so that $g(\bar{x}) < b$, and x^* is a solution of the first problem, then there exists a $\lambda^* \ge 0$ so that (x^*, λ^*) solves the second problem. When the functions $f(x)$ and $g_i(x)$ are continuously differentiable then the Kuhn-Tucker conditions for the first problem (see Exercise 8.14) are necessary and sufficient for a solution of the first problem.

8.26   show that $x^* = \begin{bmatrix} 1 & 0 \end{bmatrix}^T$ maximizes the problem

$$\text{Maximize } x_1 \text{ subject to } (x_1 - 1)^3 + x_2 \leq 0, x_1 \geq 0, \text{ and } x_2 \geq 0$$

but the Kuhn-Tucker conditions of Section 5 do not hold. Show that x^* does not satisfy the constraint qualificiation (Kuhn and Tucker [5]).

8.27   Let problem (32) have a relative maximum at x^*. Let I be the collection of indices that correspond to the active constraints of (32) relative to x^*. Determine whether Theorem 6 remains true whenever the assumptions concerning g are reduced to those g_i for which $i \in I$.

8.28   (a) Prove that the problem

$$\text{Maximize } c^T x \text{ subject to } Ax \leq b \text{ and } x \geq 0 \tag{38}$$

has a solution at x^* if and only if there exists a λ^* so that (x^*, λ^*) solves Problem (24). Assume that $A\bar{x} < b$ for some $\bar{x} \geq 0$.
(b) If x^* solves Problem (38) and w^* solves the problem

$$\text{Minimize } b^T w \text{ subject to } A^T w \geq c \text{ and } w \geq 0 \tag{39}$$

then show that $L(x, w^*) \leq L(x^*, w^*) \leq L(x^*, w)$ whenever $x \geq 0$ and $w \geq 0$.
(c) Verify that the Kuhn-Tucker conditions for Problems (38) and (39) are identical.

REFERENCES

1. M. S. Bazaraa and C. M. Shetty, *Nonlinear Programming*, John Wiley and Sons, New York (1979).
2. L. Cooper and D. Steinberg, *Introduction to Methods of Optimization*, W. B. Saunders, Philadelphia (1970).
3. S. I. Gass, *Linear Programming, Methods and Applications*, McGraw-Hill, New York (1969).
4. B. Gottfried and J. Weisman, *Introduction to Optimization Theory*, Prentice-Hall, Englewood Cliffs, N.J. (1971).
5. H. W. Kuhn and A. W. Tucker, "Nonlinear Programming" in *Proceedings of the Second Berkeley Symposium on Mathematical Statistics and Probability* (ed., J. Neyman), University of California Press, Berkeley (1951), 481–492.
6. M. Intriligator, *Mathematical Optimization and Economic Theory*, Prentice-Hall, Englewood Cliffs, N.J. (1971).
7. D. Luenberger, *Introduction to Linear and Nonlinear Programming*, Addison-Wesley, Reading, Mass. (1973).

8. W. H. Marlow, *Mathematics for Operations Research*, John Wiley and Sons, New York (1978).
9. K. Murty, *Linear and Combinatorial Programming*, John Wiley and Sons, New York (1976).
10. H. Uzawa, "The Kuhn-Tucker Theorem in Convex Programming," *Studies in Linear and Nonlinear Programming*, (eds. K. J. Arrow, L. Hurwicz and H. Uzawa), Stanford University Press, Stanford, Ca. (1958).
11. F. Valentine, *Convex Sets*, McGraw-Hill, New York (1964).
12. G. R. Walsh, *Methods of Optimization*, John Wiley and Sons, New York (1975).

9

Search Techniques for Unconstrained Optimization Problems

In this chapter we shall consider several iterative techniques for solving an unconstrained optimization problem. These techniques usually require many iterations of rather tedious computations. As a result such techniques usually require the use of a high-speed computer. A large number of these techniques exists and the variety of algorithms presented in this chapter is certainly not an exhaustive collection. The chapter is intended to be a brief introduction to iterative search techniques for unconstrained optimization problems. Thus, only a few algorithms are discussed. A best general purpose algorithm for solving such problems does not exist. Throughout the chapter, emphasis is placed on the computational aspects of the techniques. Little attention is paid to the theoretical development and the economy (rate of convergence, etc.) of the various methods. Such theoretical developments usually apply only to problems having a highly specialized structure. Nevertheless, many problems not possessing such a specialized structure can be satisfactorily solved using iterative techniques such as the ones listed in this chapter. It is left to the individual to seek out a procedure that works well with a given problem and avoids the inherent difficulties that may be associated with the problem.

1. ONE-DIMENSIONAL LINEAR SEARCH TECHNIQUES

We begin our study of search techniques by first considering linear search techniques, i.e., finding a maximum or a minimum of the objective function along a straight line. Many of the multidimensional search techniques require a sequence of unidimensional (line) searches. For simplicity we will assume for the time that f is a real valued function of a real variable x over an interval $[a_1, b_1]$. We also will assume that f is *unimodal*, i.e., f has a single

268

relative minimum x^* (or maximum), i.e., f is strictly decreasing (strictly increasing) when $x < x^*$ and f is strictly increasing (strictly decreasing) when $x > x^*$ (see Figure 1).

Three-Point Equal Interval Search

In the three-point equal interval search technique three points x_L, x_M and x_R are equally placed in the interior of the interval $[a_1, b_1]$ and the min $\{f(x_L), f(x_M), f(x_R)\}$ is computed. Then all of the original interval except the subinterval whose midpoint corresponds to min $\{f(x_L), f(x_M), f(x_R)\}$ is discarded. For illustration suppose that $\min\{f(x_L), f(x_M), f(x_R)\} = f(x_M)$. Notice that since f is unimodal, the optimal solution x^* cannot lie in the intervals $[a_1, x_L)$ and $(x_R, b_1]$. For instance, if $x^* \in (x_R, b_1]$, then the function must be strictly decreasing to the left of x^*. But $f(x_R) > f(x_M) > f(x^*)$ and $x_M < x_R < x^*$. Thus, $x^* \notin (x_R, b_1]$. Likewise, $x^* \notin [a_1, x_L)$. The process is again repeated with the new subinterval $[a_2, b_2]$, where in this case $a_2 = x_L$ and $b_2 = x_R$ (see Figure 2).

To maximize f over $[a_1, b_1]$ we can either minimize $-f$ over $[a_1, b_1]$ (recall that $-\min\{-f(x) : x \in [a_1, b_1]\} = \max\{f(x) : x \in [a_1, b_1]\}$) or we can proceed as above retaining the subinterval of $[a_1, b_1]$ whose midpoint corresponds to the max $\{f(x_L), f(x_M), f(x_R)\}$. Notice in Figure 3 that $[a_2, b_2] = [a_1, x_M]$.

In either case the decision of when to end the search is left to the individual. For example, when minimizing f we could decide to end the process after n iterations. In this case we could estimate $f(x^*)$ by $f(x^*) = \min\{f(a_n), f(x_L), f(x_M), f(x_R), f(b_n)\}$. Another possibility would be to end the search after n iterations provided either

(1) $b_n - a_n < \varepsilon_1$, $|\min\{f(x_L), f(x_M), f(x_R)\} - f(a_n)| < \varepsilon_2$, and $|\min\{f(x_L), f(x_M), f(x_R)\} - f(b_n)| < \varepsilon_2$.

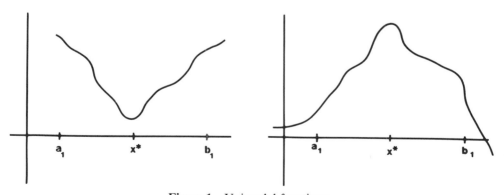

Figure 1 Unimodal functions

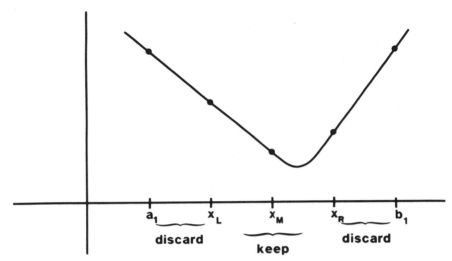

Figure 2   Discard $[a_1, x_L] \cup (x_R, b_1]$

(2) n is equal to some preassigned value (such as 100).
       The procedure is summarized as follows.
0.   Let $n = 1$.
1.   Compute $\ell_n = 0.25 (b_n - a_n)$.
2.   Let $x_L = a_n + \ell_n$, $x_M = a_n + 2\ell_n$ and $x_R = a_n + 3\ell_n$.
3.   If either (i) $b_n - a_n < \varepsilon_1, |m - f(a_n)| < \varepsilon_2$, and $|m - f(b_n)| < \varepsilon_2$, or (ii) $n = N_0$, where N_0 is some preassigned natural number, then go to (6).

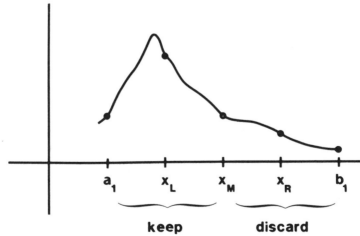

Figure 3   Discard (x_M, b_1)

4. Compute $m = \min \{f(x_L), f(x_M), f(x_R)\}$.
5. If $m = f(x_L)$, then set $a_{n+1} = a_n$ and $b_{n+1} = x_M$. Go to (1).
   If $m = f(x_R)$, then set $a_{n+1} = x_M$ and $b_{n+1} = b_n$. Go to (1).
   If $m = f(x_M)$, then set $a_{n+1} = x_L$ and $b_{n+1} = x_R$. Go to (1).
6. Stop and estimate $f(x^*)$ by $f(x^*) = \min \{f(a_n), f(x_L), f(x_M), f(x_R), f(b_n)\}$ and select x^* accordingly.

Here Step (3), which dictates when the process is to terminate, can be changed to suit the individual. In order to maximize f we can either let $m = \max \{f(x_L), f(x_M), f(x_R)\}$ in Step (4), or we can minimize $-f$.

Notice that at each iteration one function value is retained from the preceding iteration and hence does not require recalculation. Since function evaluation can be costly, this is an important feature of this technique.

EXAMPLE 1   Maximize $x(3 - x)^{5/3}$ over the interval $[0,3]$ using an equal interval search. First approximate the solution by completing 3 iterations of a three point equal interval search. Then continue the process as outlined above using $\varepsilon_i = 0.00001$, where $i = 1,2,3$.

Solution: Let $[a_1, b_1] = [0, \ 3]$. Then $\ell_1 = 0.25(3 - 0) = 0.75$. Hence, $x_L = 0.750000$, $x_M = 1.500000$, and $x_R = 2.250000$. Also,

$$f(x_L) = 0.750000(3 - 0.750000)^{5/3} = 2.897558$$

$$f(x_M) = 1.500000(3 - 1.500000)^{5/3} = 2.948334$$

$$f(x_R) = 2.250000(3 - 2.250000)^{5/3} = 1.393001$$

Since $\max \{f(x_L), \ f(x_M), \ f(x_R)\} = f(x_M)$, let $a_2 = 0.750000$ and $b_2 = 2.250000$. Then $\ell_2 = 0.25(b_2 - a_2) = 0.375$. Hence the new values for x_L, x_M and x_R are 1.125000, 1.500000, and 1.875000, respectively. Hence,

$$f(x_L) = 1.125000(3 - 1.125000)^{5/3} = 3.207411$$

$$f(x_M) = 1.500000(3 - 1.500000)^{5/3} = 2.948334$$

$$f(x_R) = 1.875000(3 - 1.875000)^{5/3} = 2.281684.$$

Now since $\max \{f(x_L), \ f(x_M), \ f(x_R)\} = f(x_L)$, let $a_3 = 0.750000$ and $b_3 = 1.500000$. Thus, $\ell_3 = (0.25)(b_3 - a_3) = 0.1875$ and the new values for x_L, x_M and x_R are 0.937500, 1.125000, and 1.312500, respectively. So that

$$f(x_L) = 0.937500(3 - 0.937500)^{5/3} = 3.133006$$

$$f(x_M) = 1.125000(3 - 1.125000)^{5/3} = 3.207411$$

$$f(x_R) = 1.312500(3 - 1.312500)^{5/3} = 3.139344.$$

Since $\max \{f(a_3), f(x_L), f(x_M), f(x_R), f(b_3)\} = 3.207411 = f(1.125000)$, we

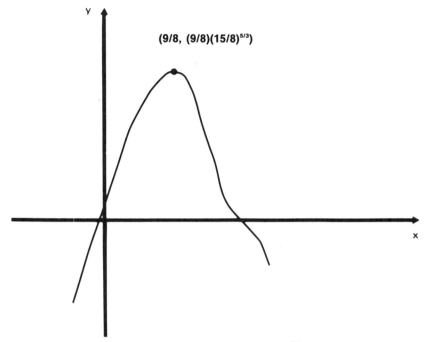

Figure 4   Maximize $x(3-x)^{5/3}$

conclude that the approximate maximum of f is 3.207411 which occurs when $x = 1.125000$.

Table 1 summarizes the results when we continue as in the outline with $\varepsilon = 0.00001$. Table 1 bears out the fact that the graph of f does not have any sharp peaks for values of x near 1.125000. The graph of $g(x) = x(3-x)^{5/3}$ is shown in Figure 4.

Other types of equal interval searches are possible (two-point, four-point, etc.). However, in [6] Kiefer has shown that the three-point search is the most "efficient."

Exercises

9.1   Write a Fortran program for the three-point equal interval search following the previously suggested outline.

9.2   Show that the following functions are unimodal over the indicated intervals.
(a) $f(x) = x(3-x)^{5/3}$ over $[0,3]$
(b) $f(x) = x^4 + x^2 - 3x$ over $[0,2]$

Table 1. Three pint equal interval search

x_L	x_M	x_R	$f(x_R)$	$f(x_M)$	$f(x_L)$	$\max F(x)$
0.750000	1.500000	2.250000	2.897558	2.948334	1.393001	2.948334
1.125000	1.500000	1.875000	3.207411	2.948334	2.281684	3.207411
0.937500	1.125000	1.312500	3.133006	3.207411	3.139344	3.207411
1.031250	1.125000	1.218750	3.189199	3.207411	3.189991	3.207411
1.078125	1.125000	1.171875	3.202907	3.207411	3.203006	3.207411
1.101563	1.125000	1.148438	3.206291	3.207411	3.206304	3.207411
1.113281	1.125000	1.136719	3.207132	3.207411	3.207134	3.207411
1.119141	1.125000	1.130859	3.207342	3.207411	3.207342	3.207411
1.122070	1.125000	1.127930	3.207394	3.207411	3.207394	3.207411
1.123535	1.125000	1.126465	3.207407	3.207411	3.207407	3.207411
1.124268	1.125000	1.125732	3.207410	3.207411	3.207410	3.207411
1.124634	1.125000	1.125366	3.207411	3.207411	3.207411	3.207411
1.124817	1.125000	1.125183	3.207411	3.207411	3.207411	3.207411
1.124908	1.125000	1.125092	3.207411	3.207411	3.207411	3.207411
1.124954	1.125000	1.125046	3.207411	3.207411	3.207411	3.207411
1.124977	1.125000	1.125023	3.207411	3.207411	3.207411	3.207411
1.124989	1.125000	1.125011	3.207411	3.207411	3.207411	3.207411
1.124994	1.125000	1.125006	3.207411	3.207411	3.207411	3.207411
1.124997	1.125000	1.125003	3.207411	3.207411	3.207411	3.207411

(c) $f(x) = x \cos x$ over $[0,\pi]$
(d) $f(x) = x \sin(1/x)$ over $[1/2\pi, 1/\pi]$
(e) $f(x) = x^4 - 6x^3 + 12x^2 - 8x + 1$ over $(0,1)$
(f) $f(x) = x\sqrt{2+x}$ over $[-2,0]$
(g) $f(x) = 4x^{1/3} - x^{4/3}$ over $[0,4]$
(h) $f(x) = 8x/(x^2+4)$ over $[-2\sqrt{3},0]$
(i) $f(x) = (x^2 + 3x + 2)/x$ over $[1/200,3]$

9.3   Use a three-point equal interval search to minimize $f(x) = x^4 + x^2 - 3x$ over $[0,2]$.

9.4   Devise a two-point equal interval search for minimizing a unimodal $f(x)$ over $[a, b]$ by selecting two points $x_2 > x_1$ in $[a, b]$ so that the interval is subdivided into three subintervals of equal length. After comparing $f(x_1)$ and $f(x_2)$ either $[a,x_1]$ or $[x_2,b]$ can be eliminated. Use your technique on the function in Exercise 9.3.

9.5   Use a three-point equal interval search to maximize $f(x) = x \cos x$ over $[0, \pi]$.

9.6   Show that $f(x_L) = f(x_R) = \min\{f(x_L), f(x_m), f(x_R)\}$ can't happen when f is unimodal.

Fibonacci Search

Consider again the problem of minimizing a function that is unimodal over the interval $[a,b]$. Let $L_1 = b - a$ and suppose that $L_1/2 < L_2 < L_1$. Then set $x_L^1 = b_1 - L_2$ and $x_R^1 = a_1 + L_2$, where $a_1 = a$ and $b_1 = b$ (see Figure 5). Notice that x_L^1 and x_R^1 are symmetrically placed about the midpoint of the interval $[a,b]$. Without loss of generality suppose that $f(x_L^1) < f(x_R^1)$. Then the minimum of f cannot occur in the interval $(x_R^1,b_1]$ and hence, $(x_R^1,b_1]$ is discarded leaving $[a_1,x_R^1]$. Let $a_2 = a_1, b_2 = x_R^1$ and $L_3 = L_1 - L_2$. Then set $x_L^2 = b_2 - L_3$ and $x_R^2 = a_2 + L_3$, where the superscript 2 in the expression x_L^2

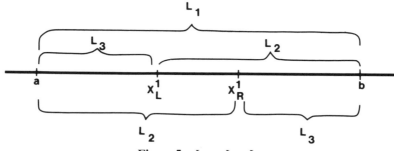

Figure 5 $L_3 = L_1 - L_2$

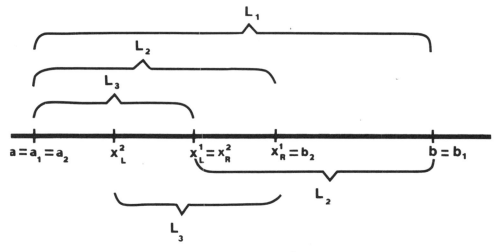

Figure 6 Discard $(x_R^1, b_1]$

denotes the second iteration rather than the square of x_L. Again notice that x_L^2 and x_R^2 are symmetrically placed (see Figure 6) about the midpoint of $[a_2, b_2]$. We shall require that the points be positioned so that $L_3 > (\frac{1}{2})L_2$ for a more general statement [see Exercise 9.7 (b)]. After $n-2$ iterations we could have a situation similar to that shown in Figure 7. Next let $L_n = L_{n-2} - L_{n-1}$ and select x_L^{n-1} and x_R^{n-1} so that they are symmetrically positioned about the midpoint $[a_{n-2}, x_R^{n-2}]$ (requiring that $L_n > L_{n-1}/2$). In Figure 8 this would mean that $x_L^{n-1} = b_{n-1} - L_n$ and $x_R^{n-1} = x_L^{n-2}$. If $\varepsilon = x_R^{n-1} - x_L^{n-1}$, then it would follow that

$$L_{n-1} = 2L_n - \varepsilon \qquad (1)$$

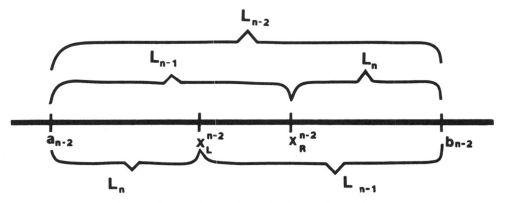

Figure 7 A Fibonacci search after $n-2$ iterations

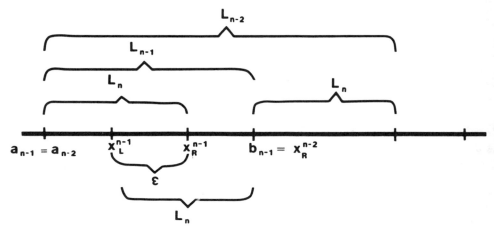

Figure 8 $L_{n-1} = 2L_n - \varepsilon$

In general after the $(n\text{-}k)$th iteration (see Figure 9), where $k = 2, 3, \ldots, n-1$, we would have

$$L_{n-(k-2)} = L_{n-k} - L_{n-(k-1)} \tag{2}$$

Thus using (1) and (2) we would have the following:

$$L_{n-1} = 2L_n - \varepsilon$$
$$L_{n-2} = L_{n-1} + L_n = 2L_n - \varepsilon + L_n = 3L_n - \varepsilon$$
$$L_{n-3} = L_{n-2} + L_{n-1} = 3L_n - \varepsilon + 2L_n - \varepsilon = 5L_n - 2\varepsilon$$
$$L_{n-4} = L_{n-3} + L_{n-2} = 5L_n - 2\varepsilon + 3L_n - \varepsilon = 8L_n - 3\varepsilon$$

$$\vdots$$

$$L_{n-k} = L_{n-(k-1)} + L_{n-(k-2)} = F_{K+1}L_n - F_{k-1}\varepsilon \tag{3}$$

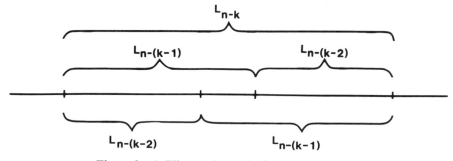

Figure 9   A Fibonacci search after $n-k$ iterations

where integral coefficients F_{k+1} and F_{k-1} are generated by the recurrence relations

$$F_0 = F_1 = 1$$

$$F_{k+1} = F_k + F_{k-1}, \quad k = 1, 2, 3, \ldots \tag{4}$$

The sequence of integers generated by (4) is called the *Fibonacci number sequence*. The first 30 Fibonacci numbers are listed in Table 2.

If we let $k = n - 1$ in equation (3), then we get the following relationship between L_1 and L_n

$$L_1 = L_{n-(n-1)} = F_{(n-1)+1} L_n - F_{n-1-1} \varepsilon = F_n L_n - F_{n-2} \varepsilon$$

or

$$L_n = \frac{L_1 + F_{n-2} \varepsilon}{F_n}$$

Likewise, if we substitute $k = n - 2$ in (3), then we have the following

$$L_2 = L_{n-(n-2)} = F_{n-2+1} L_n - F_{n-2-1} \varepsilon = F_{n-1} L_n - F_{n-3} \varepsilon$$

or

$$L_2 = F_{n-1} \left(\frac{L_1 + F_{n-2} \varepsilon}{F_n} \right) - F_{n-3} \varepsilon = \frac{F_{n-1} L_1}{F_n} + \frac{(F_{n-1} F_{n-2} - F_n F_{n-3})}{F_n} \varepsilon$$

Simplifying we have

$$F_{n-1} F_{n-2} - F_n F_{n-3} = F_{n-1} F_{n-2} - F_{n-3}(F_{n-1} + F_{n-2})$$

$$= F_{n-1} F_{n-2} - F_{n-1} F_{n-3} - F_{n-2} F_{n-3}$$

$$= F_{n-2}(F_{n-1} - F_{n-3}) - F_{n-1} F_{n-3} = F_{n-2}^2 - F_{n-1} F_{n-3}$$

But it can be shown by mathematical induction that

$$F_{n-2}^2 - F_{n-1} F_{n-3} = (-1)^n$$

Table 2. The first thirty Fibonacci numbers

n	F_n	n	F_n	n	F_n
0	1	10	89	20	10,946
1	1	11	144	21	17,711
2	2	12	233	22	28,657
3	3	13	377	23	46,368
4	5	14	610	24	75,025
5	8	15	987	25	121,393
6	13	16	1597	26	196,418
7	21	17	2584	27	317,811
8	34	18	4181	28	514,229
9	55	19	6765	29	832,040

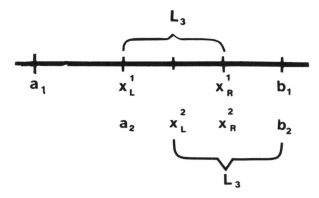

Figure 10   When $f(x_L^1) > f(x_R^1)$

Hence,

$$L_2 = \frac{F_{n-1} L_1}{F_n} + \frac{(-1)^n \varepsilon}{F_n} \tag{5}$$

Thus, for a predetermined number n of search points and a predetermined ε, the Fibonacci search technique for finding the minimum of a unimodal function over an interval $[a, b]$ can be summarized as follows:

1. Set $a_1 = a, b_1 = b$ and $L_1 = b_1 - a_1$.
2. Use equation (5) to compute L_2.
3. Let $x_L^1 = b_1 - L_2$ and $x_R^1 = a_1 + L_2$.
4. Evaluate $f(x_L^1)$ and $f(x_R^1)$.
5. Evaluate $L_3 = L_1 - L_2$.
6. If $f(x_L^1) > f(x_R^1)$, then set $a_2 = x_L^1$, $b_2 = b_1$, $x_L^2 = x_R^1$ and $x_R^2 = a_2 + L_3$. (Figure 10). Go to Step 7.

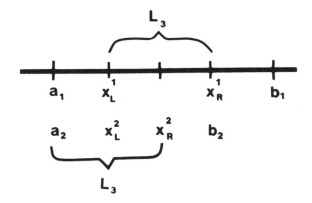

Figure 11

If $f(x_R^1) > f(x_L^1)$, then set $a_2 = a_1$, $b_2 = x_R^1$, $x_R^2 = x_L^1$ and $x_L^2 = b_2 - L_3$. (Figure 11). Go to Step 7.

7. Return to step (4) and continue for $n - 1$ trials.

Notice (Figure 8) that when the process ends the search points are ε units apart and the interval of uncertainty has length L_n, i.e., the true minimum occurs at a point not more than L_n units from the estimated solution. Moreover, the length of $L_n = L_1 + F_{n-2}\,\varepsilon/F_n$ is determined at the beginning of the problem once L_1 and ε are specified.

EXAMPLE 2   Using an eight point Fibonacci search with $\varepsilon = 0.0001$, determine the minimum $x \sin(1/x)$ over $[1/2\pi, 1/\pi]$.

Solution: First let $a_1 = 1/2\pi = 0.15915494$ and $b_1 = 1/\pi = 0.31830989$. Let

$$L_1 = b_1 - a_1 = 0.15915495$$

Using equation (5), compute

$$L_2 = \frac{F_7 L_1}{F_8} + \frac{(-1)^8 \varepsilon}{F_8} = \frac{21(0.15915495)}{34} + \frac{0.0001}{34} = 0.09830452$$

Next let

$$x_L^1 = b_1 - L_2 = 0.31830989 - 0.09830452 = 0.22000537$$

and

$$x_R^1 = a_1 + L_2 = 0.15915494 + 0.09830452 = 0.25745946$$

Hence,

$$f(x_L^1) = -0.21694297 \text{ and } f(x_R^1) = -0.17407923$$

Also, $L_3 = L_1 - L_2 = 0.06085043$. Since $f(x_L^1) < f(x_R^1)$, we set

$$a_2 = a_1 = 0.15915494$$
$$b_2 = x_R^1 = 0.25745946$$
$$x_R^2 = x_L^1 = 0.22000537$$

and
$$x_L^2 = b_2 - L_3 = 0.19660903$$

Next computing $f(x_L^2)$, $f(x_R^2)$ and L_4 we have

$$f(x_L^2) = -0.18302913$$
$$f(x_R^2) = -0.21694297$$

and
$$L_4 = L_2 - L_3 = 0.03745409$$

Since $f(x_L^2) > f(x_R^2)$, we set

$$a_3 = x_L^2 = 0.19660903$$
$$b_3 = b_2 = 0.25745946$$
$$x_L^3 = x_R^2 = 0.22000537$$

and
$$x_R^3 = a_3 + L_4 = 0.23406312$$

The process is continued for a total of seven trials. Table 3 summarizes the results of the seven trials. Notice that the two final trial points x_L^7 and x_R^7 are $\varepsilon = .0001$ apart. The optimal solution is estimated to be

$$f(0.22462463) = -0.21704508$$

and the interval of uncertainty is $L_8 = [a_7, x_R^7]$ which has length 0.00471916. The graph of $f(x) = x \sin 1/x$ is shown in Figure 12.

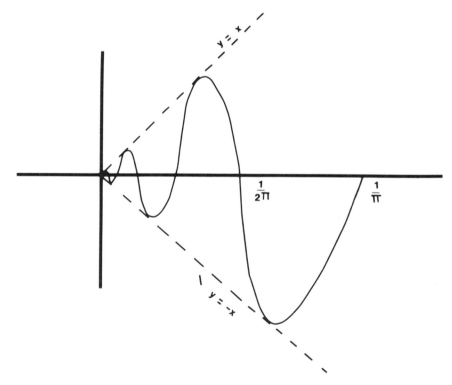

Figure 12   Minimize $x \sin(1/x)$

Table 3 Eight–point Fibonacci search

n	a_i	b_i	x_L^i	x_R^i	$f(x_L^i)$	$f(x_R^i)$	L_{i+1}
1	0.15915494	0.31830989	0.22000537	0.25745946	-0.21694237	-0.17407923	0.09830452
2	0.15915494	0.25745946	0.19660903	0.22000537	-0.18302913	-0.2169429	0.06085043
3	0.19660903	0.25745946	0.22000537	0.23406312	-0.21694297	-0.21176526	0.03745409
4	0.19660903	0.23406312	0.21066684	0.22000537	-0.21054189	-0.2169429	0.02339634
5	0.21066684	0.23406312	0.22000537	0.22472463	-0.21694297	-0.2170266	0.01405775
6	0.22000537	0.23406312	0.22472463	0.22934389	-0.2170266	-0.21527183	0.00933859
7	0.22000537	0.22934389	0.22462463	0.22472463	-0.21704508	-0.2170266	0.00471916

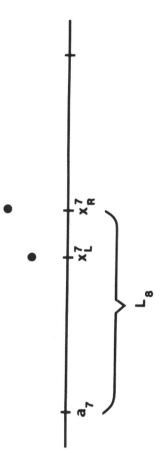

Golden-Ratio Section Search

The golden-ratio section search is a Fibonacci type of search in which the ration L_n/L_{n-1} in any iteration remains constant. Consider an iteration (Figure 13) which begins with the interval $[a,d]$. Two points b and c are placed symmetrically about the midpoint of $[a,d]$ in a way such that

$$r = \frac{c-a}{d-a} = \frac{d-b}{d-a} \qquad (6)$$

If we are trying to minimize a unimodal function f over this interval, then we evaluate f at b and c. If $f(b) < f(c)$, then we discard the interval $(c, d]$ and continue working with the interval $[a, c]$. Otherwise when $f(b) > f(c)$ we discard the interval $[a, b)$ and continue our search in $[b,d]$. In either case we require that

$$r = \frac{b-a}{c-a} = \frac{d-c}{d-b} \qquad (7)$$

Thus from equations (6) and (7) we have

$$r^2 = \frac{c-a}{d-a} \frac{b-a}{c-a} = \frac{b-a}{d-a}$$

and

$$1 - r = 1 - \frac{d-b}{d-a} = \frac{d-a-(d-b)}{d-a} = \frac{b-a}{d-a}$$

Hence,

$$r^2 = 1 - r$$
$$r^2 + r - 1 = 0$$

and

$$r = \frac{-1 \pm \sqrt{5}}{2}$$

Since $r > 0$, it follows that $r = (-1+\sqrt{5})/2 \doteq 0.6180339887$ (or $r = 0.618$).
    Thus, to minimize a unimodal function f over the interval $[a, b]$ using a golden-ratio section search we proceed as follows:
1.   Set $a_1 = a$, $b_1 = b$ $x_R^1 = a_1 + 0.618(b - a)$ and $x_L^1 = b - 0.618(b - a)$.
2.   Evaluate $f(x_R^1)$ and $f(x_L^1)$.

Figure 13   Golden-ratio section search

3. If $f(x_L^1) < f(x_R^1)$, then set $a_2 = a_1$, $b_2 = x_R^1$, $x_R^2 = x_L^1$ and $x_L^2 = b_2 - 0.618(b_2 - a_2)$. Otherwise when $f(x_R^1) < f(x_L^1)$ set $a_2 = x_L^1$, $b_2 = b_1$, $x_L^2 = x_R^1$ and $x_R^2 = a_2 + 0.618(b_2 - a_2)$.

4. Continue the process until the desired accuracy is attained or until the desired number of iterations have been completed.

Using Table 2 and computing values for F_{n-1}/F_n it appears that $\lim_{n \to \infty} F_{n-1}/F_n = 0.618$. Moreover, if we consider a Fibonacci search with an infinite number of search points then it follows from equation (5) that $L_2 = 0.618\, L_1$. Thus, the golden-ratio section search can be considered as a limiting form of a Fibonacci search in which there are an infinite number of search points.

EXAMPLE 3   Minimize $f(x) = x^4 - 6x^3 + 12x^2 - 8x + 1$ **over** $[0, 1]$ **using a thirteen iteration (fourteen-point) golden-ratio section search.**
Solution: Let $a_1 = 0$, $b_1 = 1$, $x_R^1 = 0.618$ and $x_L^1 = 1 - 0.618 = 0.382$. Then $f(x_L^1) = -0.61807599$ and $f(x_R^1) = -0.63122025$. Since $f(x_L^1) > f(x_R^1)$, set $a_2 = x_L^1 = 0.382$, $b_2 = b_1 = 1$, $x_L^2 = 0.618$ and $x_R^2 = 0.382 + 0.618(1 - 0.382) = 0.763924$. Now $f(x_L^2) = -0.63115172$ and $f(x_R^2) = -0.44273204$. Hence, set $a_3 = a_2 = 0.382$, $b_3 = x_R^2 = 0.763924$, $x_R^3 = x_L^2 = 0.618$ and $x_L^3 = b_3 -0.618(b_3 - a_3) = 0.52789497$. This is continued for thirteen iterations and the results are summarized in Table 4 and a graph of the function is shown in Figure 14.

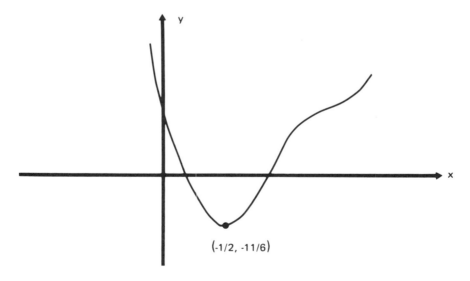

(-1/2, -11/6)

Figure 14   Graph of $f(x) = x^4 - 6x^3 + 12x^2 - 8x + 1$.

Table 4. Golden-ratio section search

a	b	x_L^i	x_R^i	$f(x_L^i)$	$f(x_R^i)$	L
0.00000000	1.00000000	0.38200000	0.61800000	-0.61807599	-0.63122025	0.61800000
0.38200000	1.00000000	0.61807600	0.76392400	-0.63115172	-0.44273204	0.38192400
0.38200000	0.76392400	0.52789497	0.61802903	-0.68408464	-0.63119408	0.23602903
0.38200000	0.61802903	0.47216309	0.52786594	-0.68392610	-0.68409165	0.14586594
0.47216309	0.61802903	0.52788388	0.56230824	-0.68408732	-0.67098210	0.09014515
0.47216309	0.56230824	0.50659854	0.52787279	-0.68730521	-0.68409000	0.05570970
0.47216309	0.52787279	0.49344420	0.50659169	-0.68730547	-0.68730562	0.03442860
0.49344420	0.52787279	0.50659592	0.51472107	-0.68730537	-0.68653752	0.02127687
0.49344420	0.51472107	0.50157196	0.50659330	-0.68748890	-0.68730552	0.01314911
0.49344420	0.50659330	0.49846716	0.51057035	-0.68748941	-0.68748892	0.00812615
0.49344420	0.51057035	0.49654839	0.49846616	-0.68744622	-0.68748940	0.00502196
0.49654839	0.50157035	0.49846677	0.49965196	-0.68748941	-0.68749945	0.00310957
0.49846677	0.50157035	0.49965234	0.50038478	-0.68749946	-0.68749933	0.00191801

Exercises

9.7   **In a Fibonacci** search show that
(a) $F_{n-2}^2 - F_{n-1} F_{n-3} = (-1)^n$
(b) $L_{k+1} = (F_{n-k}/F_{n-k+1})L_k$ (When is $F_{n-k}/F_{n-k+1} > \frac{1}{2}$?)
(c) $x_L^k = a_k + (F_{n-k-1}/F_{n-k+1})(b_k - a_k)$
(d) $x_R^k = a_k + (F_{n-k}/F_{n-k+1})(b_k - a_k)$

9.8   Write a **Fortran** program that will minimize the **function in** Example 2 with a 16-**point Fibonacci** search.

9.9   Use an **eight-point** Fibonacci search to minimize $f(x) = x^2/(x-1)^2$ over $[-1, \frac{1}{2}]$.

9.10   Use an eight-point Fibonacci search to maximize $f(x) = \sin^2 x + \cos x$ over $[0, \pi]$. Use $\varepsilon = 0.0001$ and also determine the length of the interval of uncertainty.

9.11   Write a Fortran program for a 14-point golden-ratio section search. Test your program by maximizing $f(x) = x \cos x$ over the interval $[0, \pi]$.

9.12   Write a Fortran program for the golden-ratio section search that terminates when both (1) the trial points differ by less than .0001, and (2) the function values at the trial points differ by less than 0.0001. Check your program by maximizing $f(x) = x \cos x$ over $[0, \pi]$.

9.13   Use an eight-point golden-ratio section search to minimize $f(x) = x^4 - 6x^3 + 12x^2 - 8x + 1$ over $[0, 1]$.

2.   LINEAR SEARCH TECHNIQUES BY CURVE FITTING

Another way of locating the minimum value of a unimodal function f of a single real variable is to "fit" a polynomial p to f and then locate the minimum of the polynomial. Several of these techniques, each using a quadratic polynomial, are presented in this section. The first two techniques, Powell's algorithm and the Davies, Swann and Campey algorithm (DSC), are both concerned with the fitting of a quadratic polynomial to three points.

   To begin suppose that the quadratic polynomial

$$q(x) = a_0 + a_1 x + a_2 x^2 \qquad \text{(where } a_2 > 0\text{)}$$

has the properties that $q(x_L) = f(x_L)$, $q(x_C) = f(x_C)$ and $q(x_R) = f(x_R)$ where $x_L \leq x_C \leq x_R$. In order to locate the minimum of q, we want to determine where $q'(x) = 0$. If $0 = q'(x) = a_1 + 2a_2 x$, then $x = -a_1/2a_2$. Now

$$f(x_L) = q(x_L) = a_0 + a_1 x_L + a_2 x_L^2$$
$$f(x_C) = q(x_C) = a_0 + a_1 x_C + a_2 x_C^2 \tag{8}$$

and

$$f(x_R) = q(x_R) = a_0 + a_1 x_R + a_2 x_R^2$$

Let

$$\Delta = \begin{vmatrix} 1 & x_L & x_L^2 \\ 1 & x_C & x_C^2 \\ 1 & x_R & x_R^2 \end{vmatrix}$$

Then solving equations (8) for a_1 and a_2 using Cramer's rule gives

$$a_1 = \frac{\begin{vmatrix} 1 & f(x_L) & x_L^2 \\ 1 & f(x_C) & x_C^2 \\ 1 & f(x_R) & x_R^2 \end{vmatrix}}{\Delta} = \frac{\begin{vmatrix} 1 & f(x_L) & x_L^2 \\ 0 & f(x_C)-f(x_L) & x_C^2-x_L^2 \\ 0 & f(x_R)-f(x_L) & x_R^2-x_L^2 \end{vmatrix}}{\Delta}$$

$$= \frac{(f(x_C)-f(x_L))(x_R^2-x_L^2) - (f(x_R)-f(x_L))(x_C^2-x_L^2)}{\Delta}$$

$$= \frac{f(x_L)(x_C^2-x_L^2-x_R^2+x_L^2) + f(x_C)(x_R^2-x_L^2) + f(x_R)(x_L^2-x_C^2)}{\Delta}$$

$$= \frac{f(x_L)(x_C^2-x_R^2) + f(x_C)(x_R^2-x_L^2) + f(x_R)(x_L^2-x_C^2)}{\Delta}$$

and

$$a_2 = \frac{\begin{vmatrix} 1 & x_L & f(x_L) \\ 1 & x_C & f(x_C) \\ 1 & x_R & f(x_R) \end{vmatrix}}{\Delta} = \frac{\begin{vmatrix} 1 & x_L & f(x_L) \\ 0 & x_C-x_L & f(x_C)-f(x_L) \\ 0 & x_R-x_L & f(x_R)-f(x_L) \end{vmatrix}}{\Delta}$$

$$= \frac{(x_C-x_L)(f(x_R)-f(x_L)) - (x_R-x_L)(f(x_C)-f(x_L))}{\Delta}$$

$$= \frac{f(x_L)(x_R-x_L-x_C+x_L) + f(x_C)(x_L-x_R) + f(x_R)(x_C-x_L)}{\Delta}$$

$$= \frac{f(x_L)(x_R-x_C) + f(x_C)(x_L-x_R) + f(x_R)(x_C-x_L)}{\Delta}$$

Hence,

$$x = -\frac{a_1}{2a_2} = -\frac{f(x_L)(x_C^2-x_R^2) + f(x_C)(x_R^2-x_L^2) + f(x_R)(x_L^2-x_C^2)}{2(f(x_L)(x_R-x_C) + f(x_C)(x_L-x_R) + f(x_R)(x_C-x_L))}$$

or

$$x = \frac{f(x_L)(x_C^2 - x_R^2) + f(x_C)(x_R^2 - x_L^2) + f(x_R)(x_L^2 - x_C^2)}{2(f(x_L)(x_C - x_R) + f(x_C)(x_R - x_L) + f(x_R)(x_L - x_C))} \tag{9}$$

It is left as an exercise to show that when $\Delta x = x_R - x_C = x_C - x_L$, then

$$x = x_c + \frac{(f(x_L) - f(x_R))\Delta x}{2(f(x_L - 2f(x_C) + f(x_R))} \tag{10}$$

Powell's Algorithm

Powell's algorithm uses equation (9) to approximate the value of x that minimizes f. The process starts with a point $(x, f(x))$ on the graph of f. Two other points on the graph of f are then determined and a quadratic approximation of the value of x that minimizes f is carried out. This gives four points on the graph of f. One is discarded and another approximation is again carried out. This continues until a solution is found that satisfies the desired accuracy in x and the values of $f(x)$. The algorithm is summarized as follows:

   1. Select a value x_1 and compute $f(x_1)$.

   2. Set $x_2 = x_1 + \Delta x$, where Δx is some predetermined positive number. Then compute $f(x_2)$.

   3. If $f(x_1) > f(x_2)$, then set $x_3 = x_1 + 2\Delta x$. Otherwise, if $f(x_1) < f(x_2)$, then set $x_3 = x_1 - \Delta x$.

   4. Compute $f(x_3)$.

   5. Use equation (9) to estimate the value x^* of x that minimizes $f(x)$, i.e.,

$$x^* \doteq \dot{x}^* = \frac{f(x_1)(x_2^2 - x_3^2) + f(x_2)(x_3^2 - x_1^2) + f(x_3)(x_1^2 - x_2^2)}{2(f(x_1)(x_2 - x_3) + f(x_2)(x_3 - x_1) + f(x_3)(x_1 - x_2))} \tag{11}$$

   6. Compare the values $\{\dot{x}^*, x_1, x_2, x_3\}$ and the values $\{f(\dot{x}^*), f(x_1), f(x_2), f(x_3)\}$. If the approximated solution satisfies whatever accuracy conditions are desired, then terminate the search. Otherwise, discard from the set $\{\dot{x}^*, x_1, x_2, x_3\}$ one of the values x_1, x_2, or x_3. Usually the value is discarded that has the largest function value. However, (since interpolation is preferable to extrapolation) whenever possible discard a value of x_1, x_2, or x_3 so that the minimum of the remaining three function values will not occur at the extreme right or the extreme left, when this happens we say that we have a "bracket" on the minimum. (see Figure 15).

EXAMPLE 4   Use Powell's algorithm to minimize $f(x) = (x^2 + 3x + 2)/x$ over $[1/200, 3]$

Solution: We arbitrarily set $x_1 = 0.5$ and $\Delta x = 1$. Then $x_2 = 1.5$. Next we compute $f(x_1) = 7.5$ and $f(x_2) = 5.8333333$. Since $f(x_2) < f(x_1)$, we let

$x_3 = x_1 + 2\Delta x = 2.5$. Next we compute $f(x_3) = 6.3$ and then substitute into equation (11) to get

$$x^* \doteq \dot{x}_1^* = \frac{(2.25 - 6.25)(7.5) + (6.25 - .25)(5.8333333) + (.25 - 2.25)(6.3)}{2(1.5 - 2.5)(7.5) + (2.5 - .5)(5.8333333) + (.5 - 1.5)(6.3)}$$

$$= \frac{-7.6}{2(-2.133334)} = 1.7812499$$

Next compute $f(\dot{x}_1^*) = 5.904057$. Now the max $\{f(x_1), f(x_2), f(x_3)\} = f(x_1)$, but we do not discard x_1 since $x_2 < \dot{x}_1^* < x_3$ and $f(x_2) < f(\dot{x}_1^*) < f(x_3)$. Instead we discard x_3.

To begin the second iteration we let $x_1 = .5$, $x_2 = 1.5$ and $x_3 = 1.7812499$. Then

$$x^* \doteq \dot{x}_2^* =$$

$$\frac{(2.25 - 3.1728512)(7.5) + (3.1728512 - 0.25)(5.8333333) + (0.25 - 2.25)(5.904057)}{2((1.5 - 1.7812499)(7.5) + (1.7812499 - 0.5)(5.8333333) + (0.5 - 1.5)(5.904057))}$$

$$= \frac{-1.6795328}{2(-0.5399736)} = 1.5566404$$

and

$$f(\dot{x}_1^*) = 5.8414587$$

Now max $\{f(x_1), f(x_2), f(x_3)\} = $ max $\{7.5, 5.8333333, 5.904057\} = f(x_1)$ but we do not discard x_1 since $x_2 < \dot{x}_2^* < x_3$ and $f(x_2) < f(\dot{x}_2^*) < f(x_3)$. Instead we discard x_3.

Hence, to begin the third iteration we let $x_1 = .5$, $x_2 = 1.5$, and

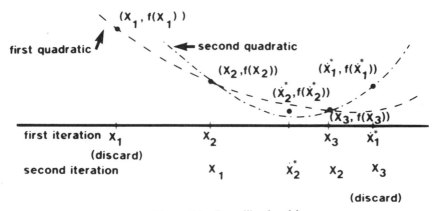

Figure 15 Powell's algorithm

$x_3 = 1.5566404$. Then

$$x^* \doteq \dot{x}_3^* =$$

$$\frac{(2.25 - 2.4231293)(7.5) + (2.4231293 - 0.25)(5.8333333) + (0.25 - 2.25)(5.8414587)}{2((1.5 - 1.5566404)(7.5) + (1.5566404 - 0.5)(5.8333333) + (0.5 - 1.5)(5.8414587))}$$

$$= \frac{-0.3047996}{2(-0.1025261)} = 1.4864493$$

and

$$f(\dot{x}_3^*) = 5.8319375$$

Since max $\{f(x_1), \ f(x_2), \ f(x_3)\} = \max \{7.5, \ 5.8333333,$ $5.8414587\} = f(x_1)$, $x_1 < x_3^* < x_2 < x_3$, $f(x_3^*) < f(x_1)$, and $f(x_3^*) < f(x_2) < f(x_3) < f(x_1)$, we discard x_3.

We will perform one more iteration with $x_1 = 0.5$, $x_2 = 1.4864493$ and $x_3 = 1.5$. Thus,

$$x^* \doteq \dot{x}_4^* =$$

$$\frac{(2.2095315 - 2.25)(7.5) + (2.25 - 0.25)(5.8319375) + (0.25 - 2.2095315)(5.8333333)}{2((1.4864493 - 1.5)(7.5) + (1.5 - 0.5)(5.8319735) + (0.5 - 1.4864493)(5.8333333))}$$

$$= \frac{-0.0702391}{2(-0.0239803)} = 1.4645167$$

and

$$f(x^*) \doteq f(\dot{x}^*) = 5.8301549$$

The process is continued until the desired accuracy in x or $f(x)$ is obtained. Actually, the optimal solution is $x^* = 1.4142136$ and $f(x^*) = 5.8284271$. A graph of the function is found in Figure 16.

The Davies, Swann, and Campey Unidimensional Algorithm (DSC Algorithm)

The DSC algorithm uses equation (10) to approximate the minimum value of a unimodal function f of a single real variable x. When using Powell's algorithm, the quadratic approximations are performed using the first three determined values of x. In the DSC algorithm to be described in this section steps of increasing size are taken until a bracket on the minimum is obtained. Then a quadratic approximation is carried out. The process may be repeated until the desired accuracy in x and the values of $f(x)$ is obtained. The

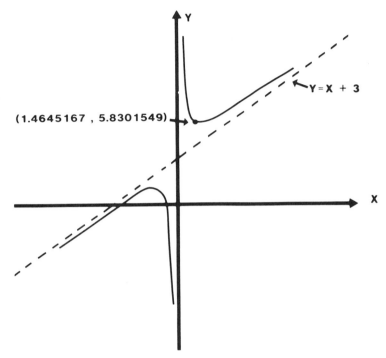

Figure 16   Minimize $(x^2 + 3x + 2)/x$.

algorithm is summarized as follows:

1. Select an initial point x_0 and a step size Δx. Compute $f(x_0)$ and $f(x_0 + \Delta x)$.

2. If $f(x_0) > f(x_0 + \Delta x)$, then let $x_k = x_{k-1} + 2^{k-1} \Delta x$ and evaluate $f(x_k)$ for $k = 1, 2, 3, \ldots$ (we already have $f(x_1)$ from Step 1) until a value of k is found such that $f(x_k) < f(x_{k+1})$. Go to Step 3i.

If $f(x_0) < f(x_0 + \Delta x)$ and $f(x_0) < f(x_0 - \Delta x)$, then go to Step 3ii.

If $f(x_0) < f(x_0 + \Delta x)$ and $f(x_0) > f(x_0 - \Delta x)$, then let $x_k = x_{k-1} - 2^{k-1} \Delta x$ and evaluate $f(x_k)$ for $k = 1, 2, 3, \ldots$ until a value of k is found such that $f(x_{k+1}) > f(x_k)$. Go to Step 3iii.

3i. Consider x_{k-1}, x_k and x_{k+1}. Let $x_{k+2} = \frac{1}{2}(x_k + x_{k+1})$. Then from the set of equally spaced points $\{x_{k-1}, x_k, x_{k+2}, x_{k+1}\}$ discard either x_{k-1} or x_{k+1} whichever is farthest from the x that corresponds to the min $\{f(x_{k-1}), f(x_k), f(x_{k+2}), f(x_{k+1})\}$. Denote the remaining values by x_L, x_C and x_R, where $x_L < x_C < x_R$. Go to Step 4 (see Figure 17a).

ii. Let $x_L = x_0 - \Delta x$, $x_C = x_0$ and $x_R = x_0 + \Delta x$. Go to step 4 (see Figure 17b).

iii. Consider x_{k-1}, x_k and x_{k+1}. Let $x_{k+2} = 1/2(x_k + x_{k+1})$. Then from the set of equally spaced points $\{x_{k+1}, x_{k+2}, x_k, x_{k-1}\}$ discard either x_{k+1} or x_{k-1} whichever is farthest from the value of x that corresponds to the $\min\{f(x_{k+1}), f(x_{k+2}), f(x_k), f(x_{k-1})\}$ Denote the remaining values by x_L, x_C and x_R, where $x_L < x_C < x_R$. Go to Step 4 (see Figure 17c).

4. Carry out a quadratic interpolation to estimate the value x^* that

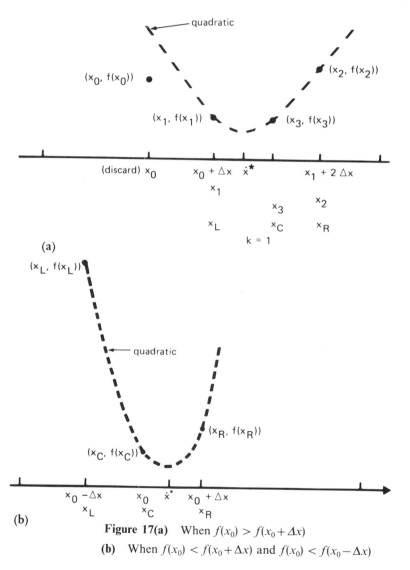

Figure 17(a)   When $f(x_0) > f(x_0+\Delta x)$

(b)   When $f(x_0) < f(x_0+\Delta x)$ and $f(x_0) < f(x_0-\Delta x)$

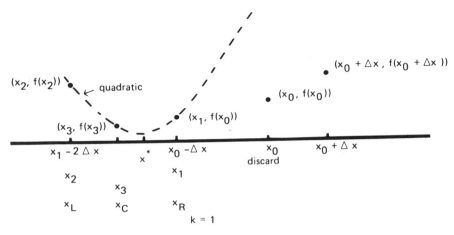

Figure 17(c)   When $f(x_0 - \Delta x) < f(x_0) < f(x_0 + \Delta x)$

minimizes f,

$$x^* \doteq \dot{x}^* = x_C + \frac{(f(x_L) - f(x_R))\Delta x}{2(f(x_L) - 2f(x_C) + f(x_R))} \qquad (12)$$

where $\Delta x = x_R - x_C = x_C - x_L$.

5. Compute $f(\dot{x}^*)$ and terminate the search if the desired accuracy has been reached. Otherwise, return to step 1 and repeat the process with the value from $\{x_L, x_C, x_R, \dot{x}^*\}$ that has the minimal function value, using a smaller value of Δx.

EXAMPLE 5 Use the DSC algorithm with $\Delta x = 0.001$ to minimize $8x/(x^2 + 4)$ over $[-4\sqrt{3}, 0]$.

Solution: We start the process by selecting some point x_0 in the interval. In this example we shall let $x_0 = -2\sqrt{3} = -3.4641016$. Since $f(x_0) = -1.7320508$ and

$$f(x_0 + \Delta x) = f(-3.4631016) = -1.7323008$$

then $f(x_0) > f(x_0 + \Delta x)$. Hence, we let $x_1 = -3.4631016$ and set $x_2 = x_1 + 2\Delta x = -3.4611016$. Then $f(x_2) = -1.7328008$. Since $f(x_2) < f(x_1)$, then set $x_3 = x_2 + 4\Delta x = -3.4571016$. Hence, $f(x_3) = -1.7338008$. Again since $f(x_3) < f(x_2)$, we let $x_4 = x_3 + 8\Delta x = -3.4491016$ and compute $f(x_4) = -1.7358007$. Since $f(x_4) < f(x_3)$, we let $x_5 = x_4 + 16\Delta x = -3.4331016$ and compute $f(x_5) = -1.739800$. Now $f(x_5) < f(x_4)$. Since this process continues for some time we shall summarize the results in Table 5. Since $f(x_{11}) > f(x_{10})$, we let

$$x_{12} = \frac{1}{2}(x_{10} + x_{11}) = -1.9290588$$

Table 5. DSC search

	x_i	$f(x_i)$
$i = 0$	-3.4641016	-1.7320508
$i = 1$	-3.4631016	-1.7323008
$i = 2$	-3.4611016	-1.7328008
$i = 3$	-3.4571016	-1.7338008
$i = 4$	-3.4491016	-1.7358007
$i = 5$	-3.4331016	-1.7398000
$i = 6$	-3.4011016	-1.7477968
$i = 7$	-3.3371016	-1.7637670
$i = 8$	-3.2091016	-1.7955109
$i = 9$	-2.9531016	-1.8571784
$i = 10$	-2.4411016	-1.9609256
$i = 11$	-1.4171016	-1.8868973

and compute that $f(x_{12}) = -1.9986980$. Since min $\{f(x_9), f(x_{10}), f(x_{11}), f(x_{12})\} = f(x_{12})$, we discard x_9. Then substituting $x_L = x_{10}$, $x_c = x_{12}$ and $x_R = x_{11}$ into equation (12) we estimate that

$$x^* \doteq \dot{x}^* = x_{12} + \frac{(x_{11} - x_{12})(f(x_{10}) - f(x_{11}))}{2(f(x_{10}) - 2f(x_{12}) + f(x_{11}))} = -2.0557504$$

and $f(\dot{x}^*) = -1.9992443$. The true optimal solution is $x^* = -2$ and $f(x^*) = -2$. A graph of the function is found in Figure 18.

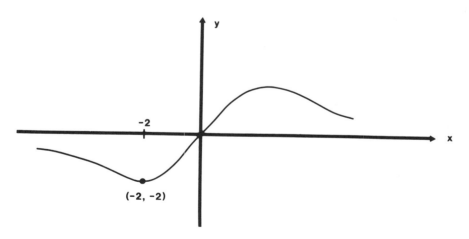

Figure 18 Graph of $f(x) = 8x/(x^2 + 4)$

Exercises

9.14   Verify equation (10), i.e., when $\Delta x = x_R - x_C = x_C - x_L$, then equation (9) simplifies to $x = x_C + [(f(x_L) - f(x_R))\Delta x]/[2(f(x_L) - 2f(x_C) + f(x_R))]$.

9.15   Use Powell's algorithm to minimize $f(x) = 3x^4 - 4x^3 - 6x^2 + 12x$ over $[-2,0]$.

9.16   Use Powell's algorithm to maximize $f(x) = \sqrt[3]{x}(x - 7)^2$ over $[0,7]$.

9.17   (a) Write a Fortran program that uses Powell's algorithm to minimize a given unimodal function.
(b) Write a Fortran program that uses Powell's algorithm to minimize a given unimodal function over a given interval $[a,b]$.
(c) Use your program to redo Example 4.
(d) Use your program to maximize $f(x) = x(3 - x)^{5/3}$ over the interval $[0,3]$. Check you answer with Example 1 in this chapter.

9.18   In both Steps 3i and 3iii of the DSC algorithm show that the points x_{k-1}, x_k, x_{k+2} and x_{k+1} are equally spaced.

9.19   (a) Write a Fortran program that uses the DSC algorithm to minimize a unimodal function.
(b) Write a Fortran program that uses the DSC algorithm to minimize a given unimodal function over a given interval. Use your program to minimize $8x/(x^2 + 4)$ over $[-4\sqrt{3},0]$. Compare your results with Example 5.

9.20   Use the DSC algorithm to minimize $x\sin(1/x)$ over $[1/2\pi,1/\pi]$. Check your results with Example 2 of this chapter.

9.21   Use the DSC algorithm to maximize $x(3 - x)^{5/3}$ over $[0,3]$. Check your results with Example 1 of this chapter.

3.   LINEAR SEARCH TECHNIQUES FOR DIFFERENTIAL FUNCTIONS BY CURVE FITTING

Many of the functions that we wish to minimize (or maximize) are differentiable in some sense. This section deals with some linear search techniques that depend upon the fitting of a quadratic function to a differentiable function.

The Method of False Position

Suppose that we have two values of x, x_L and x_R where $x_L < x_R$, and suppose also that we have computed $f(x_R)$, $f'(x_R)$ and $f'(x_L)$. Assume that $f'(x_R)$

and $f'(x_L)$ have opposite signs. Then the quadratic function

$$q(x) = f(x_R) + f'(x_R)(x - x_R) + \frac{(f'(x_L) - f'(x_R))}{2(x_L - x_R)}(x - x_R)^2 \qquad (13)$$

has the properties

$$\begin{aligned} q(x_R) &= f(x_R) \\ q'(x_R) &= f'(x_R) \\ q'(x_L) &= f'(x_L) \end{aligned} \qquad (14)$$

This is left as an exercise. A typical situation is illustrated in Figure 19. We wish to estimate the value of x^* of x that gives the minimal value of $f(x)$ by finding the value $\dot{x}^*$ of x that gives the minimal value of $q(x)$. Now $q(x)$ is minimized at the value of x that solves the equation $q'(x) = 0$. Thus, setting $q'(x) = 0$, solving for x and denoting the results by $\dot{x}^*$, we have

$$x^* \doteq \dot{x}^* = x_R - f'(x_R)\frac{x_L - x_R}{f'(x_L) - f'(x_R)} \qquad (15)$$

For greater accuracy, the process may be repeated using $\dot{x}^*$ and one of the values from $\{x_R, x_L\}$. In fact, if $f'(x_L) < 0$, $f'(x_R) > 0$ and $f'(\dot{x}^*) > 0$, then the next iteration could be performed using $\dot{x}^*$ and x_L. If $f'(\dot{x}^*) < 0$, then the iteration could be carried out using $\dot{x}^*$ and x_R. Notice that equation (15) does

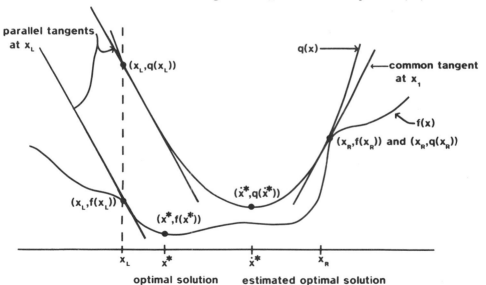

Figure 19 The method of false position.

not contain $f(x_R)$ nor $f(x_L)$. Hence, our fit could have been through either $(x_R, f(x_R))$ or $(x_L, f(x_L))$.

EXAMPLE 6   Use the method of false position to minimize $f(x) = x\sqrt{2 + x}$ over the interval $[-1.99, 0]$.
Solution: We begin the process by selecting two values from the interval, in particular $x_R = 0$ and $x_L = -1.99$. Next, since

$$f'(x) = \sqrt{2 + x} + \frac{x}{2\sqrt{2 + x}}$$

then $f'(x_R) = 1.41421356$ and $f'(x_L) = -9.85$. Using equation (15) we estimate the value of x that minimize $x\sqrt{2 + x}$ to be

$$x* \doteq \dot{x}* = 0 - (1.41421356)\frac{-1.99 - 0}{-9.85 - 1.41421356} = -.24984301$$

Also, $f(\dot{x}*) = -0.33052606$ and $f'(\dot{x}*) = 1.22850744$. Thus, in the second iteration we let $x_R = -0.24983401$ and $x_L = -1.99$. Then again using equation (15) we have a new estimation for $x*$, namely

$$x* \doteq \dot{x}* = -0.24984301 - (1.22850744)\frac{-1.99 - (-0.24984301)}{-9.85 - 1.22850744}$$
$$= -0.44281086$$

For this value $\dot{x}*$ we have that $f(\dot{x}*) = -0.55257209$ and $f'(\dot{x}*) = 1.70460814$. Hence, in the next iteration $x_L = -1.99$ and $x_R = -0.44281086$. See Table 6 for a summary of the results of 20 iterations. Actually, the minimum value of $x\sqrt{2 + x}$ is $-(4/3)/\sqrt{2/3}$ which occurs when $x = -4/3$. A graph of the function is shown in Figure 20.

Newton's Method

Minimizing a differentiable unimodal function f of a single real variable x by the method of false position requires that we work with two points at each iteration. Other techniques exist that allow us to estimate the optimal solution by using a single point. One such technique is Newton's method. For this method we require that f be twice differentiable over an interval that contains a point x_k for which $f''(x_k) > 0$. Then Taylor's formula guarantees the existence of a point c between x and x_k such that

$$f(x) = f(x_k) + f'(x_k)(x - x_k) + \frac{f''(x_k)}{2}(x - x_k)^2 + \frac{f'''(c)}{3!}(x - x_k)^3$$

This suggests that we approximate the optimal solution by finding the minimum of the quadratic function

$$q(x) = f(x_k) + f'(x_k)(x - x_k) + \frac{f''(x_k)}{2}(x - x_k)^2 \qquad (16)$$

Table 6.   The method of false position

iteration	x_L	x_R	$\dot{x}*$	$f(\dot{x}*)$
1	−1.99	0	−0.24984301	−0.33052606
2	−1.99	−0.24984301	−0.44281086	−0.55257209
3	−1.99	−0.44281086	−0.59446994	−0.70477416
4	−1.99	−0.59446994	−0.71543530	−0.81086463
5	−1.99	−0.71543530	−0.81314066	−0.88586043
6	−1.99	−0.81314066	−0.89291443	−0.93950790
7	−1.99	−0.89291443	−0.95865513	−0.97827215
8	−1.99	−0.95865513	−1.01326759	−1.00652334
9	−1.99	−1.01326759	−1.05895186	−1.02726416
10	−1.99	−1.05895186	−1.09739876	−1.04258730
11	−1.99	−1.09739876	−1.12992508	−1.05396933
12	−1.99	−1.12992508	−1.15756863	−1.06246349
13	−1.99	−1.15756863	−1.18115624	−1.06882812
14	−1.99	−1.18115624	−1.20135322	−1.07361381
15	−1.99	−1.20135322	−1.21869963	−1.07722319
16	−1.99	−1.21869963	−1.23363749	−1.07995258
17	−1.99	−1.23363749	−1.24653119	−1.08202125
18	−1.99	−1.24653119	−1.25768317	−1.08359228
19	−1.99	−1.25768317	−1.26734592	−1.08478746
20	−1.99	−1.26734592	−1.27573140	−1.08569809

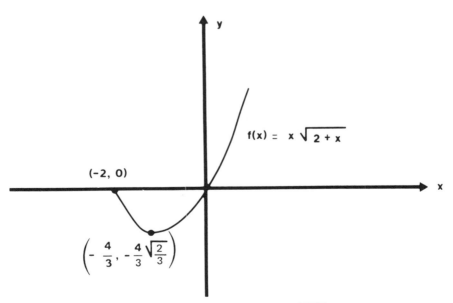

$$f(x) = x\sqrt{2+x}$$

$(-2, 0)$

$\left(-\dfrac{4}{3}, -\dfrac{4}{3}\sqrt{\dfrac{2}{3}}\right)$

Figure 20   Graph of $f(x) = x\sqrt{2+x}$

Notice that $q(x_k) = f(x_k)$, $q'(x_k) = f'(x_k)$ and $q''(x_k) = f''(x_k)$. To minimize $q(x)$ we set

$$0 = q'(x) = f'(x_k) + f''(x_k)(x - x_k)$$

and solve for x. Denoting the solution by x_{k+1} we get that

$$x_{k+1} = x_k - \frac{f'(x_k)}{f''(x_k)} \qquad (17)$$

(see Figure 21). The process works for maximization problems as well.

EXAMPLE 7   Use Newton's method to maximize $f(x) = 4x^{1/3} - x^{4/3}$ over $[0, 4]$.
Solution: Pick a point, say $x_1 = 2$, from the interval. Since

$$f'(x) = \frac{4}{3}x^{-2/3} - \frac{4}{3}x^{1/3} \quad \text{and} \quad f''(x) = -\frac{8}{9}x^{-5/3} - \frac{4}{9}x^{-2/3}$$

then $f''(2) = -2.5198421$ and $f''(2) = 0.5599649$. Hence, substituting into equation (17) we have that

$$x_2 = 2 - \frac{f'(2)}{f''(2)} = 0.5000000$$

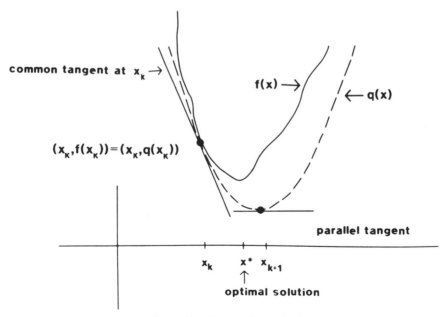

Figure 21   Newton's method

Now $f(x_2) = 2.7779518$, $f'(x_2) = 1.0582673$, and $f''(x_2) = -3.5275578$. Hence

$$x_3 = x_2 - \frac{f'(x_2)}{f''(x_2)} = 0.8000000$$

The process may be continued until the desired accuracy is obtained. Table 7 shows the results of five iterations. The actual maximum value of $f(x)$ is 3 which occurs when $x = 1$. A graph of the function is found in Figure 22.

Table 7. Iterations

x	$f(x)$
2.0000000	2.5198421
0.5000000	2.7779518
0.8000000	2.9706169
0.9714286	2.9994488
0.9994506	2.9999998
0.999998	3.0000000

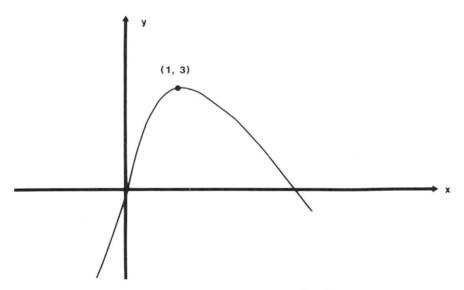

Figure 22 Graph of $f(x) = 4x^{1/3} - x^{4/3}$

Exercises

9.22   Verify that equations (14) are valid.

9.23   Use the method of false position to minimize $f(x) = (x^2 + 3x + 2)/x$ over the interval $[1/2,3]$ (see Example 4).

9.24   Write a Fortran program that will use the method of false position to minimize $f(x) = x\cos x$ over the interval $[0,2]$.

9.25   Use the method of false position to maximize $f(x) = x(3 - x)^{5/3}$ over the interval $[0,3]$ (see Example 1).

9.26   Use Newton's method to minimize $f(x) = x^4 - 6x^3 + 12x^2 - 8x + 1$ over $[0,1]$ (see Example 3).

9.27   Write a Fortran program that will use Newton's method to maximize $f(x) = x\cos x$ over $[0,\pi/2]$.

9.28   Use Newton's method to minimize $f(x) = x\sqrt{2 + x}$ over the interval $[-1.99,0]$ (see Example 7).

4.   MULTIDIMENSIONAL SEARCH TECHNIQUES

We now turn our attention to the problem of minimizing a real valued function of several real unconstrained variables. The functions in this section are assumed to have continuous first partial derivatives. Recall that in the multidimensional case the optimal solutions may be only relative minimums or relative maximums and not necessarily absolute minimums or absolute maximums. However, in most practical problems the relative extreme solutions are strict relative extreme solutions. This means that given a solution x there exists an open neighborhood about x such that along any line segment (in the neighborhood) passing through x the objective function is unimodal.

Method of Steepest Descent

Let $x = [x_1 \ldots x_n]^T \in \mathbf{R}_{n \times 1}$ and $f{:}\mathbf{R}_{n \times 1} \rightarrow \mathbf{R}$ have continuous first partial derivates. Recall that at a particular point $x_0 = [x_{0_1} \ldots x_{0_n}]^T$ the directional derivative

$$D_v f(x_0) = \lim_{t \to 0} \frac{f(x_0 + tv) - f(x_0)}{t} = \nabla f(x_0)^T v$$

gives the instantaneous rate of change of f at x_0 in the direction v. In particular, the gradient vector points in the direction of steepest ascent while the negative of the gradient vector points in the direction of steepest descent. When using the method of steepest descent to minimize the function f a linear search for the minimum of f is conducted along the line $\ell = \{x_0 - t\nabla f(x_0) : t \geq 0\}$, which extends from x_0 in the direction $-\nabla f(x_0)$. The object of the search is to determine the value of f that minimizes f along ℓ. If the function is simple enough the optimal value of f may be found by solving

$$\frac{d(f(x_0 - t\nabla f(x_0)))}{dt} = 0 \qquad \text{(optimal method of steepest descent)} \quad (18)$$

for t. However, in most of the problems to follow the value of t will be determined by one of the previous linear search techniques. Once the value, t_0, of t is determined the process starts over with

$$x_1 = x_0 - t_0\nabla f(x_0)$$

where $x_1 = [x_{1_1} \ldots x_{1_n}]^T$. Continuing in this fashion, a sequence

$$x_{k+1} = x_k - t_k\nabla f(x_k) \tag{19}$$

is generated, where t_k is the value of t that minimizes f along the line $\{x_k - t\nabla f(x_k) : t \geq 0\}$ that extends from x_k in the direction $-\nabla f(x_k)$, and $x_k = [x_{k_1} \ldots x_{k_n}]^T$.

An alternate procedure is to simply let the step size t in each step be some predetermined fixed value $t*$, i.e.,

$$x_{k+1} = x_k - t*\nabla f(x_k) \tag{20}$$

for each step k.

In either case the process is continued until some predetermined stopping condition is satisfied. Some common stopping conditions are

(1) Stop when $f(x_k) - f(x_{k+1}) < \varepsilon$, where $\varepsilon > 0$ is some predetermined number.

(2) Stop when $\nabla f(x_k) = 0$, i.e., when $(\partial f/\partial x_i)(x_k) = 0$ for all $i = 1, \ldots, n$. (Recall that $x_k = [x_{k_1} \ldots x_{k_i} \ldots x_{k_n}]^T$.) In this case it may be desirable to approximate $(\partial f/\partial x_i)(x_k)$ by

$$\frac{\partial f}{\partial x_i}(x_k) \doteq \frac{f([x_{k_1} \ldots x_{k_i} + \lambda \ldots x_{k_n}]^T) - f([x_{k_1} \ldots x_{k_i} \ldots x_{k_n}])}{\lambda}$$

(3) Stop when $\|\nabla f(x_k)\| = \sqrt{\sum_{i=1}^{n}(\partial f/\partial x_i)(x_k)} < \varepsilon$, where $\varepsilon > 0$ is a predetermined number.

(4) Stop when $\max\{(\partial f/\partial x_i)(x_k) : i = 1, \cdots, n\} < \varepsilon$, where $\varepsilon < 0$ is a predetermined number.

Usually, the process terminates at a point x_k where $\nabla f(x_k) \doteq 0$. It is possible that such a point could be a saddle point. Although this is unlikely in many practical problems it still may be desirable to check the point for this condition. This can be done by examining the Hessian matrix of f at x_k or by simply computing the numerical values of f for several points in a small neighborhood of x_k.

The method of steepest descent does have some serious drawbacks. These include

(1) Difficulty at boundary points if the region over which the function is being optimized is bounded. The process works without difficulty in the interior of the region.

(2) The rate of convergence can be very slow.

The method of steepest descent does give rise to more sophisticated techniques which attempt to overcome some or all of these difficulties.

EXAMPLE 8   Starting at the point $x_0 = [2 \quad 0]^T$ use the method of steepest descent to minimize $f(x) = x_1^2 + 7x_2^2 - 6x_1 + 56x_2 + 121$, where $x = [x_1 \quad x_2]^T$.
Solution: We will indicate three solutions to the problem by
(1)   Using the method of steepest descent with a fixed value of t.
(2)   Using the optimal method of steepest descent.
(3)   Using Powell's one dimensional line search to determine t in each iteration.

Since the rate of convergence is slow only a few iterations will be shown in each of these solutions. Thus, one should not expect a great deal of accuracy. Before beginning, notice that by completing the square it is obvious that the function f has an absolute minimum value of 0 when $x = [3 \quad -4]^T$. Also, in each iteration

$$\nabla f(x_k) = [2x_{k_1} - 6 \quad 14x_{k_2} + 56]^T, \text{ where } x_k = [x_{k_1} \quad x_{k_2}]^T$$

Solution 1:   Let $t = 0.1$ then

$$x_{k+1} = x_k - 0.1\,\nabla f(x_k)$$

for all k. If $x_0 = [2 \quad 0]^T$, then $f(x_0) = 113$. Also $\nabla f(x_0) = [-2 \quad 56]^T$. Hence,

$$x_1 = [2 \quad 0]^T - 0.1[-2 \quad 56]^T = [2.2 \quad -5.6]^T$$

Thus, $f(x_1) = 18.56$, $\nabla f(x_1) = [-1.6 \quad -22.4]^T$ and

$$x_2 = [2.2 \quad -5.6]^T - 0.1[-1.6 \quad -22.4]^T = [2.36 \quad -3.36]^T$$

Hence, $f(x_2) = 3.277$ and $\nabla f(x_2) = [-1.28 \quad 8.96]^T$. Next,

$$x_3 = [2.36 \quad -3.36]^T - 0.1[-1.28 \quad 8.96]^T = [2.488 \quad -4.256]^T$$

$f(x_3) = 0.721$ (rounding to three decimal places) and $\nabla f(x_3) = [-1.024 \quad -3.584]^T$.

One final iteration gives

$$x_4 = [2.488 \quad -4.256]^T - 0.1[-1.024 \quad -3.584]^T = [2.590 \quad -3.898]^T$$

and $f(x_4) = 0.241$.

Solution 2: Since along the ray $\{x_0 - t\nabla f(x_0) : t \geq 0\}$ the function f is a function in the variable t alone and since

$$x_0 - t\nabla f(x_0) = [2 \quad 0]^T - t[-2 \quad 56]^T = [2 + 2t \quad -56t]^T$$

then

$$f(t) = f(x_0 - t\nabla f(x_0)) = 21956t^2 - 3140t + 113$$

and

$$f'(t) = 43912t - 3140$$

Setting $f'(t) = 0$ and solving for t gives

$$t_0 = \frac{3140}{43912} = 0.0715$$

Hence,

$$x_1 = [2 \quad 0]^T - 0.0715[-2 \quad 56]^T = [2.143 \quad -4.004]^T$$

$f(x_1) = 0.735$ (rounding to three decimal places) and $\nabla f(x_1) = [-1.714 \quad -0.056]^T$. Next, since $x_1 - t\nabla f(x_1) = [2.143 \quad -4.004]^T - t[-1.714 \quad -0.056]^T = [2.143 + 1.714t \quad -4.004 + 0.056t]^T$, then along $\{x_1 - t\nabla f(x_1) : t \geq 0\}$ it follows that

$$f(t) = f(x_1 - t\nabla f(x_1)) = 2.96t^2 - 2.941t + 0.734$$

(rounding to three decimal places). Hence

$$0 = f'(t) = 5.92t - 2.941$$

implies that

$$t = \frac{2.941}{5.92} = 0.497$$

Thus,

$$x_2 = [2.143 \quad -4.004]^T - 0.497[-1.714 \quad -0.056]^T = [2.995 \quad -3.976]^T$$

$f(x_2) = 0.004$ and $\nabla f(x_2) = [-0.01 \quad 0.336]^T$.

Clearly, the convergence is faster than in the previous solution.

Solution 3: Let $t_1 = 0.1$ and $x_0 = [2 \quad 0]^T$. Consider the ray $\{x_0 - t\nabla f(x_0) : t \geq 0\}$. Let

$$x_1 = x_0 - t_1\nabla f(x_0) = [2 \quad 0]^T - 0.1[-2 \quad 56]^T = [2.2 \quad -5.6]^T$$

be a point on this ray. Then along this ray f is a function of t, $f(t) = f(x_0 - t\nabla f(x_0)) = 21956t^2 - 3140t + 113$. In fact if $t_0 = 0$ and $t_1 = 0.1$, then

$$f(t_0) = f(x_0) = 113 \quad \text{and} \quad f(t_1) = f(x_1) = 18.56$$

Since $f(t_0) > f(t_1)$, then let $x_2 = x_0 - 2(0.1)\nabla f(x_0) = [2.4 \quad -11.2]^T$ Then, $f(t_2) = f(x_2) = 363.24 > f(t_1)$, where $t_2 = 0.2$. Using equation (11) of Powell's algorithm we estimate the optimal solution along this ray to occur when

$$t* \doteq \frac{1}{2}\frac{((0.1)^2 - (0.2)^2)113 + ((0.2)^2 - 0^2)18.56 + (0^2 - (0.1)^2)363.24}{(0.1 - 0.2)113 + (0.2 - 0)18.56 + (0 - 0.1)363.24}$$

$$= 0.0715$$

Let

$$\dot{x}* = [2 \quad 0]^T - 0.0715[-2 \quad 56]^T = [2.143 \quad -4.004]^T$$

Then $f(\dot{x}*) = 0.735$. This is the end of the first iteration.

To begin the second iteration rename $x_0 = \dot{x}* = [2.143 \quad -4.004]^T$. Then $f(x_0) = 0.735$ and $\nabla f(x_0) = [-1.714 \quad -0.056]^T$. Consider the ray $\{x_0 - t\nabla f(x_0) : t \geq 0\}$. Again along this ray f is a function in t, $f(t) = f(x_0 - t\nabla f(x_0))$. If $t_0 = 0$, then $f(t_0) = 0.735$. Next let $x_1 = x_0 - (0.1)\nabla f(x_0) = [2.314 \quad -3.998]^T$ and $t_1 = 0.1$. Then $f(t_1) = f(x_1) = 0.471 < f(t_0)$. Thus, let $x_2 = x_0 - 2(0.1)\nabla f(x_0) = [2.486 \quad -3.993]^T$ and $t_2 = 0.2$. Then $f(t_2) = f(x_2) = 0.265$. Again using equation (11) of Powell's method we estimate the optimal solution along this ray to occur when

$$t* \doteq \frac{1}{2}\frac{0.735(0.01 - 0.04) + 0.471(0.04 - 0) + 0.265(0 - 0.01)}{0.735(0.1 - 0.2) + 0.471(0.2 - 0) + 0.265(0 - 0.1)} \doteq 0.505$$

Let $\dot{x}* = x_0 - t* \nabla f(x_0) = [2.143 \quad -4.004]^T - 0.505[-1.714 \quad -0.056]^T$

$$= [3.009 \quad -3.976]^T$$

then $f(\dot{x}*) = 0.0041$. The process is continued until the desired accuracy is obtained.

Exercises

9.29 Continue the iterations in Example 8 as indicated below.
(a) Do two more iterations for solution (1).

(b) Do two more iterations for solution (2).
(c) Do one more iteration for solution (3).

9.30   Maximize $f(x_1, x_2) = -x_1^2 + x_1 x_2 - x_2^2 + 3x_1 + 4$ using the following procedures.
(a) Optimal method of steepest ascent.
(b) Method of steepest ascent with a fixed value of t, $t = 0.1$
(c) Method of steepest ascent with the DSC line search algorithm.

9.31   Use the method of steepest descent and any line search algorithm to minimize $f(x_1, x_2) = x_1^2 + x_2^2 + 2x_2 - \ln x_1$.

Conjugate Direction Methods

Some of the best and most effective methods of minimizing a function $f:\mathbf{R}_{n \times 1} \to \mathbf{R}$, where f has continuous first partial derivatives, are conjugate direction methods. In the following discussion two such methods are introduced. The first method, called the conjugate gradient method, is used to minimize functions of the form $f(x) = 1/2 x^T A x + b^T x + c$, where A is an $n \times n$ positive definite symmetric matrix. The second method is known as the Fletcher – Reeves method and it extends the conjugate gradient method to more general nonquadratic functions. We begin with the following definition.

DEFINITION 1   Let A be a symmetric matrix of order n. Then two vectors v_1 and v_2 are said to be *A-orthogonal (A-conjugate)* if and only if $v_1^T A v_2 = 0$.

Notice that when $A = I$ then A-orthogonal is equivalent to orthogonal. Further, if A is positive definite, then any set $\{v_0, \cdots, v_k\}$ of $k+1$, where $k < n$, nonzero A-orthogonal vectors is a linearly independent set. To see this let $0 = \Sigma_{i=0}^{k} \alpha_i v_i$. Then $0 = A\Sigma_{i=0}^{k} \alpha_i v_i$ and hence

$$0 = v_j^T A \sum_{i=0}^{k} \alpha_i v_i = \sum_{i=0}^{k} \alpha_i v_j^T A v_i = \alpha_j v_j^T A v_j$$

Since A is positive definite it follows that $v_j^T A v_j > 0$. Hence, $\alpha_j = 0$ for each α_j. Thus, $\{v_0, v_1, \dots, v_k\}$ is a linearly independent set of vectors.
   Now consider the problem of minimizing $f(x) = (1/2)x^T A x + b^T x + c$, where A is a positive definite symmetric matrix. Notice that f is a strictly convex function and that the unique solution $x*$ to this problem is also the unique solution to the equation

$$0 = \nabla f(x) = Ax + b \qquad (21)$$

Suppose that $\{v_0, v_1, \dots, v_{n-1}\}$ is a set of nonzero A-orthogonal vectors. Since $\{v_0, v_1, \dots, v_{n-1}\}$ is a linearly independent set of n vectors it follows that $\{v_0, v_1, \ldots, v_{n-1}\}$ is a basis for $\mathbf{R}_{n \times 1}$. Hence, there exist unique scalars $\alpha_0, \alpha_1, \ldots, \alpha_{n-1}$ such that the solution $x*$ of Equation (21) can be expressed as

$$x* = \sum_{i=0}^{n-1} \alpha_i v_i$$

Thus,

$$-b = Ax* = A \sum_{i=0}^{n-1} \alpha_i v_i = \sum_{i=0}^{n} \alpha_i A v_i$$

Hence,

$$-v_j^T b = v_j^T \sum_{i=0}^{n-1} \alpha_i A v_i = \sum_{i=0}^{n-1} \alpha_i v_j^T A v_i = \alpha_j v_j^T A v_j$$

Since A is positive definite it follows that

$$\alpha_j = \frac{-v_j^T b}{v_j^T A v_j} \tag{22}$$

and

$$x* = \sum_{j=0}^{n-1} \frac{-v_j^T b}{v_j^T A v_j} v_j \tag{23}$$

It is important to notice that the α_j can be determined without actually knowing $x*$.

So far we have considered only the case in which the vectors $\{v_0, v_1, \ldots, v_{n-1}\}$ are known at the beginning of the process. In that case the optimal solution vector $x*$ can be computed using equation (23). We shall next consider an iterative procedure which will generate n A-orthogonal vectors $\{v_0, v_1, \cdots, v_{n-1}\}$ as well as a sequence of vectors $\{x_0, x_1, \cdots, x_{n-1}, x_n\}$ such that $x_n = x*$. The technique is known as the conjugate gradient method.

Conjugate Gradient Method

To minimize $f(x) = 1/2 x^T A x + b^T x + c$, where A is a positive definite symmetric matrix, using the conjugate gradient method, pick $x_0 \in \mathbf{R}_{n \times 1}$ and set $v_0 = -q_0$ where $q_0 = b + A x_0$. Then define for each $k = 0, \cdots, n$,

$$q_k = A x_k + b \text{ (here } x_k = [x_{k_1} \ldots x_{k_n}]^T)$$

$$\alpha_k = -\frac{q_k^T v_k}{v_k^T A v_k}$$

$$x_{k+1} = x_k + \alpha_k v_k \tag{24}$$

$$\beta_k = \frac{q_{k+1}^T A v_k}{v_k^T A v_k}$$

$$v_{k+1} = -q_{k+1} + \beta_k v_k$$

During each iteration of this procedure we are minimizing $f(x)$ along the ray $\{x_k + \alpha v_k : \alpha \geq 0\}$ using the optimal method of steepest descent (18). To see this consider

$$g(\alpha) = f(x_k + \alpha v_k) = \frac{1}{2}(x_k + \alpha v_k)^T A(x_k + \alpha v_k) + b^T(x_k + \alpha v_k) + c$$

$$= \frac{1}{2}x_k^T A x_k + \alpha v_k^T A x_k + \frac{\alpha^2}{2} v_k^T A v_k + b^T x_k + \alpha b^T v_k + c$$

The minimum of $g(\alpha)$ will occur when

$$0 = g'(\alpha) = D_\alpha f(x_k + \alpha v_k) = v_k^T A x_k + \alpha v_k^T A v_k + b^T v_k$$

i.e., when

$$\alpha = -\frac{v_k^T(Ax_k + b)}{v_k^T A v_k} = -\frac{v_k^T q_k}{v_k^T A v_k} = -\frac{q_k^T v_k}{v_k^T A v_k}$$

Moreover it is a simple matter to show that each v_{k+1} is A-orthogonal to its predecessor v_k. For example,

$$v_k^T A v_{k+1} = v_k^T A(-q_{k+1} + \beta_k v_k) = -v_k^T A g_{k+1} + \frac{q_{k+1}^T A v_k}{v_k^T A v_k} v_k^T A v_k = 0$$

It is left as an exercise (see Exercise 9.32) for the reader to verify that the entire set $\{v_0, v_1, \ldots, v_{n-1}\}$, generated by this procedure, is A-orthogonal.

Since the set $\{v_0, v_1, \ldots v_{n-1}\}$ is A-orthongonal, then it follows, as before, that the set is linearly independent and forms a basis for $\mathbf{R}_{n \times 1}$. Since A is a symmetric positive definite matrix it follows that there exists an $x*$ at which f has a global minimum. Thus, there exist unique scalars $\lambda_0, \lambda_1, \ldots, \lambda_{n-1}$ for which

$$x* - x_0 = \sum_{i=0}^{n-1} \lambda_i v_i$$

So,

$$v_j^T A(x* - x_0) = \sum_{i=0}^{n-1} \lambda_i v_j^T A v_i = \lambda_j v_j^T A v_j$$

and

$$\lambda_j = \frac{v_j^T A(x* - x_0)}{v_j^T A v_j} = \frac{v_j^T A(x* - x_j)}{v_j^T A v_j} + \frac{v_j^T A(x_j - x_0)}{v_j^T A v_j}$$

But $x*$, minimizes $f(x)$. Hence, $Ax* = -b$. Thus,

$$\lambda_j = \frac{v_j^T(-b - Ax_j)}{v_j^T A v_j} + \frac{v_j^T A(x_j - x_0)}{v_j^T A v_j} = -\frac{v_j^T q_j}{v_j^T A v_j} + \frac{v_j^T A(x_j - x_0)}{v_j^T A v_j}$$

and

$$\lambda_j = \alpha_j + \frac{v_j^T A(x_j - x_0)}{v_j^T A v_j} \tag{25}$$

But

$$x_j = x_0 + \alpha_0 v_0 + \cdots + \alpha_{j-1} v_{j-1}$$

and

$$x_j - x_0 = \alpha_0 v_0 + \cdots + \alpha_{j-1} v_{j-1}$$

Thus,

$$v_j^T A(x_j - x_0) = \sum_{i=0}^{j-1} \alpha_i v_j^T A v_i = 0$$

and it follows from equation (25) that each $\lambda_i = \alpha_i$. Thus,

$$x^* = x_0 + \sum_{i=0}^{n-1} \alpha_i v_i = x_n$$

It follows that the conjugate gradient method solves the problem of minimizing $f(x) = (1/2)x^T A x + b^T x + c$, where A is a symmetric positive definite matrix in n steps or less. Within n steps the sequence $\{x_{k+1}\}$ will terminate at x^*. The process ends at the step where $0 = q_k = A x_k + b$. It generates the A-orthogonal vectors as it proceeds from step to step.

EXAMPLE 9   Minimize $f(x) = 5(x_1 + 5)^2 + x_2^2$ using the conjugate gradient method. (Here, $x = [x_1 \ x_2]^T$.)
Solution: First notice

$$f(x) = 5(x_1 + 5)^2 + x_2^2 = 5x_1^2 + x_2^2 + 50x_1 + 125$$

$$= [x_1 \ \ x_2] \begin{bmatrix} 5 & 0 \\ 0 & 1 \end{bmatrix} \begin{bmatrix} x_1 \\ x_2 \end{bmatrix} + [50 \ \ 0] \begin{bmatrix} x_1 \\ x_2 \end{bmatrix} + 125$$

$$= \frac{1}{2}[x_1 \ \ x_2] \begin{bmatrix} 10 & 0 \\ 0 & 2 \end{bmatrix} \begin{bmatrix} x_1 \\ x_2 \end{bmatrix} + [50 \ \ 0] \begin{bmatrix} x_1 \\ x_2 \end{bmatrix} + 125$$

Let $x_0 = [1 \ \ 1]^T$. Then

$$q_0 = A x_0 + b = \begin{bmatrix} 10 & 0 \\ 0 & 2 \end{bmatrix} \begin{bmatrix} 1 \\ 1 \end{bmatrix} + \begin{bmatrix} 50 \\ 0 \end{bmatrix} = \begin{bmatrix} 60 \\ 2 \end{bmatrix}$$

Define $v_0 = -q_0 = \begin{bmatrix} -60 \\ -2 \end{bmatrix}$. Then

$$\alpha_0 = -\frac{q_0^T v_0}{v_0^T A v_0} = -\frac{[60 \ \ 2] \begin{bmatrix} -60 \\ -2 \end{bmatrix}}{[-60 \ \ -2] \begin{bmatrix} 10 & 0 \\ 0 & 2 \end{bmatrix} \begin{bmatrix} -60 \\ -2 \end{bmatrix}} = \frac{3604}{36008} = 0.1000888961$$

Hence,

$$x_1 = x_0 + \alpha_0 v_0 = \begin{bmatrix} 1 \\ 1 \end{bmatrix} + 0.1000888961 \begin{bmatrix} -60 \\ -2 \end{bmatrix} = \begin{bmatrix} -5.005332148 \\ 0.7998222617 \end{bmatrix}$$

$$q_1 = Ax_1 + b = \begin{bmatrix} -0.00533214 \\ 1.599644523 \end{bmatrix}$$

$$\beta_0 = \frac{q_1^T A v_0}{v_0^T A v_0} = \frac{[-0.00533214 \quad 1.599644523] \begin{bmatrix} 10 & 0 \\ 0 & 2 \end{bmatrix} \begin{bmatrix} -60 \\ -2 \end{bmatrix}}{36008} = -0.0000888495$$

and

$$v_1 = -q_1 + \beta_0 v_0 = -\begin{bmatrix} -0.00533214 \\ 1.599644523 \end{bmatrix} - 0.0000888495 \begin{bmatrix} -60 \\ -2 \end{bmatrix} = \begin{bmatrix} 0.01066311 \\ -1.599466824 \end{bmatrix}$$

Next,

$$\alpha_1 = \frac{-q_1^T v_1}{v_1^T A v_1} = -\frac{[-0.00533214 \quad 1.599644523] \begin{bmatrix} 0.0106631 \\ -1.599466824 \end{bmatrix}}{[0.01066311 \quad -1.599466824] \begin{bmatrix} 10 & 0 \\ 0 & 2 \end{bmatrix} \begin{bmatrix} 0.01066311 \\ -1.599466824 \end{bmatrix}}$$

$$= \frac{2.558635202}{5.117725261} = 0.4999555606$$

and

$$x^* \doteq x_2 = x_1 + \alpha_1 v_1 = \begin{bmatrix} -5.005332148 \\ 0.7998222617 \end{bmatrix} + 0.4999555606 \begin{bmatrix} 0.01066311 \\ -1.599466824 \end{bmatrix}$$

$$= \begin{bmatrix} -5.000001067 \\ 0.000159929 \end{bmatrix}$$

which is a good approximation to the obvious solution of

$$x^* = \begin{bmatrix} -5 \\ 0 \end{bmatrix}$$

We now consider a method of extending this technique to nonquadratic functions.

Fletcher-Reeves Method

This method while not requiring that f be quadratic does require some type of line search at each step. The method may be used for non-quadratic as well as quadratic minimization problems. The method is summarized in the following steps.

(1) Pick $x_0 \in \mathbf{R}_{n \times 1}$ and compute $v_0 = -\nabla f(x_0)$ (here, $k = 0$).

(2)   Set $x_{k+1} = x_k + \alpha_k v_k$, where α_k minimizes $f(\alpha) = f(x_k + \alpha v_k)$ along the ray $\{x_k + \alpha v_k : \alpha \geq 0\}$ extending from x_k in the direction v_k (A line search is used in this step). Here, $x_k = [x_{k_1} \ldots x_{k_n}]^T$.

(3)   Compute $f(x_{k+1})$ and $\nabla f(x_{k+1})$. The computation of $f(x_{k+1})$ is optimal.

(4)   If $k + 1 = n$, then go to Step 5. Otherwise, set

$$v_{k+1} = -\nabla f(x_{k+1}) + \frac{\nabla f(x_{k+1})^T \nabla f(x_{k+1})}{\nabla f(x_k)^T \nabla f(x_k)} v_k$$

and return to Step 2.

(5)   If a satisfactory solution has been obtained, then stop.
Otherwise set $x_0 = x_n$ and return to Step 1.

This method generates a sequence of directions $v_0, v_1, \ldots v_k$ for which v_k is a linear combination of the preceding directions $v_0, \ldots, v_{k-1}$, and the current direction of steepest descent $-\nabla f(x_k)$. When $f(x) = (1/2x)^T A x + b^T x + c$, where A is symmetric and positive definite, then the Fletcher-Reeves method minimizes f in at most n steps of one iteration and the direction vectors generated are A-orthogonal (see Exercise 9.35).

Notice that with this technique we do not need any information about the Hessian matrix of f at x_k (see Exercise 9.40).

EXAMPLE 10   Minimize $f(x_1, x_2) = 100(x_1 - x_2)^2 + (1 - x_1)^2$ using the Fletcher-Reeves method. This function is known as Rosenbrock's function. Its graph contains a deep parabolic valley.

Solution: Pick any point, say $x_0 = [-1.2 \quad 1]^T \in \mathbf{R}_{2 \times 1}$. Notice that any point $x = [x_1 \quad x_2]^T$,

$$\nabla f(x) = \begin{bmatrix} 202x_1 - 200x_2 - 2 \\ -200x_1 + 200x_2 \end{bmatrix}$$

Hence,

$$\nabla f(x_0) = \begin{bmatrix} -444.4 \\ 440 \end{bmatrix} \text{ and } v_0 = \begin{bmatrix} 444.4 \\ -440 \end{bmatrix}$$

Consider

$$x_0 + \alpha v_0 = \begin{bmatrix} -1.2 \\ 2 \end{bmatrix} + \alpha \begin{bmatrix} 444.4 \\ -440 \end{bmatrix} = \begin{bmatrix} 444.4\alpha - 1.2 \\ -440\alpha + 1 \end{bmatrix}$$

where $\alpha \geq 0$. For such points

$$f(\alpha) = f(x_0 + \alpha v_0) = 78413827.36\alpha^2 - 391091.36\alpha + 488.84$$

We want to minimize f along the ray $\{x_0 + \alpha v_0 : \alpha \geq 0\}$. This can be done with some type of line search. However, since the function along the line has the simple form shown above, we simply set the derivative of f with respect to α equal to zero and solve for α. Actually, we are setting the directional derivative $D_v f(x_0) = \lim_{\alpha \to 0} [f(x_0 + \alpha v_0) - f(x_0)] / \alpha$ equal to zero. Thus

$$0 = f'(\alpha) = 156827654.72\alpha - 391091.36$$

which gives $\alpha_0 = 0.0024938$. Therefore,

$$x_1 = \begin{bmatrix} -0.0917707 \\ -0.0972567 \end{bmatrix}$$

Moreover, $\nabla f(x_1) = \begin{bmatrix} -1.0863414 \\ -1.0972 \end{bmatrix}$ and

$$[-1.0863414 \quad -1.0972] \begin{bmatrix} -1.0863414 \\ -1.0972 \end{bmatrix} \bigg/ [-444.4 \quad 440] \begin{bmatrix} -444.4 \\ 444 \end{bmatrix}$$

$$= \frac{2.3839855}{391091.36} = 0.0000061$$

This implies that

$$v_1 = \begin{bmatrix} 1.0863414 \\ 1.0972 \end{bmatrix} + 0.0000061 \begin{bmatrix} 444.4 \\ -440 \end{bmatrix} = \begin{bmatrix} 1.0890503 \\ 1.0945179 \end{bmatrix}$$

Next consider the ray of points of the form

$$x_1 + \alpha v_1 = \begin{bmatrix} -0.0917707 \\ -0.0972567 \end{bmatrix} + \alpha \begin{bmatrix} 1.0890503 \\ 1.0945179 \end{bmatrix}$$

$$= \begin{bmatrix} 1.0890503\alpha - 0.0917707 \\ 1.0945179\alpha - 0.0972567 \end{bmatrix}$$

Along this ray

$$f(\alpha) = f(x_1 + \alpha v_1) = 1.1890201\alpha^2 - 2.3839855\alpha + 1.1949729$$

Setting the derivative $df(x_1 + \alpha v_1)/d\alpha = 0$ and solving for α gives

$$\alpha = 1.0025001$$

Hence,

$$x^* \doteq x_2 = \begin{bmatrix} -0.0917707 \\ -0.0972567 \end{bmatrix} + 1.0025001 \begin{bmatrix} 1.0890503 \\ 1.0945179 \end{bmatrix} = \begin{bmatrix} 1.0000023 \\ 0.9999979 \end{bmatrix}$$

This concludes one iteration of the Fletcher-Reeves method. For greater accuracy we could set $x_0 = x^*$ and perform another iteration. Obviously, the optimal solution occurs when $x^* = [1 \quad 1]^T$ and the minimum value is 0.

Exercises

9.32   In the conjugate gradient method show that the vectors v_0, $v_1, \ldots, v_{n-1}$ generated by equations (24) are A-orthogonal.

9.33 Use the conjugate gradient method to minimize the function

$f(x) = 2x_1^2 + x_2^2 + x_3^2 + x_1 + 2x_2 + x_1x_2$, where $x = [x_1 \quad x_2 \quad x_3]^T$.

9.34 Use the conjugate gradient method to maximize the function
$f(x_1, \quad x_2) = 6x_1 - 4x_2 - x_1^2 - 2x_2^2$.

9.35 Let $f(x) = (1/2)x^TAx + b^Tx + c$, where A is symmetric and positive definite. Verify that the direction vectors $v_0, v_1, \ldots, v_{n-1}$ that are produced in one iteration of the Fletcher-Reeves method are A-orthogonal.

9.36 Consider equations (24). For each $k \in \{1, \cdots, n-1\}$ define $\mathcal{B}_k$ to be the space spanned by $\{v_0, \ldots, v_{k-1}\}$. Show that the following statements hold:
(a) $x_k \in x_0 + \mathcal{B}_k$ for each k.
(b) $q_k \perp \mathcal{B}_k$ for each k.
(c) $\nabla f(x_k)^T(x - x_k) \geq 0$, for each $x \in x_0 + \mathcal{B}_k$.
(d) x_k minimizes f over the affine set $x_0 + \mathcal{B}_k$.
(e) x_k minimizes f over the line $\{x_{k-1} + \alpha v_{k-1} : \alpha \in \mathbf{R}\}$.

9.37 Redo Example 10 using one of the line search techniques in this chapter.

9.38 Use the Fletcher-Reeves method to maximize $f(x, x_2) = \sin(x_1 + x_2) + \sin x_1 + \sin x_2$. Start the search at the point $[x_1 \quad x_2]^T = [0 \quad 0]^T$.

9.39 Use the Fletcher-Reeves method to minimize the function
$f(x) = x_1^2 + e^{x_2^2 + x_3^2}$ starting at $x_1 = x_2 = x_3 = 2$.

9.40 (Newton's method extended to functions of several variables). Recall that a function $f:\mathbf{R}_{n \times 1} \to \mathbf{R}$, which is twice continuously differentiable in a neighborhood of a point $x_k \in \mathbf{R}_{n \times 1}$, can be approximated by

$$q(x) = f(x_k) + \nabla f(x_k)^T(x - x_k) + \frac{1}{2}(x - x_k)^TH(x_k)(x - x_k)$$

(a) Select an x_k for which $H(x_k)$ is positive definite (this ensures that q is a strictly convex function). Prove that the minimum of q occurs at x_{k+1}, where

$$x_{k+1} = x_k - H(x_k)^{-1}\nabla f(x_k)$$

(b) If f has a relative minimum at x^* and $H(x^*)$ is positive definite, then show that $H(x_k)$ is positive definite provided that x_k is sufficiently close to x^*.
(c) Use the results of parts (a) and (b) to devise an iterative procedure that generalizes Newton's method for minimizing $f(x)$. Notice that the procedure requires that each Hessian matrix, $H(x_k)$, be nonsingular and positive

definite, and that $H(x_k)^{-1}$ be computed in each iteration. (In comparison, we do not need any information about the Hessian to apply the Fletcher-Reeves method.) Notice that whenever $H(x_k) = I$, Newton's method is analogous to the method of steepest descent.

(d) Show that $H(x_k)^{-1}$ is positive definite whenever $H(x_k)$ is positive definite.

(e) Use your procedure from (c) to resolve Example 9.

9.41 (Davidon-Fletcher-Powell method.) The difficulty in applying Newton's method (see Exercise 9.40) is in maintaining a sequence $\{H(x_k)^{-1}\}$ of inverses of nonsingular positive definite Hessian matrices. The Davidon-Fletcher-Powell (DFP) method begins as a method of steepest descent. As the iterations proceed, the method resembles that of Newton. At each step, the inverse Hessian, $H^{-1}(x_k)$, is replaced with a positive definite approximation matrix H_k. The positive definite matrices H_k approach $H(x*)^{-1}$ at an optimal solution $x*$. The method is summarized below:

1. Pick a point x_0 and compute $\nabla f(x_0)$. Set $k = 0$ and $H_0 = I$.
2. If each $\|(\partial f/\partial x_i)(x_k)\| < \varepsilon$, where ε is a small acceptable positive number, then stop. Otherwise set $U_k = -H_k \nabla f(x_k)$.
3. Set $x_{k+1} = x_k + a v_k$, where a_k minimizes $f(a) = f(x_k + a v_k)$ along the ray $\{x_k + a v_k : a \geq 0\}$ (a line search can be used in this step).
4. Compute $\nabla f(x_{k+1})$, $q_k = f(x_{k+1}) - \nabla f(x_k)$, $p_k = a_k v_k$ and

$$H_{k+1} = H_k + \frac{p_k p_k^{\,T}}{p_k^T p_k} - \frac{H_k q_k q_k^T H_k}{q_k^T H_k q_k}$$

5. If $k+1 < n$, then replace k by $k+1$ and return to step (2). If $k+1 = n$, return to step (1) with x_n denoted by x_0.

(a) Use this method to resolve Example 10.
(b) Prove that H_{k+1} is positive definite whenever H_k is positive definite.
(c) Show that the DFP method applied to the problem

$$\text{Minimize } f(x) = \frac{1}{2} x^T A x + b^T x + c$$

where A is a symmetric positive definite matrix, is the conjugate gradient method. The DFP method is one of a class of similar techniques known as *quasi-Newton* procedures.

9.42 Use the Fletcher-Reeves method to minimize the following functions
(a) $f(x) = (x_1^2 - x_2)^2 + (x_1 - 1)^2$.
(b) $f(x) = 100(x_1^3 - x_2)^2 + (x_1 - 1)^2$.
(c) $f(x) = x_1 + 10x_2 + 5(x_3 - x_4)^2 + (x_2 - 2x_3)^4 + 10(x_1 - x_4)^4$.
(d) $f(x) = x_1^8 + (x_4 - 1)^2 + (e^{x_1} - x_2)^4 + 100(x_2 - x_3)^6 + \tan^4(x_3 - x_4)$.

Problems (a) and (b) are due to B. F. Whitte and W. R. Holst, while (c) is due to M. J. D. Powell, and (d) is due to E. E. Craig and A. V. Levy (see page 195 or [5]).

9.43   Use the Davidon-Fletcher-Powell method to minimize each of the problems in Exercise 9.42.

Readers who are interested in the continuation of the study of iterative search techniques for unconstrained optimization problems should consult the references. Those interested in the theoretical development are referred to [2], [3], [4], [8] and [9]. For a more detailed discussion and comparison of the economy of the various techniques the reader should refer to [4], [5], [8] and [9]. Additional techniques are discussed in all of the references.

REFERENCES

1.   L. Cooper and D. Steinberg, *Introduction to Methods of Optimization*, W. B. Saunders Company, Philadelphia (1970).
2.   J. E. Dennis, Jr. and R. B. Schnabel, *Numerical Methods for Unconstrainted Optimization and Nonlinear Equations*, Prentice-Hall, Englewood Cliffs, N.J. (1983).
3.   R. Fletcher, *Practical Methods of Optimization, Vol. 2*, John Wiley and Sons, New York (1981).
4.   B. S. Gottfried and J. Weisman, *Introduction to Optimization Theory*, Prentice-Hall, Englewood Cliffs, N.J. (1973).
5.   D. M. Himmelblau, *Applied Nonlinear Programming*, McGraw-Hill, New York (1972).
6.   J. Kiefer, Optimum Sequential Search and Approximation Methods Under Minimum Regularity Assumption, *J. Soc. Inc. and Appl. Math.*, *5*, No. 3 (1959).
7.   J. Kowalik and M. R. Osborne, *Methods for Unconstrained Optimization Problems*, American Elsevier Publishing Company, New York (1968).
8.   J. L. Kuester and J. H. Mize, *Optimization Techniques with Fortran*, McGraw-Hill, New York (1973).
9.   D. G. Luenberger, *Introduction to Linear and Nonlinear Programming*, Addison-Wesley, Reading, Mass. (1973).
10. G. R. Walsh, *Methods of Optimization*, John Wiley and Sons, New York (1975).

10

Penalty Function Methods

In this chapter, we shall be concerned with the problem of determining a solution of a constrained nonlinear programming problem. Currently, there is not a universally accepted method for dealing with such a problem. The approach in this chapter is to replace a constrained problem with one that is unconstrained. The latter problem is then solved using an iterative technique from (or similar to those found in) the last chapter. In Section 1, the constrained problem is converted into an unconstrained problem by adding a penalty function, $p(x)$, to the objective function $f(x)$. The resulting unconstrained objective function has the form $f(x) + \beta p(x)$, where $\beta > 0$. The function $p(x)$ imposes a penalty on $f(x) + \beta p(x)$ whenever x does not satisfy the constraints of the original problem. Actually, a sequence $\{f(x) + \beta_k p(x)\}$ of functions is minimized (or maximized). The solutions, $\{x_k\}$, of the sequence will usually approach, in some sense, the solution of the original problem. Normally, each x_k is not a feasible of the original problem. The process terminates whenever the required accuracy has been obtained, or whenever some solution, x_k, is generated that is a feasible solution of the original problem.

A similar approach is presented in Section 2. Again a sequence of functions $\{f(x) + (1/\beta_k)b(x)\}$ is minimized (or maximized) and the sequence of solutions $\{x_k\}$ normally tends to a solution of the original problem. The difference in Section 2 is that the solutions, x_k, are all feasible solutions of the original problem. The function $b(x)$ is called a barrier function because it imposes a penalty near the boundary of the set of feasible solutions of the original problem.

Both of these methods can possess the undesirable property of slow convergence. In the third section, the penalty function method is modified using Lagrange multipliers to obtain a more efficient method. The technique

315

is called the method of multipliers, and has emerged as an important tool for solving constrained nonlinear programming problems.

Throughout this chapter we shall normally work with problems for which the objective and constraint functions are at least continuous, and usually differentiable. Although some nondifferentiable penalty functions will be displayed, we shall normally restrict our attention to penalty functions which are differentiable. A brief survey of nondifferentiable methods can be found in [6] (also [7]).

The methods presented in this chapter deal with constraints having a very general form. In practice, more efficient techniques are often available which can capitalize on special constraint features such as linearity and bounded variables (see [7]).

Results which deal with the rate and global convergence of these techniques can be found in several of the references at the end of this chapter. Unfortunately, these results are normally based on restrictive assumptions which are necessarily placed on the functions or procedures involved. In actual practice these results do not always apply. Computational experience is often more valuable. The problem solver should always carefully assess his problem and the performance of his method of solution, and make modifications when they are warranted.

1.   INTRODUCTION

In a penalty function method an expression involving the constraints is added to the objective function. The expression is selected so that the value of the updated objective function is excessively high (or low) at a point x where the problem is infeasible.

EXAMPLE 1   Minimize $f(x_1) = x_1{}^2$ subject to $x_1 \geq 5$.
Solution: Reformulate the problem as

$$\text{Minimize } x_1{}^2 \text{ subject to } x_1 - x_2^2 = 5$$

Notice that by subtracting x_2^2 from the left side of the constraint we have converted the inequality into an equality without requiring that x_2 be nonnegative. Now define a penalty function for this problem by

$$p(x) = (5 - x_1 + x_2^2)^2$$

Then let β be a large positive number and consider the new problem (with β fixed)

$$\text{Minimize } f(x) + \beta p(x) = x_1^2 + \beta(5 - x_1 + x_2^2)^2 \text{ where } x \in \mathbf{R}_{2 \times 1} \quad (1)$$

It is intuitively obvious that any x for which $x_1 - x_2^2$ is not near to 5 will not solve (1) since $\beta p(x)$ will be large. Moreover, it appears that the larger β becomes, the closer $x_1 - x_2^2$ must be to 5 in order for x to solve (1). Actually, using the techniques of Chapter 8, we can easily establish that $x = [5(1+1/\beta)^{-1}\ \ 0]^T$ solves (1) for any $\beta > 0$, and $f(x) = 25(1+1/\beta)^{-1}$. Notice that $x \to [5\ \ 0]^T$ as $\beta \to \infty$. Clearly, the minimum value of $f(x_1)$ is 25 whenever $x_1 = 5$. Other penalty functions that could have been used include

$$p(x) = |5 - x_1 + x_2^2|^k$$

and

$$p(x) = (\max \{0,\ 5 - x_1\})^k \tag{2}$$

where k is a natural number. When the last penalty function (2) is used, it is not necessary to convert the inequality constraint to equality.

In general, one penalty function for the problem

$$\text{Minimize } f(x)$$

$$\text{subject to}$$

$$g_i(x) = b_i \text{ for } i = 1, \ldots, \ell$$
$$g_i(x) \le b_i \text{ for } i = \ell + 1, \ldots, m \tag{3}$$

is

$$p(x) = \sum_{i=1}^{\ell} |b_i - g_i(x)|^k + \sum_{i=\ell+1}^{m} (\max \{0, g_i(x) - b_i\})^k \tag{4}$$

where k is a natural number (usually in this book, $k = 2$). Notice that $p(x) \ge 0$. In fact $p(x) = 0$ if and only if x is feasible.

Problem (3) could be converted into the form

$$\text{Minimize } f(x)$$

$$\text{subject to} \tag{5}$$

$$h_i(x) = 0 \text{ for } i = 1, \ldots, m$$

by adding the square of an unrestricted variable to the left side of each inequality constraint, and then moving each b_i to the left side of each constraint. A typical penalty function for (5) is

$$p(x) = \sum_{i=1}^{m} |h_i(x)|^k \tag{6}$$

where k is a (usually even) natural number. Again notice that $p(x) \ge 0$. The remainder of this section deals with Problem (5).

To simplify the development, any variable restriction such as $x_i \ge 0$ or $a_i \le x_i \le b_i$ has been incorporated into the constraints, $g_i(x) \le b_i$ or $h_i(x) = 0$. In actual practice a variable restriction of the form $a_i \le x_i \le b_i$ can be dealt with directly by adding a term

$$\left|\frac{2x - a_i - b_i}{b_i - a_i}\right|^k \tag{7}$$

where k is a natural number, to the penalty function $p(x)$. The substitution $x_i = a_i + (b_i - a_i)\sin^2 z_i$, where z_i is unrestricted, can sometimes be used to eliminate the restriction $a_i \le x_i \le b_i$.

This means that the set $F = \{x: h_i(x) = 0 \text{ for } i = 1, \ldots, m\}$ is closed and bounded. We shall assume that the objective function f and each constraint function h_i is continuous. In most practical problems, each variable has an upper and a lower bound. Hence, we shall temporarily assume that the constraints of (5) impose such bounds on each variable. Then Problem (5) has a solution.

Now let $\{\beta_k\}$ be an increasing sequence of positive numbers that satisfy the condition that $\beta_k \to \infty$ as $k \to \infty$. For each k, consider the corresponding problem

$$\text{Minimize } P_k(k) = f(x) + \beta_k p(k)$$

where $x \in \mathbf{R}_{n \times 1}$. The following result suggests an iterative procedure for solving (5) whenever each $P_k(x)$ has a minimal value.

THEOREM 1   In Problem (5), assume that the functions $h_1, \ldots, h_m$ and f are all continuous, and that the set F of feasible solutions is nonempty, closed and bounded. Let $\{\beta_k\}$ be an increasing sequence of positive numbers that tends to infinity, and for each k let x_k denote a solution of the problem

$$\text{Minimize } P_k(x) = f(x) + \beta_k p(x) \tag{8}$$

where $p(x)$ is a differentiable penalty function of form (6). Then any accumulation point $\bar{x}$ of $\{x_k\}$ solves Problem (5). In fact, if a subsequence $\{x_{k_i}\}$ of $\{x_k\}$ converges to $\bar{x}$, then

(i) $p(\bar{x}) = \lim_{k_i \to \infty} p(x_{k_i}) = 0$

(ii) $\lim_{k_i \to \infty} \beta_{k_i} p(x_{k_i}) = 0$

(iii) $\min \{f(x): x \varepsilon F\} = \lim_{k_i \to \infty} p_{k_i}(x_{k_i})$

Proof: Since f is continuous and F is closed and bounded, then the Weierstrass theorem implies that there exists a solution x^* for Problem (5). By the definition of x_{k_i}, it follows that

$$f(x_{k_i}) + \beta_{k_i} p(x_{k_i}) \le f(x) + \beta_{k_i} p(x) \tag{9}$$

for all $x \in \mathbf{R}_{n \times 1}$. Recall that $p(x) = 0$ whenever $x \in F$. Thus, it follows from (9) that

$$f(x_{k_i}) + \beta_{k_i} p(x_{k_i}) \le \inf\{f(x) + \beta_{k_i} p(x) : x \in \mathbf{R}_{n \times 1}\} \le \inf\{f(x) + \beta_{k_i} p(x) : x \in F\}$$

$$= \inf\{f(x) : x \in F\} = f(x^*)$$

Hence,

$$0 \le \beta_{k_i} p(x_{k_i}) \le f(x^*) - f(x_{k_i}) \tag{10}$$

Since $\lim_{k_i \to \infty} (f(x^*) - f(x_{k_i}))$ exists, it follows that the

$$\sup \{\beta_{k_i} p(x_{k_i})\} \tag{11}$$

exists. Denote (11) by L. Let a subsequence $\{\beta_{n_{i_j}} p(x_{n_{i_j}})\}$ of $\{\beta_{k_i} p(x_{k_i})\}$ converge to L. Since p is continuous, it follows that $p(x_{k_{i_j}}) \to p(\bar{x})$. Suppose that $p(\bar{x}) \ne 0$. Then

$$\lim_{k_{i_j} \to \infty} \beta_{k_{i_j}} = \lim_{k_{i_j} \to \infty} \frac{\beta_{k_{i_j}} p(x_{k_{i_j}})}{p(x_{k_{i_j}})} = \frac{L}{p(\bar{x})} < \infty$$

which is a contradiction. Thus, $p(\bar{x}) = 0$ and condition (i) holds. Since $p(\bar{x}) = 0$, then $\bar{x} \in F$. From inequality (10), it follows that each $f(x_{k_i}) \le f(x^*)$ and, hence, $f(\bar{x}) \le f(x^*)$. But x^* solves (5) and $\bar{x} \in F$. Thus, $f(x^*) \le f(\bar{x})$. So, $f(x^*) = f(\bar{x})$. But then it follows from inequality (10) and $\lim_{k_i \to \infty} (f(x^*) - f(x_{k_i})) = f(x^*) - f(\bar{x}) = 0$ that

$$\lim_{k_i \to \infty} \beta_{k_i} p(x_{k_i}) = 0$$

Hence, (ii) holds. Finally,

$$\lim_{k_i \to \infty} (f(x_{k_i}) + \beta_{k_i} p(x_{k_i})) = f(\bar{x}) = f(x^*)$$

and the proof is complete.

The penalty function should force x_k to be close to F when β_k is large. Thus, in light of the assumption that each variable of (5) is bounded, it is not unreasonable to expect the sequence $\{x_k\}$ to usually have accumulation points. It is possible, however, to construct problems for which the corresponding sequence $\{x_k\}$ does not possess any accumulation points [see Exercise 10.3b], or in which some of the functions $p_k(x)$ possess unbounded minimums [see Exercise 10.3c].

Theorem 1 suggests a strategy for solving Problem (5). Namely, select (or select a method for generating) an increasing sequence $\{\beta_k\}$ that tends to infinity. For each k, minimize $p(x) = f(x) + \beta_k p(x)$. When the hypothesis of Theorem 1 is satisfied, a converging subsequence of the solutions x_k will usually approach a solution of (5). It is a simple exercise to show that x_k solves (5) whenever $p(x_k) = 0$ (see Exercise 10.1). Thus, the process continues until either an x_k is generated for which $p(x_k) = 0$, or until the desired accuracy has been obtained. For example we could proceed as follows:

(1)   Set $k = 1$ and $\beta_1 > 0$.
(2)   Minimize $f(x) + \beta_k p(x)$. Denote the vector that solves this problem by x_k.
(3)   If $p(x_k) = 0$ or if the desired accuracy has been obtained, then stop. Otherwise, let $\beta_{k+1} = c\beta_k$, where c is a positive constant, and return to Step (2).

Usually the point x_k is used as an initial point in any search technique for minimizing $p_{k+1}(x)$.

As β_k becomes very large it is often necessary to use small step sizes in a line search algorithm (that uses a definite step size) in order to improve the value of $P_k(x_k)$. This is because with large values of β_k more attention is paid to decreasing the term $p(x)$, in $\beta_k p(x)$, than to decreasing $f(x)$. The result can be slow convergence or premature termination. Modifications of the method exist which try to alleviate this difficulty. We shall consider one such method in Section 3.

It should be emphasized that the penalty function method generates a sequence $\{x_k\}$ of nonfeasible points, i.e., $x_k \notin F$, which ultimately leads to a solution of the problem. The technique can run into difficulties whenever the objective function is undefined at points outside the set of feasible solutions (see Exercise 10.8 of Section 2).

Methods exist for handling problems having form (3) without converting to form (5). A complete discussion can be found in [7].

EXAMPLE 2   Use a penalty function to minimize $x_1 + x_2$ subject to $x_1^2 + x_2^2 \leq 1$.

Solution: First rewrite the problem as

$$\text{Minimize } x_1 + x_2 \text{ subject to } x_1^2 + x_2^2 + x_3^2 = 1$$

Then set $\beta_1 = 1$ and solve the sequence of problems

$$\text{Minimize } P_k(x) = x_1 + x_2 + \beta_k(x_1^2 + x_2^2 + x_3^2 + x_3^3 - 1)^2$$

where $\beta_k = 2\beta_{k-1}$, using Newton's method in the direction of steepest descent starting at the nonfeasible point $x_0 = [2 \ 0 \ 0]^T$. An easy computation gives

$$\nabla p_1(x) = \begin{bmatrix} 1 - 4x_1(1 - x_1^2 - x_2^2 - x_3^2) \\ 1 - 4x_2(1 - x_1^2 - x_2^2 - x_3^2) \\ -4x_3(1 - x_1^2 - x_2^2 - x_3^2) \end{bmatrix}$$

In particular, $\nabla P_1(x_0) = [25 \quad 1 \quad 0]^T$. We want to minimize $P_1(x)$ along the ray that extends from x_0 in the direction $-\nabla P_1(x_0)$, i.e., along the ray $\{x = x_0 - \alpha \nabla P_1(x_0) : \alpha \geq 0\} = \{x = [2 - 25\alpha \quad -\alpha \quad 0]^T : \alpha \geq 0\}$. Along this ray $P_1(x)$ becomes

$$P_1(\alpha) = (-626\alpha^2 + 100\alpha - 3)^2 - 26\alpha + 2$$

Using Newton's method, a minimum value of $P_1(\alpha)$ occurs approximately when $\alpha* = 0.046946$, i.e. at the point

$$x_0^* = x_0 - \alpha* \nabla P_1(x_0) = [0.82635 \quad -0.04695 \quad 0]^T$$

Next $P_1(x)$ is minimized along the ray that extends from x_0^* in the direction $-\nabla P_1(x_0^*)$. Again using Newton's method, the solution is approximately $[0.85389 \quad -0.75805 \quad 0]^T$. Continuing in this fashion, the minimum value of $P_1(x)$ can be shown to occur near the point

$$x_1 = [-0.75434 \quad -0.77123 \quad 0]^T$$

Next let $\beta_2 = 2\beta_1$ and minimize $P_2(x)$ beginning at the nonfeasible point x_1. The solution is

$$x_2 = [0.73645 \quad -0.73655 \quad 0]^T$$

Table 1 summarizes the results of fifteen such iterations. Notice that the values of $f(x_k)$ and $P_k(x_k)$ are nondecreasing while the values of $p(x_k)$ are nonincreasing.

Exercises

10.1 For the terms in Theorem 1 verify the following
(a) The sequence $\{p(x_k)\}$ is nonincreasing (Hint: Consider the inequalities $P_{k+1}(x_k) \geq P_{k+1}(x_{k+1})$ and $P_k(x_{k+1}) \geq P_k(x_k)$.)
(b) The sequence $\{f(x_k)\}$ is nondecreasing.
(c) The sequence $\{P_k(x_k)\}$ is nondecreasing.
(d) Verify that x_{k_i} solves (3) whenever $p(x_{i_j}) = 0$.

10.2 If Problem (5) has a unique solution $x*$, then prove that any limit of a convergent subsequence of $\{x_k\}$ must be $x*$. Hence prove that the sequence $\{x_k\}$ converges to $x*$.

10.3 (a) Construct an example of a constrained optimization problem in which some of the variables are constrained and the solutions $\{x_k\}$ of the corresponding penalty function problems do not converge to a solution of the original problem.
(b) Construct a problem of form (5) for which the sequence $\{x_k\}$ of solutions of the related penalty function problems does not possess any converging subsequences.

Table 1 Penalty function method

k	β_k	x_k	$P_k(x_k)$	$p(x_k)$	$f(x_k)$
1	1	$[-0.75434 \ -0.77123 \ 0]^T$	-1.49873	2.6838×10^{-2}	-1.52557
2	2	$[-0.73645 \ -0.73655 \ 0]^T$	-1.45860	7.2019×10^{-3}	-1.473
3	4	$[-0.73650 \ -0.73650 \ 0]^T$	-1.44419	7.2019×10^{-3}	-1.473
4	8	$[-0.72224 \ -0.72224 \ 0]^T$	-1.42951	1.8715×10^{-3}	-1.44448
5	16	$[-0.71479 \ -0.71479 \ 0]^T$	-1.42194	4.7739×10^{-4}	-1.42958
6	32	$[-0.71098 \ -0.71098 \ 0]^T$	-1.41810	1.2067×10^{-4}	-1.42196
7	64	$[-0.70905 \ -0.70905 \ 0]^T$	-1.41616	3.0291×10^{-5}	-1.4181
8	128	$[-0.70808 \ -0.70808 \ 0]^T$	-1.41519	7.5876×10^{-6}	-1.41616
9	256	$[-0.70759 \ -0.70759 \ 0]^T$	-1.41470	1.8692×10^{-6}	-1.41518
10	512	$[-0.70735 \ -0.70735 \ 0]^T$	-1.41446	4.7339×10^{-7}	-1.4147
11	1024	$[-0.70723 \ -0.70723 \ 0]^T$	-1.41434	1.2148×10^{-7}	-1.41446
12	2048	$[-0.70717 \ -0.70717 \ 0]^T$	-1.41427	3.1973×10^{-8}	-1.41434
13	4096	$[-0.70714 \ -0.70714 \ 0]^T$	-1.41424	8.8266×10^{-9}	-1.41428
14	8192	$[-0.70712 \ -0.70712 \ 0]^T$	-1.41423	1.3972×10^{-9}	-1.41424
15	16384	$[-0.70711 \ -0.70711 \ 0]^T$	-1.41422	8.281×10^{-11}	-1.41422

(c) Construct a problem of form (5) for which all of the functions $P_k(x)$ possess unbounded minimums.

10.4 Determine the characteristics of a more general type of penalty function for which the results of Theorem 1 remain valid.

10.5 Solve the following problems:
(a) Minimize $x_1^2 + x_2^2$ subject to $x_1 + x_2 \leq 1$, $x_1 x_2 \geq 1$ and each $x_i \geq 0$.
(b) Minimize $x_1^2 + (2x_1 - x_2)^3$ subject to $x_1 = x_2^3$.
(c) Minimize $x_1 + \ln x_2$ subject to $x_1^2 + x_2^2 = 9$ and $x_2 \geq 1$.
(d) Minimize $7(x_1 - 3x_2)^2 + (x_1 - 6)^2$ subject to $x_1^2 + x_2 = 10$.
(e) Minimize $x_1 x_2 x_3$ subject to $x_1^2 + x_2^2 + x_3^2 \leq 4$, $x_1 x_2 \geq 1$, and all $x_i \geq 0$.
(f) Minimize $x_1 x_2 + \ln x_3$ subject to $x_1^2 + x_2^2 + x_3^2 \leq 100$, $x_2 + x_3^4 = 4$, $x_1 \geq 0$, and $x_3 \geq 1$.
(g) Minimize $f(x) = (x_1 - 2)^2 + (x_2 - 1)^2$ subject to $x_1 - 2x_2 = -1$, $(x_1^2/4 + x_2^2 \leq 1$.
(h) Minimize $f(x) = 100(x_2 - x_1^2)^2 + (1 - x_1)^2 + 90(x_4 - x_3^2)^2 + (1 - x_3)^2 + 10.1((x_2 - 1)^2 + (x_4 - 1)^2) + 19.8(x_2 - 1)(x_4 - 1)$ subject to each $-10 \leq x_i \leq 10$ (C. F. Wood, referenced in [10]).

2.   BARRIER FUNCTION METHODS

In this section we shall consider only those problems which have the form

$$\text{Minimize } f(x) \text{ subject to } g_i(x) \leq 0 \text{ for } i = 1, \dots, m \qquad (13)$$

Notice that Problem (13) does not contain any equality constraints. We shall assume that the function f is continuous over the set $F = \{x: \text{each } g_i(x) \leq 0\}$, and that the functions $g_i, \dots, g_m$ are continuous over $\mathbf{R}_{n \times 1}$. Moreover, we shall assume that F has a nonempty interior and that each boundary point of F is an accumulation point of the interior of F. This means that each boundary point of F can be approached via the interior of F.

Barrier function methods are similar to penalty function methods in that a barrier function is added to the objective function, and the resulting function is minimized. The difference is that the solutions are interior points of F (rather than points exterior to F). The purpose of the barrier function is to prevent the solutions from leaving the interior of F.

Some common barrier functions for Problem (13) are

$$b(x) = -\sum_{i=1}^{m} \frac{1}{g_i(x)} \qquad (14)$$

and

$$b(x) = \sum_{i=1}^{m} \ln |g_i(x)| \qquad (15)$$

Notice that $b(x)$ is, in either case, continuous throughout the interior of F. Moreover, $b(x) \to \infty$ as x approaches the boundary of F via the interior of F. Rather than solve (13), we intend to solve the following problem:

$$\text{Minimize } f(x) + \frac{1}{\beta} b(x) \text{ subject to each } g_i(x) < 0 \qquad (16)$$

where $\beta > 0$.

EXAMPLE 3   Minimize x subject to $x \geq 5$.
Solution: A barrier function of type (15) will be used to simplify computations. In particular, let $\beta > 0$ and solve the problem

$$\text{Minimize } x - \frac{1}{\beta} \ln|5 - x| \text{ subject to } x > 5 \qquad (17)$$

Problem (17) can be solved in any standard fashion. The minimum value of the objective function occurs when

$$x = 5 + \frac{1}{\beta}$$

and is equal to

$$5 + \frac{1}{\beta} - \frac{1}{\beta} \ln \frac{1}{\beta} \qquad (18)$$

Notice that for each $\beta > 0$, x is larger than 5 and approaches 5 as $\beta \to \infty$. Since $\lim_{\beta \to \infty} -(1/\beta)\ln(1/\beta) = 0$, it also follows that (18) approaches the minimum value of 5 for f as $\beta \to \infty$.

  Since most practical problems have bounded variables, we shall again assume that the set of feasible solutions, F, of problem (13) is bounded. It should be noted that the bounds may or may not be explicitly displayed in the constraints $g_i(x) \leq 0$. Nevertheless, with this assumption F is both closed and bounded.
  We are now ready to state a theorem which will give rise to a procedure for solving problem (13) that is similar to the method in Section 1. The proof is also similar to that of Theorem 1 and is outlined in Exercise 9.6.

THEOREM 2   In Problem (13) assume that f is continuous over F, and that each g_i is continuous over $\mathbf{R}_{n \times 1}$. Assume also that F is nonempty, closed and bounded; that each boundary point of F is an accumulation point of the interior of F; that $\{\beta_k\}$ is an increasing sequence of positive numbers for which $\lim_{k \to \infty} \beta_k = \infty$; and that for each k there exists an x_k which solves the problem

$$\text{Minimize } B_k(x) \text{ subject to each } g_i(x) < 0$$

where $B_k(x) = f(x) + (1/\beta_k) b(x)$, and $b(x)$ has form (14). Then

$$\text{Min } \{f(x) : g_i(x) \le 0 \text{ for all } i\} = \lim_{k \to \infty} B_k(x_k)$$

Moreover, if $\bar{x}$ is a limit point of any converging subsequence of $\{x_k\}$, then $\bar{x}$ solves Problem (13). Finally, $\lim_{k \to \infty} (1/\beta_k) b(x_k) = 0$.

This result suggests an iterative procedure for solving (13). Namely, select or determine a method for generating an increasing sequence of positive numbers $\{\beta_k\}$ that tends to infinity. Then for each k, solve the problem

$$\text{Minimize } f(x) + \frac{1}{\beta_k} b(x) \text{ subject to } g_i(x) < 0 \text{ for } i = 1, \dots, m \qquad (19)$$

Denote the solution by x_k. The values $f(x_k) + (1/\beta_k) b(x)$ will usually approach the minimum value of f. The process terminates when the desired accuracy is obtained.

Since the process works in the interior of F, it is necessary to determine an initial point x_0 for which each $g_i(x_0) < 0$ (The iterative procedure for minimizing $f(x) + (1/\beta_1) p(x)$ would start at x_0.) Finding such a point can itself be a difficult problem, and is a serious drawback of the method. As with penalty functions, barrier function methods can be slow to converge as the boundary is approached.

Ideally, the barrier function prevents the search from leaving the interior of F by becoming infinite as the boundary of F is approached. Hence, in the ideal situation one would not have to worry about any constraint, $g_i(x) < 0$, of (19). However in practice, a line search might step over the boundary into the exterior of F. When this happens it is possible to experience a decrease in $b(x)$ and, hence, $B(x)$. For instance, in Example 3 let $x_k = 5.01$, $\Delta x = 0.02$, $\beta = \frac{1}{2}$, and $x_{k+1} = x_k - \Delta x = 4.99$. Then $B(5.01) = 205.01$ and $B(4.99) = 195.01$, where $B(x) = x - 2(5 - x)^{-1}$. In general, one must determine if $g_i(x_k) < 0$ for $i = 1, \dots, m$ for each x_k that is generated.

In summary a barrier function method for solving (13) could have a form that is similar to the following:

Step 1 Determine (or compute) a point x_0 that satisfies the inequalities $g_i(x) < 0$ for $i = 1, \dots, m$.

Step 2 Determine a sequence of increasing positive numbers $\{\beta_k\}$ that tend to infinity. Let $k = 1$.

Step 3 Determine a solution x_k of the problem

$$\text{Minimize } f(x) - \frac{1}{\beta_k} \sum_{i=1}^{m} \frac{1}{g_i(x)} \text{ subject to } g_i(x) < 0 \text{ for } i = 1, \dots, m$$

Check to be sure that $g_i(x_k) < 0$ for each i before continuing. (If some $g_i(x_k) > 0$, you will have to determine a different solution before continuing). Start your search with x_{k-1}.

Step 4 If the required accuracy has been obtained, then stop. Otherwise replace k with $k+1$ and return to Step 3.

EXAMPLE 4   Use a barrier function method to solve the problem

$$\text{Minimize } f(x) = x_1 + x_2 \text{ subject to } x_1^2 + x_2^2 \le 1$$

Solution: The solution can be obtained using Newton's method in the direction of steepest descent to solve the sequence of problems

$$\text{Minimize } B_k(x) = f(x) + \frac{1}{\beta_k} b(x)$$

where $b(x) = -(x_1^2 + x_2^2 - 1)^{-1}$, $\beta_1 = 1$, and $\beta_k = 2\beta_{k-1}$. The results are summarized in Table 2.

Exercises

10.6 Let $\{\beta_k\}$ be an increasing sequence of positive numbers that satisfies the condition that $\beta_k \to \infty$ as $k \to \infty$. Assume that for each β_k there exists an x_k that solves the problem

$$\text{Minimize } f(x) - \frac{1}{\beta_k} b(x) \text{ subject to each } g_i(x) < 0 \qquad (19)$$

where $b(x)$ has the form (14) (see Theorem 2).

Assume also that (19) has an optimal solution. Then prove the following statements:

(a) Min $\{f(x) : g_i(x) \le 0, \, i = 1, \dots, m\} \le \inf_k \{f(x_k) - \frac{1}{\beta_k} b(x_k)\}$.

(b) The sequence $\{f(x_k)\}$ is nondecreasing.

(c) The sequence $\{b(x_k)\}$ is nonincreasing.

(d) The sequence $\{B(x_k)\}$ is nondecreasing.

(e) Min $\{f(x) : g_i(x) \le 0, i = 1, \dots, m\} = \inf_k B(x_k)$.

(f) If $\bar{x}$ is a limit point of any converging subsequence of $\{x_k\}$, then $\bar{x}$ solves (17).

(g) $\lim\limits_{k \to \infty} \frac{1}{\beta_k} b(x_k) = 0$.

10.7 Solve the following problems using barrier function methods.

(a) Minimize $7(x_1 - 3x_2)^2 + (x_1 - 6)^2$ subject to $x_1^2 + x_2 = 10$.

Table 2 Barrier function method

k	β_k	x_k	$B_k(x_k)$	$f(x_k)$	$B(x_k)$
1	1	$[-0.31810 \quad -0.31810]^T$	-0.61752	-0.63620	1.25372
2	2	$[-0.41964 \quad -0.41964]^T$	-0.06744	-0.83928	1.54368
3	4	$[-0.50000 \quad -0.50000]^T$	-0.500000	-1.00000	2.00000
4	8	$[-0.55947 \quad -0.55947]^T$	-0.78470	-1.11894	2.67389
5	16	$[-0.60233 \quad -0.60233]^T$	-0.97689	-1.20466	3.64435
.	.	.	.	.	.
.	.	.	.	.	.
20	524288	$[-0.70629 \quad -0.70629]^T$	-1.41175	-1.41258	433.11043
21	1048576	$[-0.70653 \quad -0.70653]^T$	-1.41248	-1.41306	613.22606
.	.	.	.	.	.
.	.	.	.	.	.
36	2^{35}	$[-0.70711 \quad -0.70711]^T$	-1.41422	-1.41422	109890.11

(b) Minimize $2(x_1 - 4) + x_2^4$ subject to

$$x_1 + x_2 \leq 8$$
$$x_1 - x_2^2 \geq 2$$
$$x_2 \geq 0$$

(c) Minimize $(x_1 - 2)^2 + (x_2 + 3)^2$ subject to

$$x_1 - x_2 \geq 1$$
$$x_1 \geq x_2$$
$$x_1 \leq 10$$

10.8 Minimize $f(x) = \sum_{i=1}^{10} \left((\ln(x_i - 2))^2 + (\ln(10 - x_i))^2 \right) - \left(\prod_{i=i}^{10} x_i \right)^{0.2}$

subject to $2.001 < x_i < 9.999$ for $i = 1, \dots, 10$,

where $\Pi_{i=1}^{10} x_i = x_1 x_2 \cdots x_{10}$. The objective function f is undefined outside of F (see Pavianni [12] or Himmelblau [10]).

10.9 Consider Problem (3) of Section 1. Let $\{\beta_k\}$ be a sequence of increasing positive numbers that tends to infinity. Consider the *mixed penalty-barrier* problem

$$\text{Minimize } M_k(x) = f(x) + \beta_k \sum_{i=1}^{\ell} (b_i - g_i(x))^2 - \frac{1}{\beta_k} \sum_{i=\ell+1}^{m} \frac{1}{g_i(x) - b_i}$$

subject to

$$g_i(x) < b_i \text{ for } i = \ell + 1, \dots, m$$

Let F be the set of feasible solutions of (3) and $\bar{F}_k$ be the set of feasible solutions of the above mixed penalty-barrier problem.

(a) For each k, let x_k be a solution of the above mixed penalty-barrier problem. Prove that under suitable restrictions

$$\text{Min } \{f(x) : x \, \varepsilon \, F\} = \inf_k M_k(x_k)$$

(b) Use this method to solve Exercise 10.5g.

3.   THE QUADRATIC PENALTY FUNCTION METHOD

In this section we shall return to the problem

$$\text{Minimize } f(x) \text{ subject to } h_i(x) = 0 \quad i = 1, \dots, m \qquad (5)$$

where $f, h_1, \dots, h_m$ are now continuously differentiable. We shall again assume that the set, F, of feasible solutions of (5) is nonempty. The continuity of the h_i ensures that F is closed. Since most practical problems have bounds imposed on their variables, we shall again assume that F is also bounded. As before, the Weierstrass theorem guarantees the existence of a solution, $x*$, of problem (5).

In Chapter 8, a method for determining $x*$ was suggested. Namely, compute vectors $x*$ and $\lambda*$ that satisfy

$$0 = \frac{\partial L}{\partial x}(x*, \lambda*) = \nabla f(x*)^T + \lambda*^T \frac{\partial h}{\partial x}(x*)$$

and   (20)

$$0 = \frac{\partial L}{\partial x}(x*, \lambda*) = h(x*)$$

where $L(x, \lambda) = f(x) + \lambda^T h(x)$ and $h(x) = [h_1(x) \ldots h_m(x)]^T$. Unfortunately, the system of equations (20) is difficult to solve.

In Section 1 of this chapter, Problem (5) was solved by solving a sequence of penalty function problems having form (8). The major advantage in this approach is that the problems (8) are unconstrained. Some disadvantages of the method include premature termination and slow convergence. As mentioned earlier these conditions can be brought about as the penalty parameters, β_k, become large. The success of the method for a particular problem may depend on the selection of the β_k.

In 1969, a new method emerged which combined the method of penalty functions with the Lagrange method of solving (20) (see Hestennes [8] and Powell [13]). This new class of methods could be called the class of *Lagrangian-penalty function methods*. However, it is often referred to as the class of *multiplier methods*. We shall restrict our attention to a particular method called the *quadratic penalty function method*. Our approach will be somewhat intuitive. A detailed development is tedious and beyond the nature of this text. Readers interested in greater detail should consult one of following references: Bertsekas [3], Hestenes [9] or Fletcher [6].

Consider a solution $x*$ of (5). Let $\lambda*$ be the corresponding vector of Lagrange multipliers for which equations (20) hold. Notice that whenever $x \in F$, then

$$L(x*,\lambda*) = f(x*) \leq f(x) = f(x) + \lambda*^T h(x) = L(x,\lambda*)$$

Thus, min $\{L(x, \lambda*) : x \in F\} = L(x*, \lambda*)$ and

$$\min \{f(x) : x \in F\} = \min\{L(x, \lambda*) : x \in F\}   \tag{21}$$

This suggests that rather than solve (5), we could solve the problem on the right side of (21), possibly using a penalty function method. That is

$$\text{Minimize } f(x) + \lambda *^T h(x) + \frac{\beta}{2} \sum_{i=1}^{m} (h_i(x))^2 \qquad (22)$$

where $\beta > 0$. Of course the problem is that $\lambda *$ is not known at the onset of the problem. The next result suggest an alternate strategy consisting of solving a sequence of problems of the form

$$\text{Minimize } f(x) + \lambda_k^T h(x) + \frac{\beta_k}{2} \sum_{i=1}^{m} (h_i(x))^2$$

where $\lambda_k \varepsilon \mathbf{R}_{m \times 1}$. The proof is similar to that of Theorem 1.

THEOREM 3   In problem (5) assume that the functions $h_1, \ldots h_m$ and f are all continuous, and that the set F of feasible solutions is nonempty, closed and bounded. Let $\{\lambda_k\}$ be a sequence of bounded vectors in $\mathbf{R}_{m \times 1}$, $\{\beta_k\}$ be a sequence of increasing positive numbers that tends to infinity, and let x_k solve the problem

$$\text{Minimize } L_k(x, \lambda_k) = f(x) + \lambda_k^T h(x) + \frac{\beta_k}{2} p(x) \qquad (23)$$

where $p(x) = \Sigma_{i=1}^{m} (h_i(x))^2$. Then every accumulation point of $\{x_k\}$ solves (5). In particular, if a subsequence $\{x_{k_i}\}$ of $\{x_k\}$ converges to $\bar{x}$, then

$$\text{(i) } p(\bar{x}) = \lim_{k_i \to \infty} p(x_{k_i}) = 0$$

$$\text{(ii) } \lim_{k_i \to \infty} \frac{\beta_{k_i}}{2} p(x_{k_i}) = 0$$

$$\text{(iii) } \min \{f(x) : x \varepsilon F\} = \lim_{k_i \to \infty} L_{k_i}(x, \lambda_{k_i})$$

Proof: Suppose that the subsequence $\{x_{k_i}\}$ of $\{x_k\}$ converges to $\bar{x}$. Since the corresponding subsequence, $\{\lambda_{k_i}\}$, of $\{\lambda_k\}$ is bounded it possesses accumulation points. Let $\bar{\lambda}$ be one of these accumulation points. Then three is a subsequence of $\{\lambda_{k_i}\}$ that converges to $\bar{\lambda}$. The corresponding subsequence of $\{x_{k_i}\}$ still converges to $\bar{x}$. To simplify notation, we shall denote both of these converging subsequences by $\{\lambda_{k_i}\}$ and $\{x_{k_i}\}$, respectively. Thus, $\lambda_{k_i} \to \bar{\lambda}$ and $x_{k_i} \to \bar{x}$ as $k_i \to \infty$.

Let $x*$ denote a solution of Problem (5). From the definition of x_{k_i}, it follows that

$$L_{k_i}(x_{k_i}, \lambda_{k_i}) \leq L_{k_i}(x, \lambda_{k_i})$$

for all $x \in \mathbf{R}_{n \times 1}$. Since $h(x) = 0$ and $p(x) = 0$ when $x \in F$, it follows from this last inequality that

$$L_{k_i}(x_{k_i}, \lambda_{k_i}) \leq \inf\{L_{k_i}(x, \lambda_{k_i}) : x \in \mathbf{R}_{n \times 1}\} \leq \inf \{L_{k_i}(x, \lambda_{k_i}) : x \in F\}$$
$$= \inf\{f(x) : x \in F\} = f(x*)$$

Hence,

$$0 \le (\frac{\beta_{k_i}}{2})\, p(x_{k_i}) \le f(x^*) - f(x_{k_i}) - \lambda_{k_i}{}^T h(x_{k_i}) \qquad (24)$$

Since the limit of the far right side of expression (24) exists, it follows that the

$$M = \sup\left\{ (\frac{\beta_{k_i}}{2}) p(x_{k_i}) \right\}$$

exists. This forces $p(\bar{x}) = \lim_{k_i \to \infty} p(x_{k_i}) = 0$ [Exercise 10.15a]. Thus, $h(\bar{x}) = 0$ and $\bar{x} \in F$. So, $f(x^*) \le f(\bar{x})$. Since the sequence $\{\lambda_{k_i}\}$ is bounded, it follows from (24) that $0 \le f(x^*) - f(\bar{x})$. Thus, $f(\bar{x}) \le f(x^*)$. therefore, $f(x^*) = f(\bar{x})$ and $M = 0$. It is left as an exercise to show that (ii) holds [Exercise 10.15 (b)]. Thus,

$$f(\bar{x}) = \lim_{k_i \to \infty} L_{k_i}(x_{k_i}, \lambda_{k_i})$$

The above proof gives no indication of how the sequence $\{\lambda_k\}$ could possibly be generated. One approach is outlined in the following discussion.

Suppose that $h_1, \ldots, h_m$ and f are twice continuously differentiable. Let $\{\beta_k\}$ be an increasing sequence of positive numbers that tends to infinity, and let $\{\varepsilon_k\}$ be a decreasing sequence of nonnegative numbers that tends to 0. Select a vector $\lambda_1 \in \mathbf{R}_{m \times 1}$ and determine a solution x_1 of the problem

$$\text{Minimize } L_1(x, \lambda_1) = f(x) + \lambda_1^T h(x) + \frac{\beta_1}{2} p(x)$$

such that

$$\left\| \frac{\partial}{\partial x} L_1(x_1, \lambda_1) \right\| \le \varepsilon_1$$

Then define λ_2 by the expression

$$\lambda_2 = \lambda_1 + \beta_1 h(x_1)$$

Next determined solution x_2 of the problem

$$\text{Minimize } L_2(x, \lambda_2) = f(x) + \lambda_2^T h(x) + \frac{\beta_2}{2} p(x)$$

such that

$$\left\| \frac{\partial}{\partial x} L_2(x_2, \lambda_2) \right\| \le \varepsilon_2$$

Continuing in this fashion two sequences $\{\lambda_k\}$ and $\{x_k\}$ are generated so that

$$\lambda_{k+1} = \lambda_k + \beta_k h(x_k) \qquad (25)$$

and

$$\|\frac{\partial}{\partial x}L_k(x_k,\lambda_k)\| \le \varepsilon_k$$

Notice that

$$(\frac{\partial}{\partial x}L_k(x_k,\lambda_k))^T = \nabla f(x_k)^T + \lambda_k^T\frac{\partial h}{\partial x}(x_k) + \beta_k h(x_k)^T\frac{\partial h}{\partial x}(x_k) \tag{26}$$

$$= \nabla f(x_k)^T + (\lambda_k^T + \beta_k h(x_k)^T)\frac{\partial h}{\partial x}(x_k)$$

Now suppose that a subsequence $\{x_{k_i}\}$ of $\{x_k\}$ converges to a vector $\bar{x}$. Assume that $\det(\partial h/\partial x)(\bar{x}) \ne 0$. Then when x_{k_i} is near $\bar{x}$, it follows that $\det[(\partial h/\partial x)(x_k)] \ne 0$ and $(\partial h/\partial x)(x_k)$ is nonsingular. It follows from (26) that

$$(\frac{\partial}{\partial x}L_{k_i}(x_{k_i},\lambda_{k_i}))^T(\frac{\partial h}{\partial x}(x_{k_i}))^{-1} = \nabla f(x_{k_i})^T(\frac{\partial h}{\partial x}(x_{k_i}))^{-1} + (\lambda_{k_i}^T + \beta_{k_i}h(x_{k_i})^T) \tag{27}$$

But

$$\frac{\partial}{\partial x}(L_{k_i}(x_{k_i},\lambda_{k_i})^T \to 0 \text{ as } k_i \to \infty$$

So, the left side of (27) tends to 0 as $k_i \to \infty$. It follows that

$$\lambda_{k_i}^T + \beta_{k_i}h(x_{k_i})^T \to \bar{\lambda}$$

where

$$\bar{\lambda} = -\nabla f(\bar{x})^T(\frac{\partial h}{\partial x}(\bar{x}))^{-1} \tag{28}$$

When the sequence $\{\lambda_{k_i}\}$ is bounded, then it is a simple exercise (see Exercise 9. 10) to show that $h(\bar{x}) = 0$ and $\bar{x}$ solves (5). Notice that

$$\nabla f(\bar{x})^T + \bar{\lambda}^T\frac{\partial h}{\partial x}(\bar{x}) = 0$$

Hence, $\bar{x}$ and $\bar{\lambda}$ satisfy equations (20). This suggest the following iterative procedure for solving (5).

Select an increasing sequence $\{\beta_k\}$ of positive numbers that tends to infinity, and a sequence $\{\varepsilon_k\}$ of nonnegative numbers that tends to 0. Let $k = 1$.

Step 1 Determine a solution x_k of the problem

$$\text{Minimize } f(x) + \lambda_k^T h(x) + (\frac{\beta_k}{2})p(x)$$

that satisfies

$$\left\| \frac{\partial}{\partial x} L_k(x_k, \lambda_k) \right\| \leq \varepsilon_k$$

If the required accuracy has been obtained, then stop. Otherwise to go step 2.

Step 2 Compute $\lambda_{k+1} = \lambda_k + \beta_k h(x_k)$. Replace k by $k+1$ and return to step 1.

Normally, x_k is used as the starting point for any line search method of minimizing $L_{k+1}(x, \lambda_{k+1})$. Practical experience indicates that defining the sequence $\{\beta_k\}$ by $\beta_{k+1} = c\beta_k$, where $4 \leq c \leq 10$, often gives satisfactory convergence [3]. If possible the first vector λ_1 in $\{\lambda_k\}$ should be selected as close to $\lambda*$ as possible, where $x*$ solves (5) and the pair $(x*, \lambda*)$ satisfies (20).

Examples are easily constructed in which f has a global minimum over a compact set F of feasible solutions, but each $L_k(x, \lambda_k)$ has no minimum (see Exercise 10.14). When this occurs it is often helpful to use a penalty function other than $p(x) = \sum_{i=1}^{m} (h_i(x))^2$. A complete discussion of non-quadratic penalty functions can be found in Chapter 5 of [3].

The class of multiplier methods has better convergence properties than does the class of penalty function methods. With multiplier methods, it is not always necessary to increase β_k to infinity in order to obtain convergence [3]. Thus, the slow convergence and premature termination, which can occur with a penalty function method as β_k becomes large, can often be avoided. The result is a method that converges much more rapidly than does a penalty function method. These multiplier methods currently represent some of the more promising techniques for solving constrained optimization problems. New research and developments can be expected.

EXAMPLE 5   Use the quadratic penalty function method to minimize $f(x) = x_1 + x_2$ subject to $x_1^2 + x_2^2 \leq 1$.
Solution: First rewrite the above problem as

$$\text{Minimize } f(x) = x_1 + x_2 \text{ subject to } 1 - x_1^2 - x_2^2 - x_3^2 = 0$$

Next set $\lambda_1 = 1$, $\beta_1 = 1$, $\varepsilon_1 = 0.001$, $\beta_{k+1} = 4\beta_k$, $\lambda_{k+1} = \lambda_k + \beta_k h(x_k)$, where $h(x) = 1 - x_1^2 - x_2^2 - x_3^2$, and $\varepsilon_{k+1} = \varepsilon_k/10$. Then solve the sequence of problems

$$\text{Minimize } L_k(x, \lambda_k) = f(x) + \lambda_k(1 - x_1^2 - x_2^2 - x_3^2) + \left(\frac{\beta_k}{2}\right)(1 - x_1^2 - x_2^2 - x_3^2)^2$$

using Newton's method in the direction of steepest descent beginning with the point $[2 \quad 0 \quad 0]^T$ when $k = 1$. The results are summarized in Table 3.

Table 3 Multiplier method

k	β_k	λ_k	x_k	$f(x_k)$	ε_k	$\left\|\frac{\partial L}{\partial x}(x_k,\lambda_k)\right\|$
1	1	1	$[-1.10776 \quad -1.10654 \quad 0]^T$	-2.21430	10^{-3}	7.933×10^{-4}
2	4	-0.4516	$[-0.72766 \quad -0.72761 \quad 0]^T$	-1.45527	10^{-4}	4.817×10^{-5}
3	16	-0.6872	$[-0.70754 \quad -0.70754 \quad 0]^T$	-1.41508	10^{-5}	1.615×10^{-6}
4	64	-0.7067	$[-0.70711 \quad -0.70711 \quad 0]^T$	-1.41422	10^{-6}	1.267×10^{-7}
5	256	-0.7071	$[-0.70711 \quad -0.70711 \quad 0]^T$	-1.41422	10^{-7}	5.508×10^{-10}

Exercises

10.10 Verify that the vector $\bar{x}$ that is associated with the vector $\bar{\lambda}$ in (28) solves (5) whenever the sequence $\{\lambda_{k_i}\}$ is bounded.

10.11 Solve each of the problems in Exercise 10.5 using the method of multipliers.

10.12 Let $c > 1$ and $0 < \gamma < 1$. Define the sequence $\{\beta_k\}$ by

$$\beta_{k+1} = \begin{cases} c\beta_k, \text{if } \|h(x_k)\| > \gamma \| h(x_{k-1})\| \\ \beta_k, \text{if } \|h(x_k)\| \leq \gamma \| h(x_{k-1})\| \end{cases}$$

(a) Prove that the sequence $\{\beta_k\}$ is bounded.
(b) Let $c = 10$ and $\gamma = 0.25$ and solve each of the problems in Exercise 9.5 using the method of multipliers and this method for generating the sequence $\{\beta_k\}$.

10.13 Define the sequence $\{\lambda_k\}$ by

$$\lambda_{k+1} = \lambda_k + \beta_k h(x_k) + \Delta\lambda_k$$

where $\Delta\lambda_k$ and some Δx_k solves the system

$$\begin{bmatrix} H & \frac{\partial h}{\partial x}(x_k)^T \\ \frac{\partial h}{\partial x}(x_k) & 0 \end{bmatrix} \begin{bmatrix} \Delta x_k \\ \Delta\lambda_k \end{bmatrix} = - \begin{bmatrix} \nabla f(x_k) + \frac{\partial h}{\partial x}(x*)^T(\lambda_k + \beta_k h(x_k)) \\ h(x_k) \end{bmatrix}$$

where H is the Hessian of $L(x, \lambda_k + \beta_k h(x_k)) = f(x) + (\lambda_k + \beta_k h(x_k))^T h(x)$. This method of generating the sequence $\{\lambda_k\}$ sometimes gives a better rate of convergence than (25). Use this approach to resolve Example 5.

10.14 Construct an example of a problem of form (5) for which f has a minimum, the set F is compact, and each $L_k(x, \lambda_i)$ is unbounded from below. Construct your example so that you can determine a penalty function $\bar{p}(x)$ (which is not the quadratic penalty function $p(x) = \sum_{i=1}^m (h_i(x))^2$) for which

$$\bar{L}_k(x, \lambda_k) = f(x) + \lambda^T h(x) + \frac{\beta_k}{2} p(x)$$

has a minimum value for every k.

10.15 (a) Verify that $\lim_{k_i \to \infty} p(x_{k_i}) = 0$ in the proof of Theorem 3.
      (b) Verify that (ii) holds in the proof of Theorem 3.
(Hint, consider $\lim_{n \to \infty} \sup\{(\beta_{k_i}/2)p(x_{k_i}): k_i \geq n\}$.)

REFERENCES

1. M. S. Bazaraa and C. M. Shetty, *Nonlinear Programming*, John Wiley and Sons, New York (1979).
2. D. P. Bertsekas, Augmented Lagrangian and Differentiable Exact Penalty Methods, in *Nonlinear Optimization 1981* (ed. M. J. D. Powell), Academic Press, New York (1981).
3. D. P. Bertsekas, *Constrained Optimization and Lagrange Multiplier Methods*, Academic Press, New York (1982).
4. J. Bracken and G. P. McCormick, *Selected Applications of Nonlinear Programming*, John Wiley and Sons, New York (1968).
5. A. R. Conn, Penalty Function Methods, in *Nonlinear Optimization 1981*, (ed. M. J. D. Powell), Academic Press, New York (1981).
6. R. Fletcher, *Practical Methods of Optimization, Vol. 2*, John Wiley and Sons, New York (1981).
7. R. Fletcher, Methods for Nonlinear Constraints, in *Nonlinear Optimization 1981* (ed. M. J. D. Powell), Academic Press, New York (1982).
8. M. R. Hestenes, Multiplier and Gradient Methods, *Journal of Optimization Theory and Applications, 4*, (1969), 303 – 320.
9. M. R. Hestenes, *Optimization Theory*, John Wiley and Sons, New York (1975).
10. D. M. Himmelblau, *Applied Nonlinear Programming*, McGraw-Hill, New York, (1972).
11. D. G. Luenberger, *Introduction to Linear and Nonlinear Programming*, Addison-Wesley, Reading, Mass. (1973).
12. D. A. Pavianni, Ph. D. dissertation, University of Texas, Austin (1968).
13. M. J. D. Powell, A Method for Nonlinear Constraints in Minimization Problems, in *Optimization* (ed. R. Fletcher), Academic Press, New York (1969).
14. G. R. Walsh, *Methods of Optimization*, John Wiley and Sons, New York (1975).

Index

A-conjugate, 305
Admissible basis, 47, 85
Affine
    combination, 51
    function, 217
    hull, 51
    set, 55
Affinely independent, 61
Almost completely basic feasible
        solution, 137
A-orthogonal, 305
Arc, 176
    forward, 181
    reverse, 181, 199
    rooted, 184–185
Artificial
    cost coefficients, 99–109
    variables, 99–116
Augmented matrix, 42

Back substitution, 44
Barrier function, 323–328
Basic, 47
    feasible solution, 76, 85
    solution, 85, 137
    variable, 137

Basis, 32
    matrix, 85
Block multiplication, 35

Capacitated, 178
Chain, 181
Circuit, 181
Column
    matrix, 29
    rank, 34
    space, 34
Complementary, 136
    basic feasible solution, 137
    cone, 137
    pivoting algorithm, 137–140
    slackness, 135–136
Concave function, 217
Conditions
    first necessary, 248
    Kuhn-Tucker, 256–259
    sufficient, 248, 259
Conical
    combination, 51
    hull, 51
Conjugate
    direction methods, 305